Geoenvironmental Engineering

Contaminated Soils, Pollutant Fate, and Mitigation

Geoenvironmental Engineering

Contaminated Soils, Pollutant Fate, and Mitigation

Raymond N. Yong

CRC Press
Boca Raton London New York Washington, D.C.

Library of Congress Cataloging-in-Publication Data

Yong, R.N. (Raymond Nen)
Geoenvironmental engineering: contaminated soils, pollutant fate and mitigation / Raymond N. Yong
p. cm.
Includes bibliographical references and index.
ISBN 0-8493-8289-0
1. Soil pollution.
2. Soil remediation. 3. Environmental geotechnology. I. Title.

TD878.Y65 2000
628.5′5—dc21 00-055652
CIP

Direct all inquiries to CRC Press LLC, 2000 N.W. Corporate Blvd., Boca Raton, Florida 33431.

International Standard Book Number 0-8493-8289-0
Library of Congress Card Number 00-055652
Printed in the United States of America 1 2 3 4 5 6 7 8 9 0
Printed on acid-free paper

Preface

The treatment of contaminated land to eliminate or reduce the presence of pollutants in the contaminated site has received (and will continue to receive) considerable attention from the practicing profession. Extensive research and development are still underway in respect to the delivery of more effective (and economic) means for site decontamination. The ongoing results can be seen in the availability of new and innovative techniques for complete or partial removal of pollutants, fixing pollutants within the soil substrate such that these remain immobile (forever?), reducing the toxicity of those pollutants in place, and a whole host of other schemes — all designed to eliminate or reduce the threat to human health and the environment posed by the pollutants. These constitute very important subjects that are being discussed and published by those professionals dealing with technology for site remediation. In this book, we are concerned with the development of a better understanding of the many basic issues that surround the control of pollutant fate in contaminated sites.

In the continuing effort to improve our understanding and appreciation of the various bonding and partitioning processes between pollutants and soil fractions, it has become increasingly clear that the processes that control the fate of pollutants should be taken into account if we are to structure effective remediation programs. The intent of this book is to provide the groundwork for a keener appreciation of some of the key factors that need to be considered when we seek to determine the fate of pollutants in soils. No attempt is made to provide all the detailed substantive data and results. Instead, the material presented is designed to remind the reader of the various factors, interactions, and mechanisms deemed to be important in the bonding and partitioning processes. As such, the treatment given in the first three chapters seeks to address the nature of soil and the soil-water system — after first examining the problems associated with contaminated lands.

It has long been known that we cannot overlook the influence of the surface characteristics and properties of the various soil fractions that make up a "regular piece of soil" if we are to understand why some soils retain more pollutants and why other soils do not. The soil-water system is considered as a separate subject for discussion (Chapter 3) because of the importance of soil structure and its relation to the pollutant partitioning process. This is further explored in Chapter 4 where the interactions between the soil fractions and pollutants are examined — particularly in respect to the resultant partitioning of the pollutants. We have taken care throughout the book to remind the reader that we consider pollutants to be contaminants that are classified as "threats" to human health and the environment.

The partitioning, fate, and persistence of pollutants are examined in Chapters 5 and 6. Heavy metals are used as a focus for discussions in Chapter 5 concerning inorganic contaminants because of their ubiquitous presence in contaminated sites. Much of the material presented in the chapter applies to other inorganic contaminants (pollutants and non-pollutants). The various processes that contribute to the transformation and degradation of organic chemical pollutants are discussed in Chapter 6 — with attention to the persistence of the organic chemicals and the associated changes in their properties. Since removal of these pollutants must require attention to their properties, and since these properties will change because of the various transformations, it becomes necessary to be aware of those processes in control of the situation. This lays the basis for Chapters 7 and 8, which examine the interactions between pollutants and soil fractions from the viewpoint of "pollutant-removal" — as remediation or pollution mitigation schemes.

It has been difficult from the beginning to determine the level of basic information and theories needed to support the discussions presented, especially in those chapters dealing with the fundamental mechanisms and processes. Undoubtedly, there will most probably be "too much" and "too little" background support information/theory in the various chapters.

The author has benefitted considerably from all the interactions with his colleagues and students. In particular, considerable benefit has been obtained from the various research studies conducted by his post-graduate students. This has been a mutual learning process. It has not been possible to list more than a few individual theses and published works by the various students and learned authorities in the texts of the various chapters. Instead, a selected reading list is given at the end of the book to provide the reader with some guidance into the more detailed aspects of the problem. Any omission of specific research studies or published works must be considered as inadvertent. This is most highly regretted.

Finally, the author wishes to acknowledge the very significant support and encouragement given by his wife, Florence, in this endeavour.

Raymond N. Yong
March 2000

Contents

CHAPTER 1

Contaminated Land

1.1 GROUND CONTAMINATION

The term *contaminated land* bears significant connotations in many jurisdictions and countries. In these areas, *contaminated land* is a special designation assigned to a land site where ground pollution has been detected. Furthermore, these pollutants are more than likely considered to be serious threats to the environment and human health. The characterization of the seriousness of the various threats posed by the contaminated land is not always easily performed. This is because agreement on the degree of risk and risk factors is not always obtained or uniformly established. To a very large extent, this is due to a lack of understanding or awareness of: (a) the nature and distribution of the pollutants in the contaminated ground, and (b) the nature, magnitude, and seriousness of the various threats posed by the pollutants.

To better appreciate the various environmental and health threat problems arising from the pollutants residing on the land surface and in the subsurface of contaminated lands, we need to consider the nature of the land environment. Contamination of the ground can lead to severe consequences. Considering *pollutants* as those contaminants deemed to be threats to human health and the environment, it is important for us to be aware of the fate of the pollutants in the soil strata underlying the ground (land) surface. For simplicity in representation, the underlying soil strata will be generally identified as the *substrate* or *substrate material*. Figure 1.1 shows a schematic view of the potential pathways to biotic receptors for which pollutants in a contaminated land site might travel. The degree of threat (risk) posed by pollutants travelling along these pathways, and the processes affecting the fate of the pollutants on these pathways, will be some of the many key factors that will determine the course of action required to minimize or eliminate the threat. Threat minimization or elimination requires consideration for removal of the pollutants, containment of the pollutants, reduction of toxicity of the pollutants, and pollution mitigation — amongst the many action choices available. One of the key factors is *risk management*, i.e., the management of the pollutant threat such that the threat is reduced to acceptable risk limits as prescribed by regulations and accepted practice.

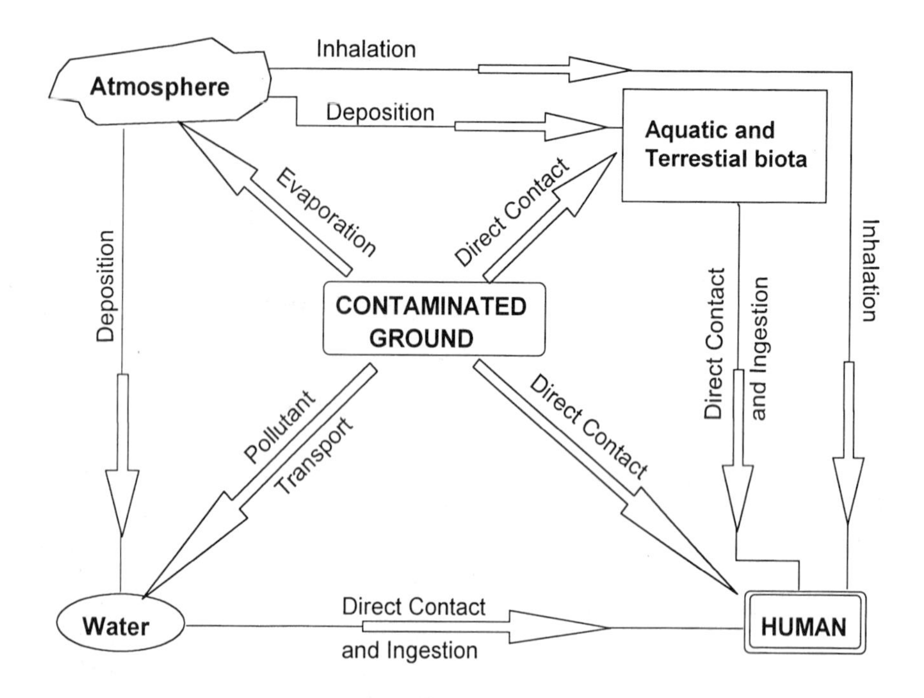

Figure 1.1 Pathways from contaminated ground to biotic receptors.

1.1.1 Elements of the Problem

The fundamental aim of the material presented in this book is to develop a better understanding of the various elements of the problem generally defined as *ground contamination*. In the diagram given as Figure 1.1, the impact of the contaminated ground is felt in many ways — as demonstrated in the diagram. What we require as basic knowledge is the nature and distribution of the pollutants in the contaminated ground. This is necessary if we are to determine whether these pollutants pose threats to the immediate environment and the various biotic species that live therein. The basic elements of the ground contamination problem are given in Figure 1.2. Some of the key pieces of information required include:

- Nature (species) of the various pollutants present in the substrate;
- Distribution and partitioning of the pollutants in the substrate;
- Potential for mobility or "change" in composition (transformation) and concentration of the pollutants;
- Role of the substrate material in respect to pollutant "bonding," distribution, transformation, and mobility — i.e., fate of pollutants;
- Toxicity of the pollutants;
- How and/or when the pollutants will become environmentally mobile; and
- Basic elements required to design and implement remediation of the contaminated ground.

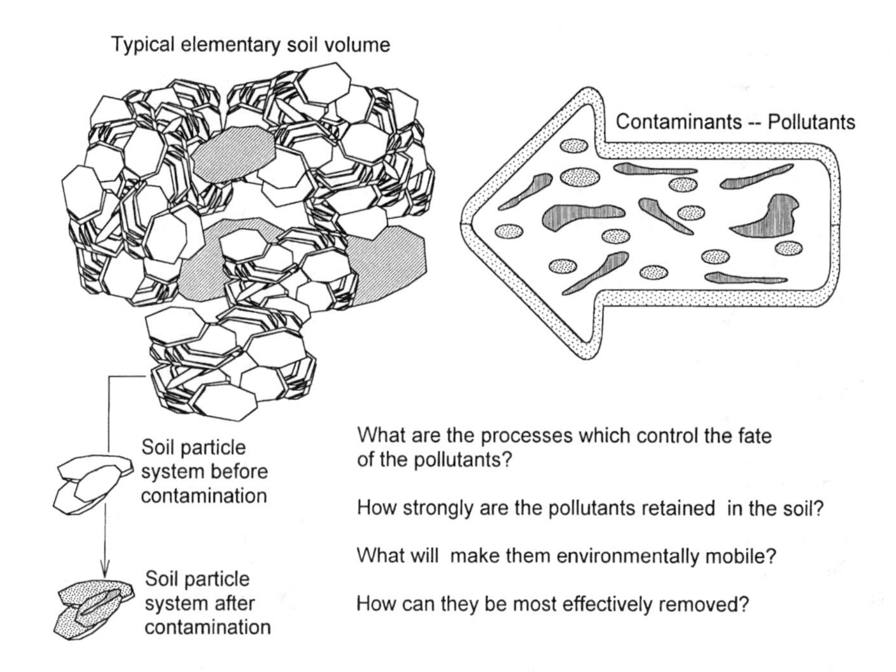

Figure 1.2 Pollutant-soil interaction problem.

The discussion material developed herein is designed to provide the basic elements which constitute the pollutant-soil system in the substrate. The fundamental question is: "What are the processes that control the persistence and fate of pollutants?" Why do we seek to learn about these processes? Because:

- This would provide us with knowledge of the durability of the bonding relationships between pollutants and soil solids, i.e., strength of bonds formed between the pollutants and the soil solids;
- Management of the contaminants (pollutants and non-pollutants) in the contaminated ground would be more effectively implemented; and
- Remediation (removal of pollutants in the contaminated ground) methods and technology and pollution mitigation can be properly developed and effectively implemented.

We assume that the principal constituent in the substrate is a soil-water system. Accordingly, the material and discussion items presented point toward the fundamental features, properties, and characteristics of pollutants and soil fractions, which determine the fate of pollutants in a soil. This type and level of knowledge is required if we are to develop the necessary procedures and tools for remediation of contaminated lands. The various items that define the degree of "toxicity" of a pollutant and associated issues are not within the scope of this book, and therefore will not be addressed. The reader is advised to consult the appropriate textbooks on toxicology and ecotoxicology to obtain the proper information on this subject.

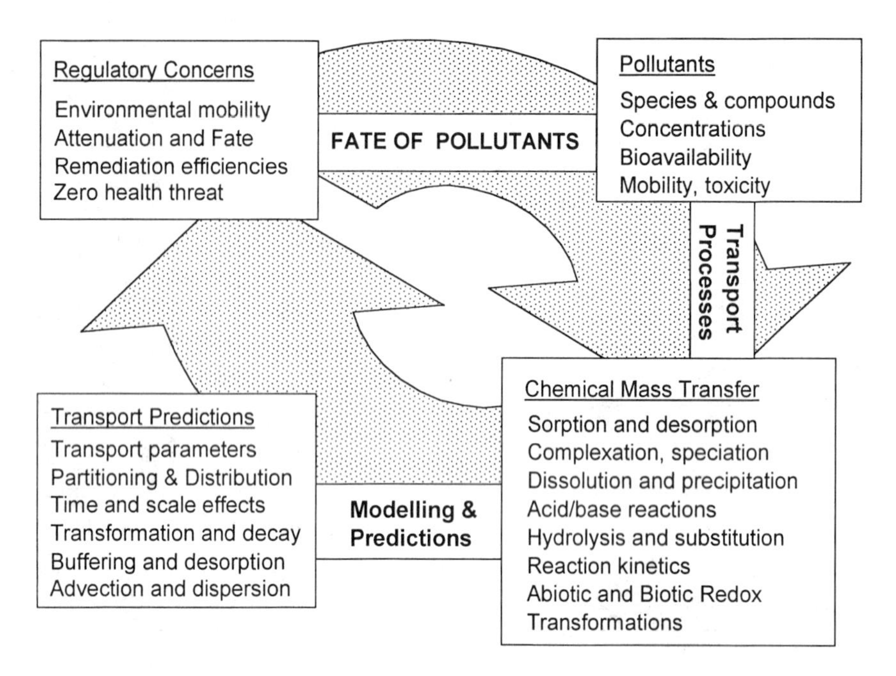

Figure 1.3 Illustration of the many factors and issues requiring attention in the interactions controlling the fate of pollutants.

The simple diagrammatic sketch shown in Figure 1.3 illustrates many of the issues that need to be considered in the assessment or evaluation of the fate of pollutants in a contaminated site. Most of the factors, properties, and parameters are considered in detail throughout this book. Whilst regulatory agencies (shown as "Regulatory Concerns" in the figure) are generally seen as the driving force behind the many sets of activities mounted to determine the fate of the pollutants, this is not a necessary requirement. The detailed listing of the various factors, properties, parameters, characteristics, etc. shown in Figure 1.3 is a "shopping list." A good working knowledge of many of the items in the shopping list would serve to provide a better understanding of the problems associated with contaminated ground — and the means whereby effective remediation techniques can be developed.

1.2 THE LAND ENVIRONMENT

In the context of geoenvironmental engineering practice, the term *land environment* is used to mean the physical landform and substrate, including the receiving waters contained therein. Four particular categories of land environmental problems are noted:

1. Problems or catastrophic disasters attributed to natural circumstances and events, such as earthquakes, floods, landslides, etc.
2. Problems associated with anthropogenic activities not directly related to waste production and management, e.g., construction activities, deforestation leading to

catastrophic erosion of slopes and decrease of watersheds, removal of thermal cover in permafrost regions leading to permafrost degradation, etc.

3. Problems or disasters arising as a result of anthropogenic activities associated with production of waste substances and waste containment, e.g., hazardous substance spills, leaking underground hazardous substance storage facilities, land mismanagement of hazardous wastes, pollution of streams and rivers, polluted sediments and sites, and other activities associated with exploitation of undeveloped land and resources and development of infrastructure.
4. Pollution of ground and receiving waters from non-point sources due to, for example, activities associated with agricultural and forestry practices such as the use of fertilizers and pesticides, or waste products from various livestock operations.

In this book, we are concerned with the problems directly associated with the last two categories, i.e., categories 3 and 4, with particular emphasis on the development of a better appreciation of the fate of pollutants, and the basic elements required for implementation of remediation technology. These are problems which arise directly from (and because of) anthropogenic activities, e.g., process streams and products, waste handling, storage, containment, discharge, etc. The particular instances or examples that immediately come to mind include discharge of waste streams into receiving waters, solid waste handling and disposal, land farming of organic wastes, lagoons for storage of sludges, underground storage tanks that may (or have) deteriorated, buried pipelines, and the whole host of historically polluted sites. The common factor to all these instances or examples is potential pollution of the *land environment*, resulting in threats to both human health and the environment.

The 1992 United Nations Conference on Environment and Development (UNCED) in Rio de Janeiro adopted 27 principles, listed as the *Rio Declaration on Environment and Development*. More than half of these principles deal directly with: (a) the need to establish and maintain a sustainable environmental resource base, and (b) the requirements to ensure protection of the environment. Principle 4 of the Declaration states, for example:

> In order to achieve sustainable development, environmental protection shall constitute an integral part of the development process and cannot be considered in isolation from it

and Principles 15 and 17 state:

> In order to protect the environment, the precautionary approach shall be widely applied by States according to their capabilities. Where there are threats of serious or irreversible damage, lack of full scientific certainty shall not be used as a reason for postponing cost-effective measures to prevent environmental degradation. Environmental impact assessment, as a national instrument, shall be undertaken for proposed activities that are likely to have a significant adverse impact on the environment and are subject to a decision of a competent national authority.

The principles cited above remind us of the need to continue seeking more information and knowledge concerning the impact of pollutants in the environment.

Agenda 21, the non-binding program of action for environmentally safe economic growth issued by UNCED, addresses various environmental protection programs and also the very difficult issues of toxic and hazardous wastes. Fundamental to the implementation and achievement of sustainable development are: (a) environmentally responsible land disposal and management of waste; (b) rehabilitation of contaminated ground; and (c) development of measures to ensure protection of the environment and its resources.

Some of the many activities that are required to ensure that the land environment is protected and that sustainable development can occur include:

- Construction of civil facilities that would ensure protection of the immediate land environment, e.g., preservation of surface cover, erosion control, frost heave, slope protection, levees, flood protection and control, etc.
- Design and construction of land disposal facilities for all kinds of waste products, including domestic, municipal, industrial, nuclear, agricultural, mining, etc.
- Management of land disposal facilities, including closure, monitoring, assessment of ongoing performance, maintenance, correction, etc.
- Site evaluation, selection, assessment, preparation, etc., including environmental audits and impact assessments for civil facilities and waste disposal facilities.
- Remediation (decontamination?), reclamation and rehabilitation of contaminated soils, sites, sediments, and underground facilities (underground storage facilities) including all affected substrate material (soil and rock), contaminated sediments, etc.
- Leachate management and groundwater, surface water, and watershed protection.
- Risk assessment and management with respect to waste handling and disposal, and also with respect to contaminated sites, remediation, and other activities associated with problems and catastrophic disasters in land environmental problems 1 and 2.

A very dramatic example of the need for ensuring proper environmental controls on management of waste and ground contamination can be deduced by studying the nature or elements of the basic problem underlying the development of many of the principles articulated in the *Rio Declaration*, e.g.,

1. **Population growth** — The global population in 1998 was estimated to be somewhat in excess of 5 billion. At the present rate of growth, by year 2050, conservative estimates give a global population ranging anywhere from 10 to 15 billion. At least 85% of the growth in global population will be in the developing countries.
2. **Depletion of productivity of agricultural lands** — Uncontrolled urban and industrial expansion, conversion of agricultural lands for other purposes, desertification, and loss of productivity all combine to reduce agricultural capability.
3. **Watershed management** — Urban and industrial expansion, poor land utilization and management, timber cutting, other forest and resource development activities, etc. have contributed to depletion of watersheds.
4. **Waste management practices** — The pressures of uncontrolled urbanization and industrial growth have contributed to minimal environmental waste management practices in many countries, increasing the overall threat to the maintenance of a sustainable environmental resource base.

The preceding issues pose challenges that can be identified as follows:

A. Our environmental resources are currently strained and already in default in many key areas to feed the present 5 billion population. It is acknowledged that we are in fact borrowing from future generations, and that if present practice is not changed, it is difficult to anticipate how one will be able to provide the various consumables for a two to three-fold increase in population within the next 50 years!
B. Waste generated by industries and consumers will continue to increase. Disposal, and the 4 R's (reduce, recover, recycle, and reuse) are by no means keeping pace with growth of waste. The methods of waste reduction, containment, and disposal have to be improved if environmental resources are to be conserved.
C. Increasing GNP and increasing population will require greater attention to products generated (waste and otherwise). Agricultural productivity and other environmental resources must be increased to meet growing demand. Water supply for many parts becomes very critical, even under today's needs and circumstances.
D. Of all the available water (global), approximately 95% is saline and unusable for drinking or other purposes except through desalination procedures. The remaining 5% of all available water is non-saline water, and is distributed as shown in Figure 1.4. We note that 0.2% of non-saline water is attributed to lakes and rivers, 31.4% is resident as snow and ice, and the remaining 68.4% appears as groundwater.

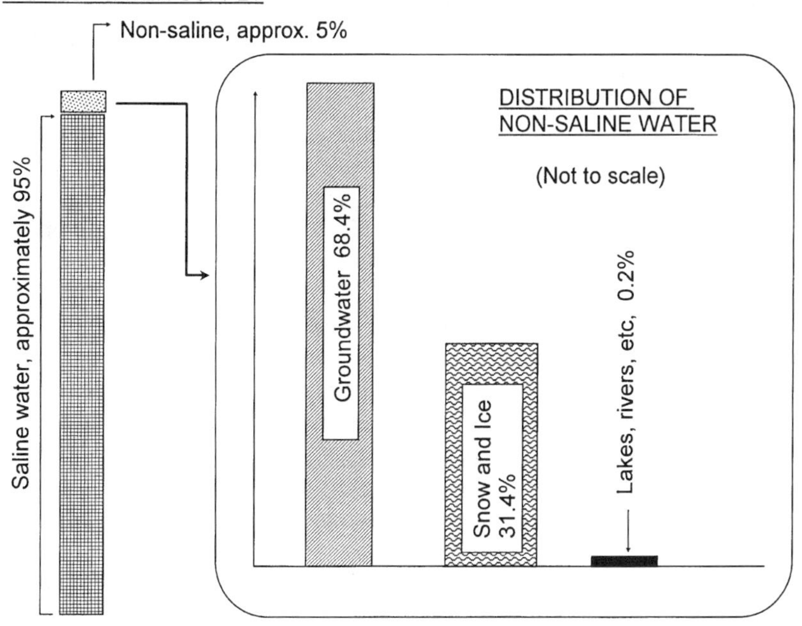

Figure 1.4 World water supply. Distribtion of non-saline water is shown in the right-hand box. (Data from Environment Canada fact sheet.)

Whereas there is an obvious need to ensure that the surface receiving waters do not become polluted from various sources of contamination, the need for protection of groundwater is not always obvious. This is because it is an unseen resource. Recognizing the large potential resource, and recognizing that all waste contaminants contained in or on the ground have the potential for migration downward into the aquifer, it is obvious that the preservation of groundwater quality becomes paramount. Because of threats to the environmental resources that are already being strained to meet present global population needs, it is now no longer acceptable for further development of societies, cities, industries, and infrastructure to be undertaken without environmental accountability, protection, and controls. This is particularly true if we are to provide responsible land management — for waste handling and natural resource management.

1.3 LAND ENVIRONMENT SENSITIVITY AND TOLERANCE

In this book, we are concerned with the various problems caused by contamination of the ground by pollutants that find their way onto and into the ground. By that we mean contamination of the soils, groundwater, and all other materials located on and under the ground surface. The term *pollutant* is used to indicate that the contaminant under discussion or investigation is deemed to be a potential threat to human health and the environment. The term *contaminant* is used in general considerations of ground contamination. In general, we mean the substrate underlying the ground surface when we refer to *ground* as a general view of the land environment. The sources of pollution have been discussed in general in the preceding section as arising from anthropogenic activities. Other sources of pollution include natural sources, e.g., arsenic poisoning of groundwater or aquifers as a result of arsenic release from source materials such as arsenopyrites in the substrate under oxidizing conditions. The more direct sources of pollution due to anthropogenic activities include: (a) byproducts of goods produced and services rendered; (b) inadvertent spills and deliberate dumping; (c) landfills; (d) underground storage tanks; (e) fertilizers and pesticides used in agricultural and forestry practices; and (f) management of animal wastes on farms. The last two sources are generally considered as non-point sources. The presence of pollutants in the substrate poses a threat not only to the immediate environment, but also to human health and other biotic species resident within the particular ecosystem. Developing the safeguards and technology for protection of public health and the environment requires an understanding of the pollutants in the contaminated ground, and also the various processes responsible for the fate of those pollutants.

For *environmentally safe pollutant management*, we must consider the nature of the health threat posed not only by the presence of the pollutants in the ground, but also the exposure route (see Figure 1.1), the acceptable daily intake (ADI) of the toxicant. In ground contamination, the particular health protection issues that are considered important — other than the nature and presence of the pollutants — are those which arise from pollutant transport in the substrate, the toxicity of the pollutant, and exposure routes (pathways).

The term *environmental impact*, which is often cited as a requirement in the assessment of performance or viability of many significant civil facilities, is now a commonplace term in the review of many kinds of activities associated with engineering activities and facilities. However, it is not unusual to encounter difficulties in establishing the details of environmental impacts for a particular activity or facility. This is because the extent of the environment that is impacted by the activity or facility is most often not easily established. This makes the application of good geoenvironmental engineering practice for control and management of the impact very difficult.

1.3.1 Environmental Impact Policy

Establishment of an environmental impact policy requires one to determine what constitutes an impact and the object, item, or activity, that is *impacted*. The U.S. National Environmental Policy Act (NEPA, 1969, PL91-190) provides a good starting point for assessing the problem of environmental impacts and consequences. This policy has been generally used as a guide by many countries and agencies in formulating their own sets of guidelines, procedures, and criteria. One observes considerable harmony between the statements issued in respect to the purposes of the Act and the Rio Declaration discussed in Section 1.1, viz:

> To declare a national policy which will encourage productive and enjoyable harmony between man and his environment; to promote efforts which will prevent or eliminate damage to the environment and biosphere and stimulate the health and welfare of man...

Amongst the six specific goals identified in Section 102 of the Act, goals 3, 4, and 6 are perhaps the most readily identifiable vis-a-vis the Rio Declaration and the underlying responsibilities that confront geoenvironmental engineering:

> ...(3) attain the widest range of beneficial uses of the environment without degradation, risk to health or safety, or other undesirable and unintended consequences; (4) preserve important historic, cultural, and natural aspects of our national heritage, and maintain, wherever possible, an environment which supports diversity and variety of individual choice; ...(6) enhance the quality of renewable resources and approach the maximum attainable recycling of depletable resources.

The many factors that need to be considered in implementing the policy and goals can be grouped into three categories as shown in Figure 1.5. These include *Environment*, *Ecology*, and *Aesthetics and Human Interests*. A listing of some of the more important factors is shown in the diagram. The *Environment* category includes many other items other than the ones listed, i.e., water, air, land, and noise. The choice and number of items to be listed will depend on the activity or project being scrutinized under the terms of the governing environmental policy, i.e., policy in force.

1.3.2 Environmental Inventory, Audit, Assessment, and Impact Statement

There are many terms used to describe the nature and outcome of work designed to describe the environment and the various impacts. In the simplest form, we

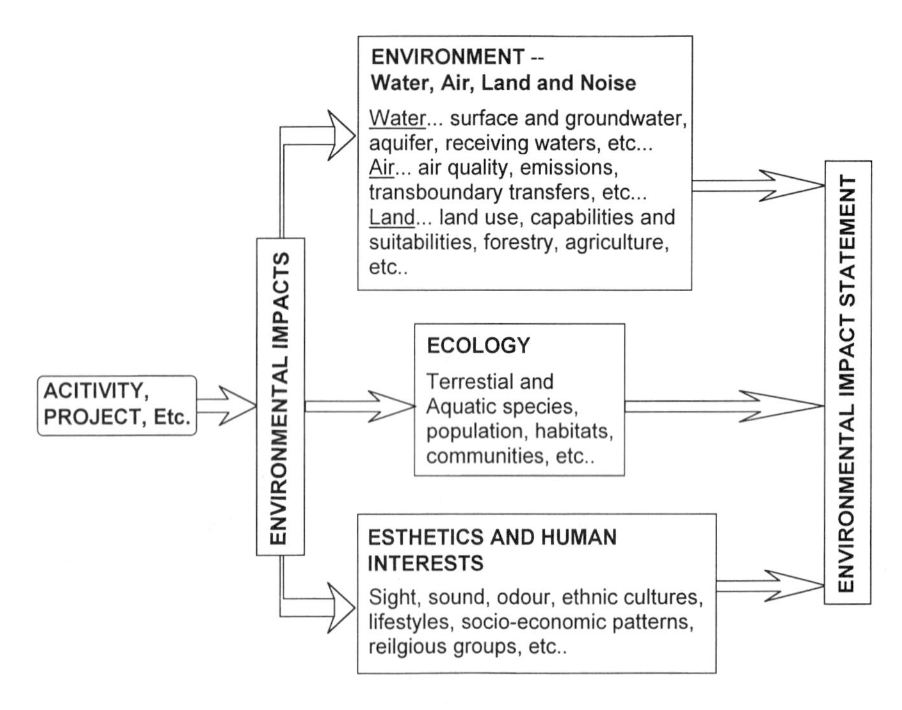

Figure 1.5 Environmental impacts and Impact Statement.

consider an *environmental inventory* to be the base-line descriptor of the state of the various constituents that constitute the environment within the region of interest. It is important to note that the region of interest not only encompasses the specific area where the activity or project is located, but also the surrounding areas. The various items given in Figure 1.5 and the sub-items (not listed in the figure) supporting the listed items form the basis for a "checklist" upon which the environmental inventory is built. In essence, the checklist consists of descriptors for the basic physical-chemical, ecological-biological and cultural-socioeconomic environments. Alternatively, the checklist descriptors can be classified as natural and man-made environments. Under this scheme, the *natural environment* will include the physical-chemical and ecological-biological descriptor environments, whilst the *man-made environment* includes the cultural-socioeconomic environments. For ease in communication, we can define these as the *descriptor environments*. The purpose of an environmental inventory is to establish or define these descriptor environments as they exist, prior to implementation of the proposed activity or project. In that sense, environmental inventory is essentially an information/data gathering process which is designed to describe the existing state of the various items identified within the specified region. No judgment is made concerning the merits (or otherwise) of the items described. One of the key features of the environmental inventory is its incorporation in the environmental impact statement. By this procedure, the adverse or beneficial impacts from the proposed activity or project can be rationally evaluated.

Environmental audit concerns itself with the determination of compliance with existing environmental laws and regulations. It is a systematic exercise which is generally initiated in response to an audit requirement for a specific activity, project, or charge. In a sense, the environmental audit is conducted to determine if there are any transgressions, and/or if the various environmental issues and items meet all the environmental, zoning, health, safety, and city (region) requirements. A good example of this is the determination of site compliance with existing environmental policies/laws, zoning statutes/requirements, and health and safety standards/requirements. Most often, an environmental inventory is used as the starting point for the audit. Environmental impact statements are necessary tools for the audit procedure.

Environmental Impact Assessment (EIA) is to a very large extent one of the most difficult of the environmental examination processes to implement since this requires considerable foresight in spotting potential problems. Conducting an environmental assessment of a particular project to be constructed/implemented or a specific contemplated set of activities requires one to predict and/or anticipate the changes that would occur to various inventory items due to those external activities. Direct impact due to external activities is only one of the routes for environmental impact. Indirect impact or secondary routes (for impact) need to be considered. A good example is the "triggering" of landslides due to initial slope erosion of surface cover. The erosion of surface cover results from, for example, deforestation. This activity (removal of tree and surface cover) alters the surface hydrological pattern, which in turn will produce surface erosional effects. These effects can provide the triggering tools for landslides in many sensitive areas.

It is necessary to determine not only the order of importance of the environmental inventory items being impacted, but also the magnitude of the adverse (or beneficial) effects, and to assign an order of importance or significance to these effects. Health impacts of projects associated with anthropogenic activities need to be considered in the decision-making process. The proposition contained in WHO (1987) concerning environmental health impact assessment (EHIA) should be considered seriously in environmental impact assessments. Because of the potential for highly subjective forecasting of adverse, beneficial, or even neutral effects, prior experience and actual case study records are used wherever possible. Mathematical/computer modelling is used to aid the assessment process. Proponents of expert systems consider environmental assessment to be a most suitable application of this method of scrutiny. Because so many issues need to be considered in assessing environmental factors and changes, and because different projects/activities are not necessarily similar in circumstances, conditions, requirements, sites, outcome, etc., there is no total and comprehensive set of checklists that can be issued to cover all the concerns.

Environmental Impact Statement (EIS) is the name given to the document that is written in response to specific charges, guidelines, mandates, etc. issued by a specific regulatory agency for a particular contemplated project or set of activities. This statement is meant to summarize the outcome of the EIA conducted for the contemplated project or set of activities, and includes the *environmental inventory* as the base-line state. Adverse environmental impacts that cannot be avoided, minimized, or totally mitigated need to be included in the EIS, together with alternatives as proposed actions that would counter the impacts. In some jurisdictions, there is a requirement to define irreversible and reversible commitment of resources, and also the impact on sustainable resources. The chart given in Figure 1.5 demonstrates the set of events necessary for the production of the EIS. Most often, environmental audit is not part of the EIS, although some will argue that an environmental audit should also be included in the EIS. Environmental impact, therefore, is seen to be a very elusive term, conditioned by the circumstances dictated by specific projects and activities.

1.4 LAND SUITABILITY AND USE

The significant factors that contribute to the status of the land environment from the perspective of ground contamination concerns are shown in Figure 1.6. *Land use*, i.e., the manner in which a land is utilized, is dictated by several factors and forces, not the least of which is the capability of the land to respond to the requirements associated with land utilization. By that, we mean: (a) the particular land usage (i.e., prevailing land use) will not degrade the quality of the land environment, and (b) environmental sustainability and protection of land-use capability are maintained.

Insofar as the land environment is concerned, environmental impact associated with anthropogenic activities can be in the form of changes in the quantity or quality

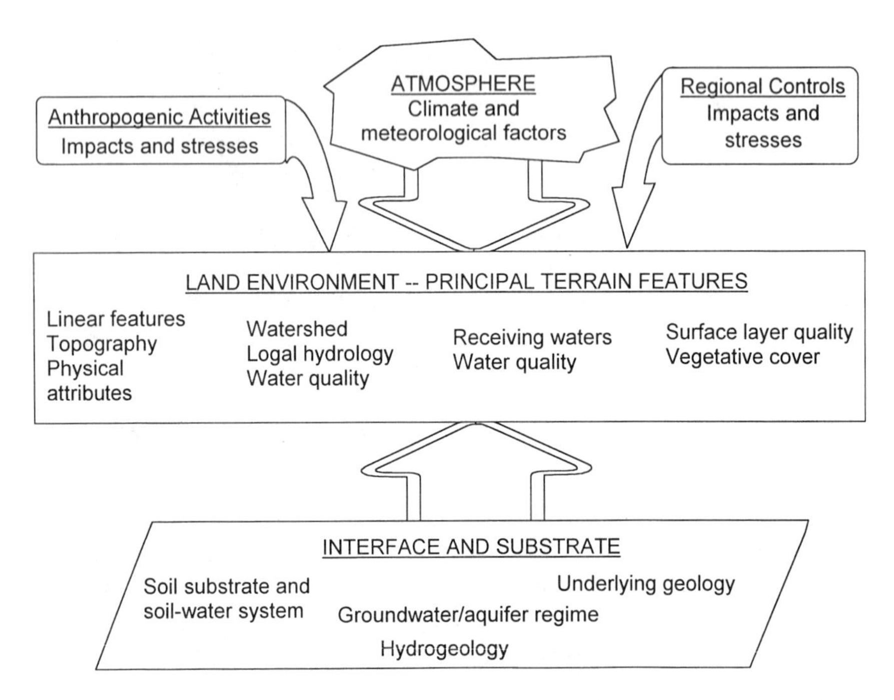

Figure 1.6 Features and factors considered in ground contamination, in the context of a land environment.

of the various features that constitute the land environment. It is fair to say that not all anthropogenic activities will result in adverse impacts on the environment. In the case of soil as a resource material for agricultural purposes, for example, we can identify both beneficial and adverse effects in the following summary list:

- **Beneficial changes and/or effects** — Use of mineral fertilizers to increase fertility; creation of crumb structure and alteration of soil moisture to improve irrigation and drainage; use of organic manure; pH manipulation; and addition of new soil as a means of rejuvenating the soil.
- **Adverse changes and/or effects** — Use of herbicides; over-removal of nutrients; compaction; alteration of soil microclimate; soil pollution.

The degree of environmental impact due to pollutants in a contaminated ground site is dependent on: (a) the nature and distribution of the pollutants; (b) the various physical, geological, and environmental features of the site; and (c) existent land use. Other than the natural setting that has not been exposed to any anthropogenic activities, the various types of land use range from natural forested regions and simple grazing land at the one end, to recreational use and urban land use at the other. Each type of land use imposes different demands and requirements from the land. The ideal situation in land utilization matches land suitability with land development consistent with environmental sensitivity and sustainability requirements. In the first order characterization for land status and quality given in Figure 1.7, we are interested in determining: (a) the many participating factors that contribute to

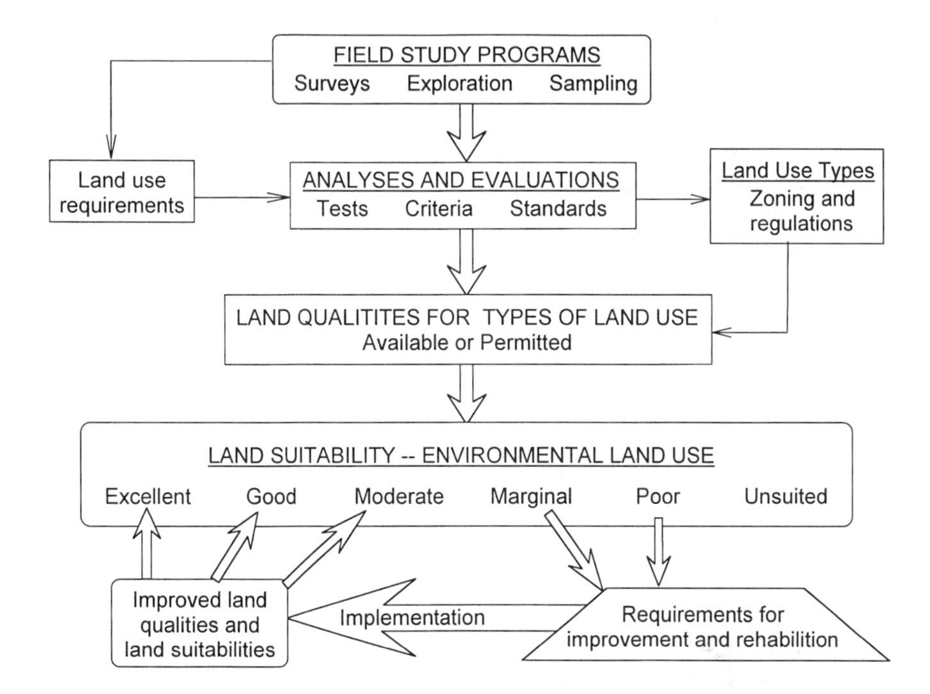

Figure 1.7 Land quality and land suitability.

the land environment of interest, and (b) the requirements for improvement and rehabilitation to increase land suitability. The latter procedure applies not only to improvement of marginal lands, but also to rehabilitation of contaminated ground.

A general 4-step procedure is used to address environmental impacts as they pertain to ground quality (i.e., soil quality). This ranges from Step 1, which identifies the impacts of the proposed activity on soil and land quality, to Step 4, which requires that the mitigation measures be identified and detailed. We define *soil quality* herein to mean the physical, chemical, and biological well-being of the soil. The 4-step procedure is thus given as follows:

- **Step 1** — Identify impacts to soil and land quality from the planned project and activities associated with the project. The types of impacts need to be detailed, e.g., physical, chemical, biological, etc. A simple example of a physical impact would be an activity that results in the loss of ground cover on a slope. As will be seen in Step 3, this could result in changes in surface hydrology and could also result in erosion of the slope.
- **Step 2** — Obtain the data base which describes the soil and land quality in the pre-project stage. Obtain and/or define the pertinent standards and criteria which protect (govern) the pre-project soil/land quality. Unless otherwise specified, the general assumption is that the quality of the soil/land (i.e., land quality and use) must be returned to the pre-project state. There are occasions, however, when regulatory bodies might decide that the pre-project state might need to be improved. This, however, is an issue of land planning or land management, and is outside the scope of this book.
- **Step 3** — Assess the impact on soil/land quality due to the planned project and its associated activities, and the significance of the impact. If we follow the example given in Step 1, it is not difficult to envisage that adverse physical impact of removal of groundcover on a slope could lead to erosion of the slope. If we had chosen an example of chemical impact in Step 1, e.g., leachate discharge in a holding pond or landfill, we would identify pollution of the soil/land by organic chemicals or heavy metals. Furthermore, we would need to specify the significance of such pollutants in the ground, e.g., potential threats to the environment and human health.
- **Step 4** — Specify or identify the measures needed to mitigate the adverse impacts. Preventative and corrective measures, actions, technology, etc., should be identified. Where possible, preventative measures should be the favoured sets of action.

1.4.1 Groundwater

Groundwater is an integral part of land use considerations. Groundwater availability and quality are principal factors in characterization of groundwater resource as part of land use studies. Causes and sources of groundwater contamination include wastewater discharges, injection wells, leachates from landfills and surface stockpiles, open dumps and illegal dumping, underground storage tanks, pipelines, irrigation practices, production wells, use of pesticides and herbicides, urban runoff, mining activities, etc. The partial listing of causes and sources shows that, by and large, the most likely sources of groundwater contamination are from anthropogenic activities. The same 4-step procedure used previously in assessing and addressing the environmental impact on soil is used in addressing the problem of impacts on groundwater quality.

Exploitation of groundwater as a resource requires continuous attention to the quality of water withdrawn from the aquifers. Classification of aquifers in respect to potential use is a means for ensuring the maintenance of the quality of the aquifer — i.e., the quality of the water. We can classify aquifers by the degree of protection required to maintain the quality of the aquifer, or by their use. In the former scheme, a semi-regulatory 3-level approach is utilized. Classifying a *Class I aquifer* as the sole-source drinking water requires the maximum protection. Criteria and regulation have to be established to protect the aquifer. A *Class II aquifer* would require moderate protection and the appropriate criteria and regulations written to provide the necessary protection. We can consider a *Class III aquifer* as a limited use aquifer because of the relatively poor quality of the product, thus requiring minimum protection. By and large, protection requirements revolve around curtailment of anthropogenic activities in the region which will impact the aquifers.

Another way of classifying aquifers is through use of the abstracted water. We can establish at least 6 different categories of aquifers as follows: a *Category 1 aquifer* provides the sole source of drinking water, and is of the highest quality. Presumably, this category accords with the Class I aquifer described above. A *Category 2 aquifer* is where the abstracted water can only be used as drinking water after some minimal treatment. When more extensive treatment is required, this will classify the aquifer as a *Category 3 aquifer*. Aquifers classified as *Categories 4*, *5* and *6* provide abstracted water suitable for agricultural use (*Category 4*), industrial use (*Category 5*), and for mining and energy development (*Category 6*).

1.5 WASTES AND WASTE STREAMS

The discharge of wastes (waste materials and waste streams) into the land environment means that pollutants will be introduced into the land environment, resulting thereby in land pollution. The terms *contaminants* and *land pollution* can have several meanings, depending upon the perspective of the reader/observer, and upon the context and application of the terms. The nature and extent of the threat posed by the pollutants will not only depend upon the nature and distribution of the pollutants, but also on the target that is threatened. This ranges from biotic receptors at the one end of the spectrum to the physical land environment (physical features and natural environmental resources) at the other end. The reader should consult the many specialized textbooks and learned articles that deal with these subjects. The consideration and treatment of these factors and issues are beyond the scope of this book.

Not all contaminants in a contaminated ground are threats to either the environment or human health. A simple example of a good waste product might be putrescibles, which when successfully composted will function as an organic fertilizer. In contrast, a bad waste product can be categorized as harmful, e.g., hazardous and/or toxic. The determination of whether a waste material or product is harmful to public health and/or the environment is the purview of various other disciplines more competent to deal with issues of public and environmental healths and which specialize in such concerns, e.g., environmental scientists, public health scientists, epidemiologists, biologists, botanists, zoologists, animal scientists, marine scientists, toxicologists, ecotoxicologists, phytotoxicologists, etc.

What constitutes a waste material is not always clear. Defining what constitutes a waste material becomes particularly tricky when we attempt to distinguish amongst the different kinds of wastes. Some of the methods used to define the waste materials include:

1. The medium to which they are released, i.e., air, water, or land;
2. Their physical characteristics, i.e., whether they are gaseous, liquid, or solid;
3. Types of risk or problems that they create;
4. Types of hazard that they pose, i.e., ignitable, corrosive, reactive, or toxic; and
5. Their origin, e.g., mine tailings, municipal waste, or industrial waste.

The methods of classification given as categories 1 and 2 can provide ready estimates of wastes generated. They do not, however, provide information on the various processes (i.e., origins) that produce the waste products or waste streams. These methods provide the basic information for management and control of the waste discharge and final resting place for the waste. Information in regard to the source of the waste is important if one is concerned with the development and implementation of technology designed to obtain reduction and recycle of waste materials at source.

The wastes included in categories 3 and 4 are of direct concern to regulatory agencies because of the dangers they pose to the public. Because of the concerns, these wastes are subject to regulatory control, and are generally known as *regulated wastes*. General minimal requirements in regulating waste handling and discharge include an integrated tracking system, which tracks the waste from generating source to final disposal. Decisions in respect to what constitutes regulated wastes are obviously critical issues since strict management of such wastes is required. Many countries and jurisdictions have generated lists of substances and pollutants that have been judged to pose threats to human health. These lists have several names, e.g., priority substances, dangerous goods, etc.

Contaminants contained in waste materials generated by activities associated with industry, agriculture, mining, cities, forestry, etc. contain both pollutants and non-pollutants, i.e., substances that are by themselves pollutants and substances that are non-pollutants (e.g., putrescibles). An example of what constitutes a waste material can be seen in the definition given in the U.S. Resource Conservation Recovery Act (1976) [RCRA 1976]. Section 1004(27) defines a "solid waste" as:

> ...any GARBAGE, REFUSE, SLUDGE from a waste treatment plant, water supply treatment plant or air pollution control facility, and other discarded material including solid, liquid, semi-solid, or contained gaseous materials resulting from industrial, commercial, mining and agricultural activities, and from community activities, but does not include solid or dissolved material in domestic sewage or irrigation return flows...

1.5.1 Characterization of Hazardous and Toxic Wastes

The example previously given in Figure 1.1 showed the various forms and manner in which land disposal of waste materials can impact public health. To ensure the safety of the public, regulatory agencies are charged with the responsibility for enacting and enforcing regulations designed to eliminate or minimize the threats. This procedure requires the development of considerable information regarding the nature of the threats and the health- and environment-threatening effects. Thus, whereas these agencies must have sufficient toxicological, phytotoxicological, and epidemiological information relative to the impact of these toxic substances and their respective pathways to the various biogenic receptors, there is never sufficient information to permit one to render the necessary sets of judgments with any degree of certainty. The problem is not restricted to waste materials but also with respect to various kinds of substances assessed for health hazard threats.

The terms *hazardous* and *toxic* used in characterizing the threat from pollutants in waste materials are quite often wrongly used or used interchangeably. Why do we need to distinguish between hazardous and toxic wastes? Characterizing wastes as *hazardous* or *toxic* is necessary if regulations designed to control the management of such waste materials are to be developed and enforced. There is common agreement on the need for regulatory control and harmful waste materials amongst all countries and agencies. However, because of how waste materials are viewed by different regulatory agencies in many different countries, an apparent wide discrepancy exists on the types of wastes and substances contained in the waste materials that fall within these categories. The discrepancy or disagreement is not in regard to the spirit of the classification scheme.

By and large, hazardous and toxic wastes are considered to be regulated wastes, i.e., they are subject to stringent regulatory controls. These controls can cover all the aspects of waste generation from collection, handling, recycle, and reuse, to storage and final disposal. From the viewpoint of the waste generator, the process in obtaining regulatory approval in management of such wastes, from cradle to grave, can be very tedious and expensive. It is therefore to the benefit of the potential waste generator to seek ways to minimize the generation of hazardous or toxic waste. The incentive to detoxify and neutralize such waste materials is considerable, and hence programs designed to reduce waste production, or even to use alternative technology in production of goods, are very attractive propositions. In contrast to the problems attendant with the control of regulated waste, waste materials that are not regulated (i.e., non-regulated waste) do not face the same set of stringent regulatory controls vis-a-vis management and disposal, and are obviously less expensive to manage.

Most regulatory agencies agree on the broad requirements for classification of hazardous wastes. The RCRA (Sec. 1004) definition of a "Hazardous Waste" states that it is

> ...a solid waste or combination of solid wastes, which because of its quantity, concentration, or physical, chemical or infectious characteristics may: (a) cause, or significantly contribute to an increase in mortality or an increase in serious irreversible, or incapacitating reversible, illness, or (b) pose a substantial present or potential

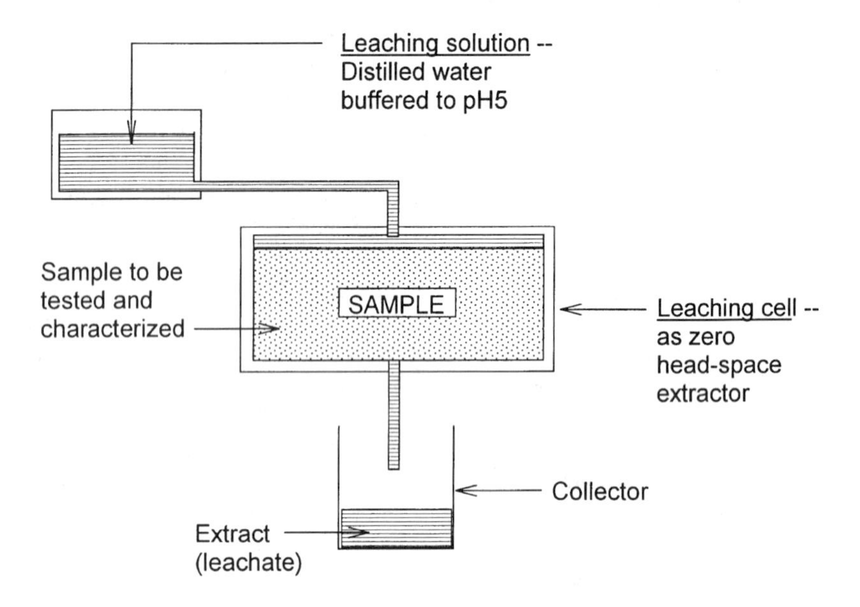

Figure 1.8 Typical leaching test for chemical assessment of leachate.

hazard to human health or the environment when improperly treated, stored, transported, or disposed of, or otherwise managed.

Wastes and/or waste materials are considered hazardous if they exhibit any of the following characteristics:

- **Ignitability** — potential for fire hazard during storage, transport, or disposal;
- **Corrosivity** — potential for corrosion of materials in contact with candidate waste, resulting in environmental and health threats;
- **Reactivity** — potential for adverse chemical reactions;
- **Toxicity** — as identified from the Toxicity Characteristics Leaching Procedure (TCLP). The procedure for the TCLP is given in Figure 1.8 and the criteria for toxicity classification are given in Table 1.1.

In the TCLP, because of the concern over the loss of volatile organics in the extract (leachate), the extraction is performed with a zero head-space extractor to ensure that such a loss does not occur, and the list of compounds that must be evaluated is shown in Table 1.1. In the final analysis, the basic philosophy in classifying a substance as toxic or non-toxic, for most classification schemes, is seen to be determined on the basis of analyses of the extract obtained from leaching tests, and the concentrations of the detected contaminants in the extract. The choice of the target contaminants and their permissible concentrations may vary amongst countries.

Table 1.1 TCLP Compounds and Regulatory Levels in Extract

Compound	Level (mg/L)	Compound	Level (mg/L)
Arsenic	5.0	Hexachloro-1,3-butadiene	0.5
Barium	100.0	Hexachloroethane	3.0
Benzene	0.5	Lead	5.0
Cadmium	1.0	Lindane	0.4
Carbon tetrachloride	0.5	Mercury	0.2
Chlordane	0.03	Methoxychlor	10.0
Chlorobenzene	100.0	Methyl ethyl ketone	200.0
Chloroform	6.0	Nitrobenzene	2.0
Chromium	5.0	Pentachlorophenol	100.0
o-Cresol	200.0	Pyridine	5.0
m-Cresol	200.0	Selenium	1.0
p-Cresol	200.0	Silver	5.0
1,4-Dichlorobenzene	7.5	Tetchloroethylene	0.7
1,2-Dichloroethane	0.5	Toxaphene	0.5
1,1-Dichloroethylene	0.7	Trichloroethylene	0.5
2,4-Dinitrotoluene	0.13	2,4,5-Trichlorophenol	400.0
Endrin	0.2	2,4,6-Trichlorophenol	2.0
Heptachlor	0.008	2,4,5-TP (Silvex)	1.0
Hexachlorobenzene	0.13	Vinyl chloride	0.2

Based on the general guidelines for classification of hazardous materials, we can see that we can have a hazardous waste that may not be toxic, but is hazardous because it is corrosive, and/or flammable, and/or reactive. On the other hand, a toxic substance will always be classified as a hazardous material because it is one of the determining characteristics of a hazardous waste. A general definition of a toxic material follows: *...A toxic waste (substance) is one which has the ability to cause serious injury or death to biotic receptors (humans, animals, etc.).*

1.5.2 Land Disposal of Non-hazardous and Hazardous Wastes

The term *waste disposal* generally refers to the discharge of waste forms into the atmosphere, receiving waters, and land. Land disposal of waste is by far the most common form of waste disposal practised by almost all countries. This refers to all kinds of waste materials, from municipal to industrial wastes (non-hazardous and hazardous wastes). In some countries, transformational operations necessary for recovery, reuse, and recycling are considered integral to the disposal of wastes. In other countries, the transformational aspects may be covered as separate issues that need to be implemented in conjunction with waste disposal.

Land disposal of hazardous waste subsequent to all the necessary procedures undertaken to neutralize and detoxify the waste requires the physical containment of the waste material in secure impoundment systems. These are generally waste containment systems constructed as waste landfills or landfarming facilities designed to prevent or control waste leachate contamination of ground (i.e., ground pollution). Design and construction requirements for the containment facilities must undergo considerable public and regulatory scrutiny to ensure that they meet environmental impact requirements, and that these are constructed in accordance with design

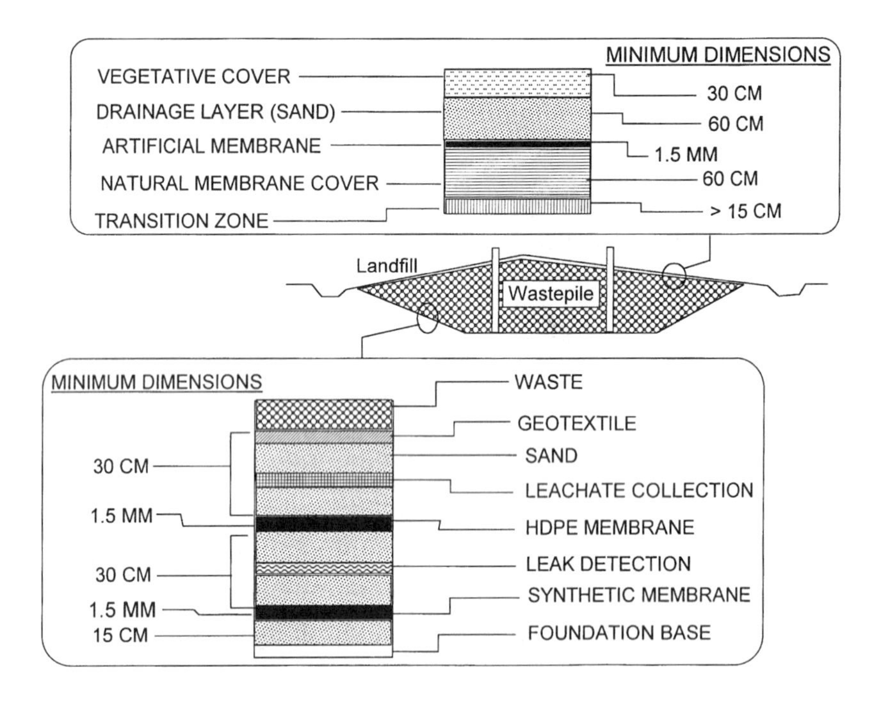

Figure 1.9 General top and bottom liner systems for hazardous waste landfill.

specifications or expected performance criteria. A typical landfill containment using engineered liner systems is shown in Figure 1.9. There are obviously many types of liner and barrier systems that can be used to provide safe containment of the wastes and the leachates generated. The design and construction of these systems constitute the subject material of many textbooks dedicated to these subjects, and is not within the scope of this book.

The importance of available (published) criteria, guidelines, etc. as targets for control of the design, management, and performance of the system cannot be overstated. As a rule, these criteria, standards, etc. are issued by various regulatory agencies and professional bodies, except that in the case of problems associated with the environment it is generally the governmental regulatory agencies that are responsible for the issuance of the guidelines, standards, etc., inasmuch as the protection of public health from environmental threats is a government responsibility. Interpretation of the requirements for protection of public health and the environment can result in differing sets of design criteria for impoundment systems. Thus for example, the suggested design for the bottom barrier/liner system for containment of municipal solid wastes shown in Figure 1.10 is seen to be somewhat different for various countries. Reconciliation between accepted practice and regulatory requirements is an absolute requirement.

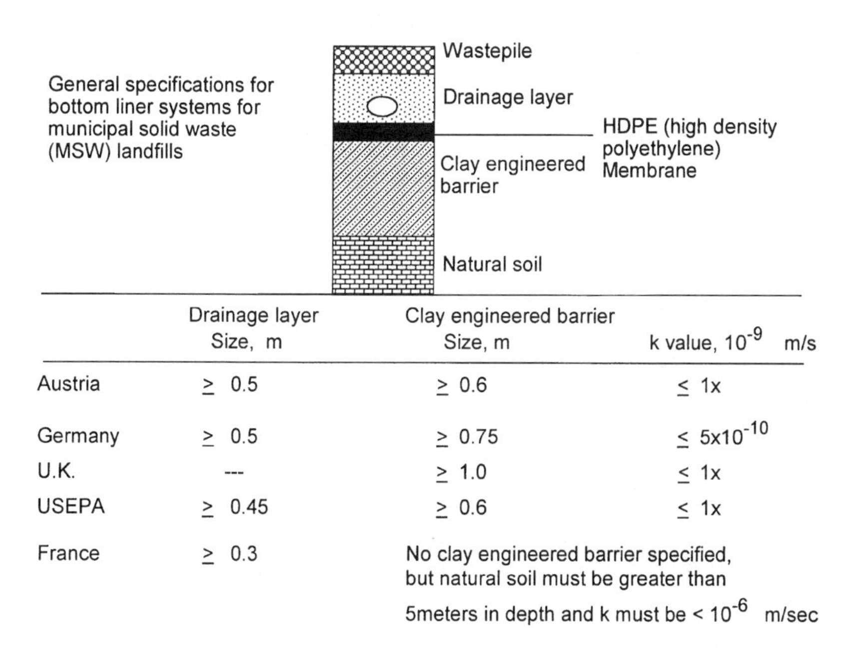

	Drainage layer Size, m	Clay engineered barrier Size, m	k value, 10^{-9} m/s
Austria	$\geq$ 0.5	$\geq$ 0.6	$\leq$ 1x
Germany	$\geq$ 0.5	$\geq$ 0.75	$\leq$ 5×10^{-10}
U.K.	---	$\geq$ 1.0	$\leq$ 1x
USEPA	$\geq$ 0.45	$\geq$ 0.6	$\leq$ 1x
France	$\geq$ 0.3	No clay engineered barrier specified, but natural soil must be greater than 5meters in depth and k must be < 10^{-6} m/sec	

Figure 1.10 Suggested minimum dimensions and properties for bottom liner system for municipal solid waste landfills. (Adapted from Manassero et al., 1997).

The particular instances of environmental and health threats posed by the disposal of waste in the ground, and other problems associated with landfarming, illicit dumping, underground storage tanks, etc., are clear examples of the need for environmental regulations designed to deal specifically with waste containment and management for the protection of public health and the environment. Environmentally safe land management of waste means that the potential health threats posed by the waste material in its present and future form, and all other products issuing therefrom (e.g., leachates), must be below limiting concentrations of toxicants. Since water is the primary carrier for contaminants in the substrate, the barrier system which separates the waste material and/or waste leachates in a landfill from the natural substrate material and groundwater constitutes the technical element that can be controlled through the use of technical design/construction specifications and through regulatory requirements and specifications. In the case of ground contamination occurring from spills and leaking underground storage tanks, we do not normally expect constructed liner-barrier systems to be present, at which time the ground substrate must perform as a contaminant barrier system if the contaminants (plume) are to be controlled and groundwater protection is to be achieved. To achieve maximum protection of public health and the environment, control of potential contamination is exercised through regulations governing illicit dumping and storage tank specifications.

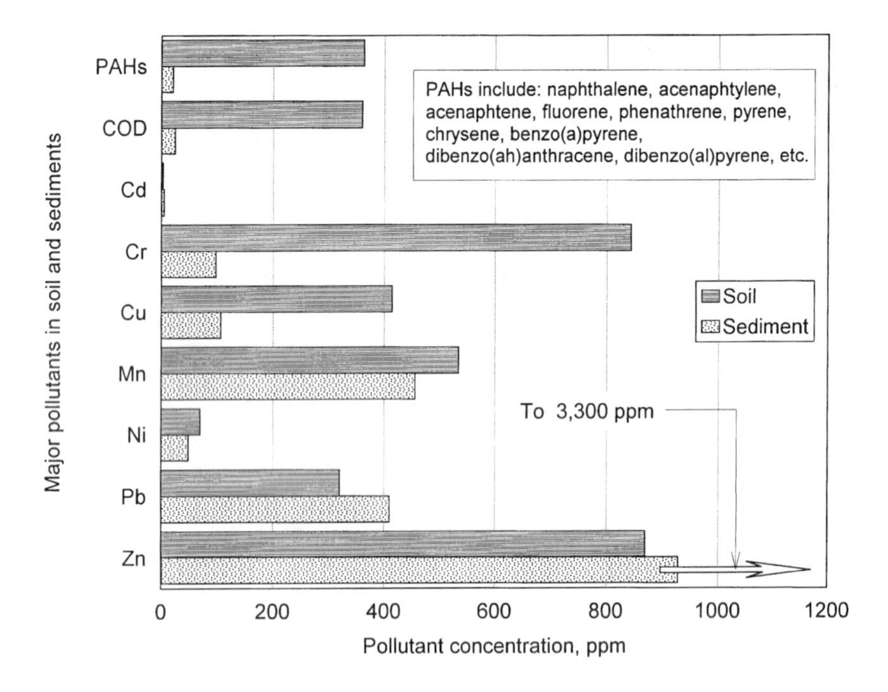

Figure 1.11 Concentrations of some major pollutants found in the soil samples taken from a contaminated site, and in the sediments in the nearby harbour. (Data from Mulligan, 1998).

1.6 CONCLUDING REMARKS

Contamination of the ground element of the land environment (i.e., ground contamination) via processes that include leaching of stored and landfilled waste products, inadvertent spills, and illicit dumping is not an uncommon phenomenon. The countless numbers of historically contaminated sites exist not only because of the processes identified above, but to a very large extent because our awareness and knowledge of environmental concerns and sustainability have not been as acute as one would have liked them to be. For example, using data reported by Mulligan (1998), Figure 1.11 shows some of the major pollutants in a hydrocarbon and metal contaminated site near the harbour of a major city. Also shown in Figure 1.11 is the distribution of the same pollutants in the sediment in the harbour. Many of the concentrations of the pollutants shown in the figure are in excess of permissible limits of some jurisdictions.

Regardless of the origin of the contaminants and pollutants in the contaminated ground, the existence of such a site requires evaluation of the potential threat to human health and the environment. A simple procedure in the evaluation of such sites is given in Figure 1.12. The various pieces of supporting information needed for the many stages of assessment of threat to the environment include not only the

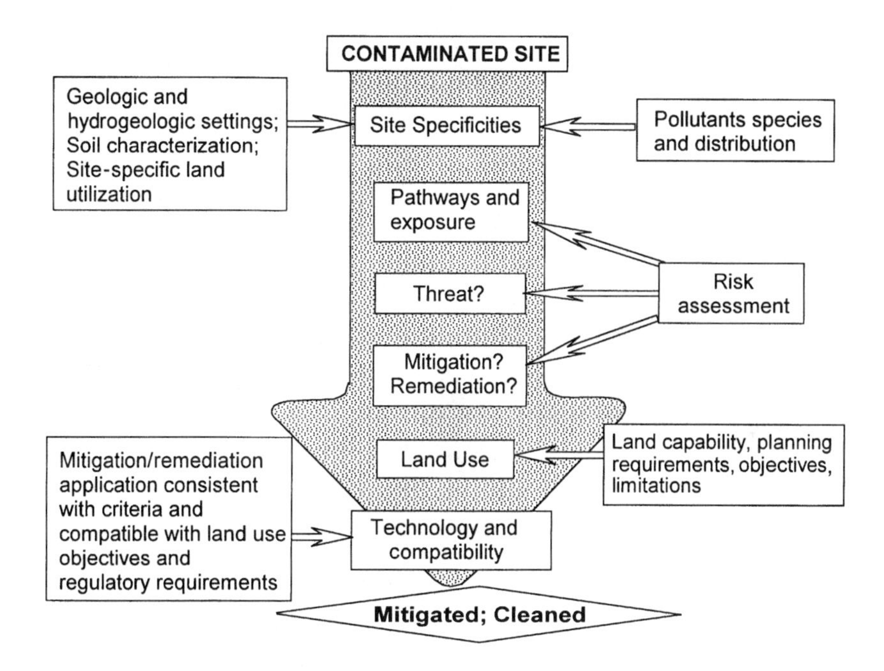

Figure 1.12 Simple protocol for rehabilitation of a contaminated site.

site specificities, but also the direct and indirect pathways. The toxic substances (i.e., *toxicants)* in the contaminated site are health threats because of the resultant effects on biotic receptors when exposure to these toxicants occur. Inorganic contaminants commonly required for identification in assessment of water quality are sometimes used as the checklist for identification of inorganic pollutants. These include: Al, Sb, As, Asbestos, Ba, Be, Cd, Cl, Cr, Co, Cu, Cn, Fl, I, Fe, Pb, Mg, Mn, Hg, Mo, Nitrite, Nitrate, P, K, Se, Ag, Na, Sulphate, Ti, V, and Zn.

The health threats posed by organic chemicals are also difficult to fully assess insofar as exposure effects are concerned. We need to take into account not only the exposure level, but also the duration of exposure. Thus, one classifies a substance as acutely toxic if it produces a lethal or sub-lethal effect within a short time frame, whereas a chronic toxic substance is viewed as requiring a relatively longer period of time to manifest itself as a health threat. The classification or ranking of toxic substances with respect to level of toxicity differs between different countries, not only with respect to terminology used, but also with respect to criteria used to distinguish between the various levels of toxicity. The types of wastes generated, and the nature of the contaminants found in waste streams are so varied and complex that it is not often possible to predict the exact composition of the waste material that will be discharged into the environment or found as pollutants in the substrate. However, we have the ability to determine and understand many of the basic elements of waste materials and their product leachates, and can develop an understanding of how they interact with soil substrate material through research and further study.

CHAPTER 2

Nature of Soils

2.1 SOIL MATERIALS IN THE LAND ENVIRONMENT

The soil materials of interest (and concern) in the study of the pollutant fate in contamination of the land environment are the soil substrate and the sediments formed at the bottom of receiving waters (lakes, rivers, etc.). We have defined *pollutants* (Section 1.5) as those contaminants judged to be threats to the environment and public health, and will continue to use the term in this sense. Pollutants are toxicants. We will continue to use the term *contaminants* in much of the material contained in this chapter since this is a general term which includes pollutants in the general grouping of contaminants. The term pollutant will be used to highlight the specific concern under discussion. *Contaminated land* is used to refer to a land area that contains contaminants (including pollutants). In this chapter, we will be interested in those properties and characteristics of the soil materials that provide the significant sets of reactions and interactions between these soil materials and contaminants. It is these reactions and interactions that control the fate of pollutants. Furthermore, it is these same reactions and interactions we must address if we are to structure successful and effective remediation programs to clean up the contaminated ground. We should also be interested in the performance of these soil materials when they are used as contaminant attenuating barriers to the transport of contaminants.

Whilst our primary interest is focused on the buffering and attenuation capabilities of the soil material since they control the transport and fate of the pollutants, we will need to make mention of problems of contaminant presence in the soil on its short- and long-term mechanical stability. This recognizes one of the prime areas of concern in the use of soil materials as contaminant containment barriers — the degradation of the physical (mechanical) and chemical properties of the material when it is subjected to all the forces developed from chemical interactions. The results of creep tests reported by Yong et al. (1985) where a natural clay soil under creep loading was subjected to leaching by 0.025 N $Na_2SiO_3 \cdot 9H_2O$ after 12,615 minutes of leaching are shown in Figure 2.1. The axial creep strain of the

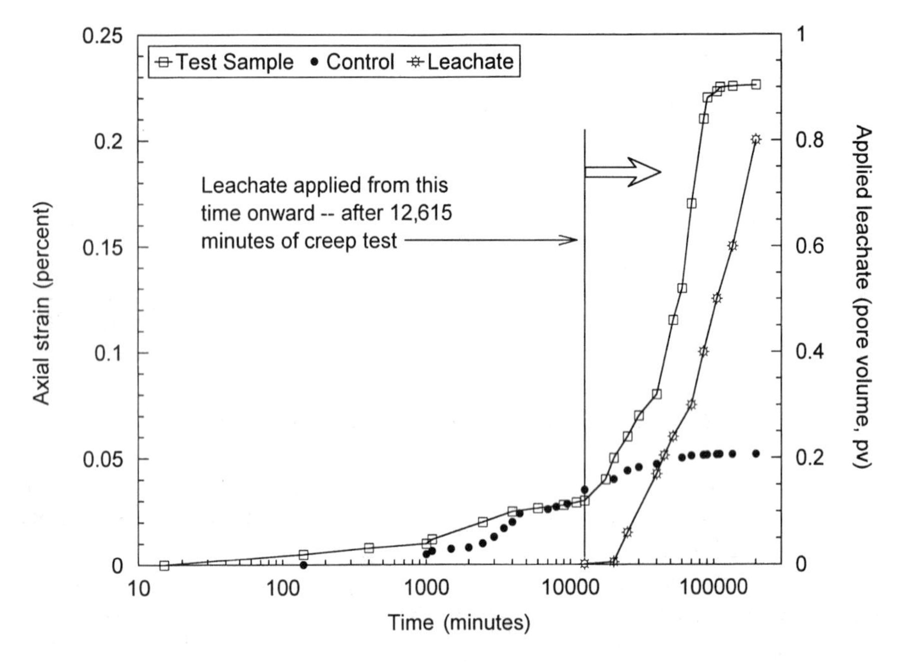

Figure 2.1 Effect of pore fluid chemistry replacement on creep of a natural clay sample. Sample leached with 0.025 *N* $Na_2SiO_3 \cdot 9H_2O$. (Adapted from Yong et al., 1985.)

control unleached sample is shown as black dots in the figure, and the amount of leachate introduced during the leaching process is given in terms of the pore volume, pv. The *pore volume parameter* is the ratio of the volume of influent leachate (leachant) divided by the pore volume of the sample. This is a dimensionless quantity, and is commonly used in leaching tests as a parameter that describes the volume of influent leachate because it permits one to view test data on a normalized basis. There are both good and bad aspects to this method of data viewing. The good aspect lies in the ability to compare leaching performance with different soils and different leachants. The bad aspects are mostly concerned with the inability to fully appreciate the time required to reach the breakthrough point. A solution to this problem is to use both kinds of data expression, pore volumes and direct time-leaching expressions, such as those used for the results of leaching and creep tests shown in Figure 2.1.

The creep test results shown in Figure 2.1 indicate that introduction of the leachate dramatically increases the magnitude of the creep (strain). The total creep strain is almost five times the strain of the control (unleached) sample. Higher applied creep loads will show higher creep strains and greater differences in creep strain due to leaching effects. The changes in the mechanical properties due to the interactions developed between the leachant and the soil fractions can be studied using techniques that seek to determine the energy characteristic of the soil (see Section 3.6).

In the case of sediments, we can consider the primary sediment material to be composed of soil material obtained from erosion processes (from land surfaces)

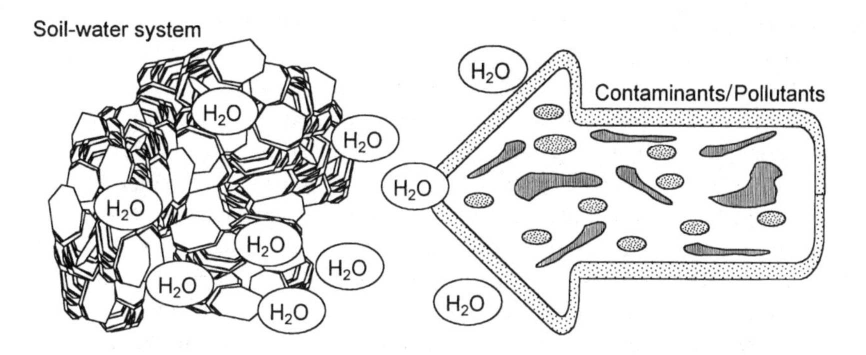

1. What are the interactions between soil fractions and water molecules?
2. How do they affect the interactions between soil fractions and contaminants?
3. What are the interactions between soil fractions and contaminants/pollutants?
4. How do they result in sorption and bonding between contaminants/pollutants and soil fractions?
5. How do these interactions control persistence and fate of the pollutants?
6. How do these relate to the "natural attenuation capability" of the soils?

Figure 2.2 Interactions amongst soil fractions, water molecules, and contaminants/pollutants.

deposited in various ways, e.g., erosion of embankments, runoffs, air particulates settling onto water bodies, clay and silt loads transported in streams and rivers, etc. The principal feature involves water, either as a carrier or as a medium within which sedimentation of all of the soil particulates occurs to form the sediment bed.

The basic interest in soil materials and contaminants is in respect to the attenuation processes resulting from the interactions and reactions between these soil materials and the contaminants. These processes result in the accumulation of the contaminants and are directly related to the *surface properties* of the soil solids. By that, we mean the properties of surfaces of those soil solids that interact directly with the contaminants. We need to understand how the interactions between contaminants and soil fractions (i.e., the various types of soil solids) result in sorption or partitioning of the contaminants by the soil fractions. This is illustrated in the simple sketch in Figure 2.2 which shows interactions between: (a) water and the soil fraction; (b) contaminants and the soil fractions; (c) contaminants and water; and (d) interactions amongst all three. The basic questions posed in Figure 2.2 follow directly from the questions posed previously in Figure 1.2. These seek to determine why and how sorption of contaminants by the soil solids (i.e., removal of contaminants from the aqueous phase of the soil-water system onto the soil solids) occur. In particular, the questions address the central issue of the relationships between soil properties and contaminants which are pertinent to the sorption or partitioning processes.

Because the bonding between contaminants and soil solids is established at the interacting surfaces of both contaminants and soil solids, i.e., interface, we need to know what specific characteristics of the surfaces are involved to establish bonding between the various kinds of contaminants and the soil solids' surfaces. These will characterize the *contaminant holding capability* of the soil (i.e., the capability of the soil fractions to sorb contaminants). The more detailed considerations of contaminant-soil interaction are given in Chapters 4, 5, and 6, when the transport, fate, and persistence of pollutants in the substrate are examined. At that time we will be interested in the basic details that define the bonds established in relation to the properties of the surfaces of both pollutants and soil solids. We would also be interested to determine the control or influence of the immediate environmental factors, such as temperature, pH, and Eh on the fate of the contaminants.

The question "Why do we need to know about contaminant bonding to soil solids?" can be addressed by citing three very simple tasks: (a) assessment of the "storage" capacity (for pollutants); (b) determination of the potential for "mobilization" or release of sorbed pollutants from the contaminated ground into the immediate surroundings; and (c) development of a strategy for removal of the sorbed pollutants from the soil fractions and from the contaminated site that would be most effective (i.e., compatible with the manner in which the pollutants are held within the substrate system).

2.1.1 Pollutant Retention and/or Retardation by Subsurface Soil Material

One of the more significant problems to be encountered in assessment of the potential for pollutant plume migration is the sorption and chemical buffering capacity of the soil substrate. The example of a waste landfill shown in Figure 2.3 illustrates the problem. A soil-engineered barrier has been used, in the example shown, to prevent waste leachate from penetrating the supporting substrate material. In most instances, prudent engineered soil-barrier design requires consideration of potential leachate breakthrough and formation of a pollutant plume. The resultant pollutant plume and its transport through the soil substrate must be examined to determine whether it poses a threat to the aquifer and to the immediate surroundings. One of the key factors in this process of examination is the *natural attenuation capability* of the soil substrate and/or the *managed attenuation capability* of the engineered barrier system. In the context of pollutant transport in soils, the term *natural attenuation capability* is used to refer to those properties of a soil which would provide for "dilution of the pollutants in the pollutant plume by natural soil-contaminant (soil-pollutant) accumulation processes." Similarly, the term *managed attenuation capability* refers to those properties of an engineered soil system that serve to accumulate the contaminants. This means that a reduction in the concentration of pollutants in the pollutant plume occurs because of pollutant transport processes in the soil.

It is often impossible to discriminate between the amounts of diluted concentration of pollutants obtained between *attenuation-dilution* and *water content-dilution* processes (Figure 2.4). However, the importance in being able to distinguish between the two pollutant-dilution processes is evident. In the attenuation-dilution process,

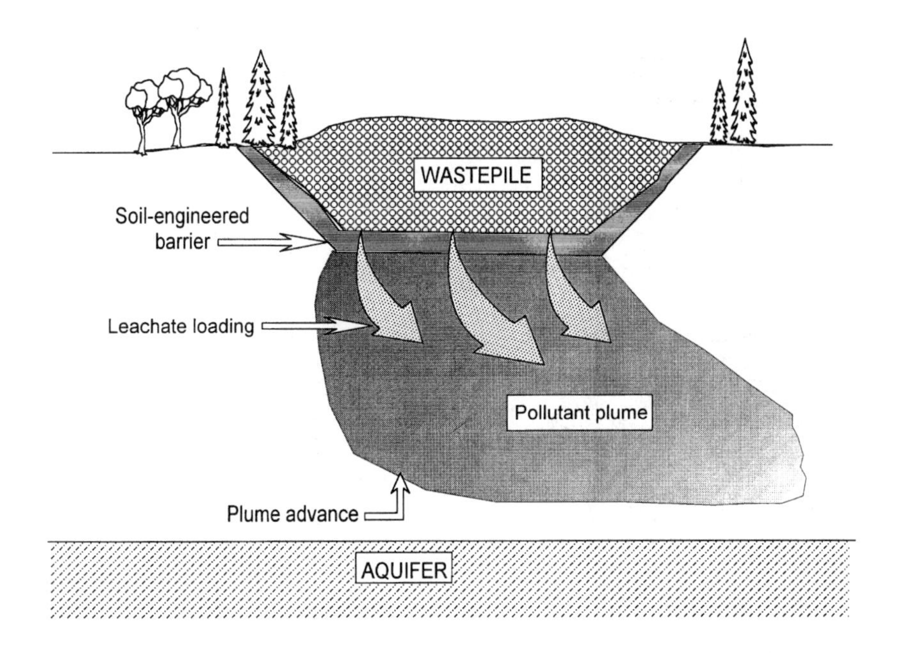

Figure 2.3 Pollutant plume and natural attenuation capability of soil substrate.

we are asking the substrate soil material to retain the pollutants within the soil medium — thereby reducing the concentration of pollutants in the pollutant plume as it continues to propagate in the soil. In the water content-dilution process, the pollutant concentrations in the pollutant plume are diluted (reduced) simply through the addition of water. Additional water contents in soil materials can quite often lead to unwelcome changes in the mechanical and physical properties of the soil.

In natural attenuation processes, both *retention* and *retardation* occur as mechanisms of pollutant accumulation and pollutant dilution in the soil system. In the former (retention), we expect the pollutants to be more or less permanently (irreversibly) held by the soil system so that no future re-release of these contaminants will occur. This means to say that irreversible sorption of pollutants by the soil fractions occurs. In the latter (retardation), we are in effect delaying the transmission of the full load of pollutants. The process is essentially one which will, in time, transmit the total pollutants in contaminant loading. The distinction between the two is shown in Figure 2.5. The various processes involved will be discussed in further detail in Chapters 5, 6, and 7.

2.2 SOIL MATERIALS

Soils are derived from the weathering of rocks, and are either transported by various agents (e.g., glacial activity, wind, water, anthropogenic activity, etc.) to new locations, or remain in place as weathered soil material. The inorganic part of the

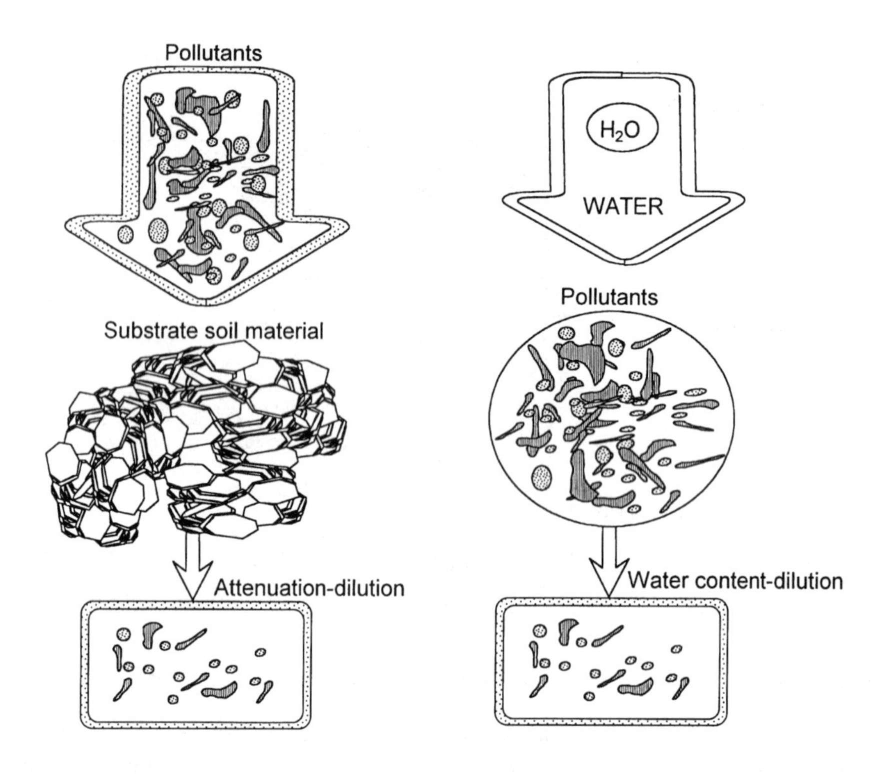

Figure 2.4 Attenuation-dilution and water content-dilution of pollutants in the substrate.

soil consists of primary and secondary minerals. These most often can be conveniently grouped into the more familiar soil and geotechnical engineering particle-size classification of gravels, sands, silts, and clays. Because the size-classification schemes pay attention only to particle size, the term *clay* used in the size-classification scheme to designate a class of soil fractions can be misleading. It is not uncommon to find references in the literature referring to clay as that size fraction of soils with particles of less than 2 microns effective diameter. Whilst this categorization of clay in relation to particle size may be popularly accepted in many instances, it can be highly misleading when we need to refer to clay as a mineral.

In this book, we should use the term *clay-sized* to indicate a particle size distinction in the characterization of the soil material. Since we need to pay attention to the surface characteristics of the soil fractions, particle size distinction does not provide us with sufficient information concerning the manner in which the fractions will interact with water and contaminants. Clay as a soil material consists of clay-sized particles (sometimes referred to as *clay particles* or *clay soil*) and *clay minerals*, with the latter being composed largely of alumina silicates which can range from highly crystalline to amorphous. Insofar as considerations of soil contamination are concerned, the surface properties of interest of the soil materials are the clay minerals, amorphous materials, soil organic materials, the various oxides, and the carbonates.

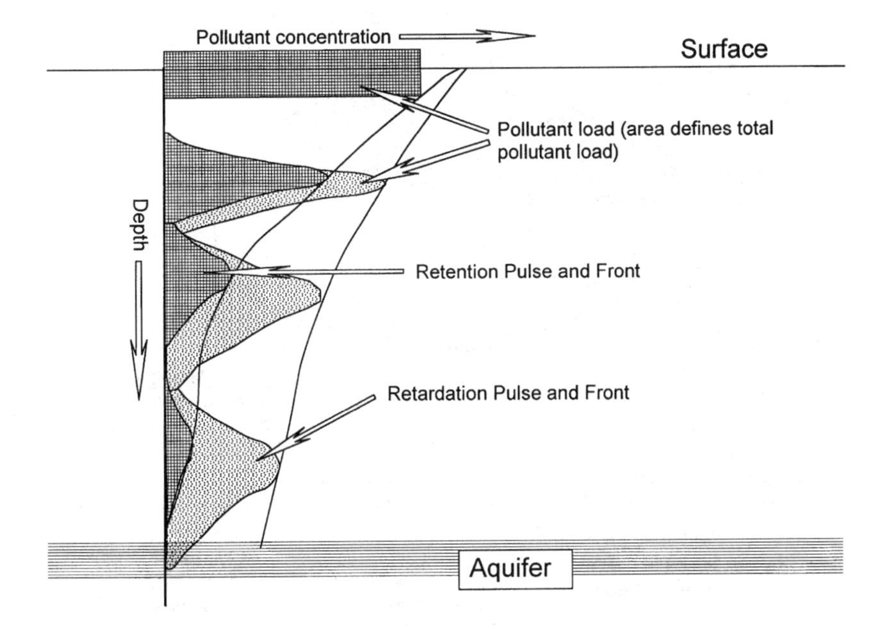

Figure 2.5 Retention and retardation pulses of pollutant load.

Strictly speaking, clay should refer to clay minerals, which are the result of chemical weathering of rocks and usually not present as large particles. Clay minerals are alumino-silicates, i.e., oxides of aluminum and silicon with smaller amounts of metal ions substituted within the crystal. Where a distinction between the two uses of the term clay is not obvious from the context, the terms clay size and clay mineral should be used.

Most clay minerals are weakly crystalline; the crystal size is smaller and there is more substitution, e.g., of H^+ for K^+, than in primary minerals. Amorphous alumina silicates are common weathering products of volcanic ash, or of crystalline material under intense leaching. On the other hand, the organic component of soils ranges from relatively unaltered plant tissues to highly humified material that is stable in soils and may be several thousands years old. This humus fraction is bonded to mineral soil surfaces to form the material that determines surface soil characteristics.

Surface soils are formed by alteration of inorganic and organic parent materials. The characteristic differences between soils and rocks that are important in the transport, persistence, and fate of contaminants include:

- Higher content of active organic constituents;
- Higher surface area and larger electric charge;
- More active biological and biochemical processes;
- Greater porosity and hence more rapid fluxes of materials; and
- More frequent changes in water content, i.e., wetting and drying. These differences are larger the closer one gets to the soil-atmosphere surface.

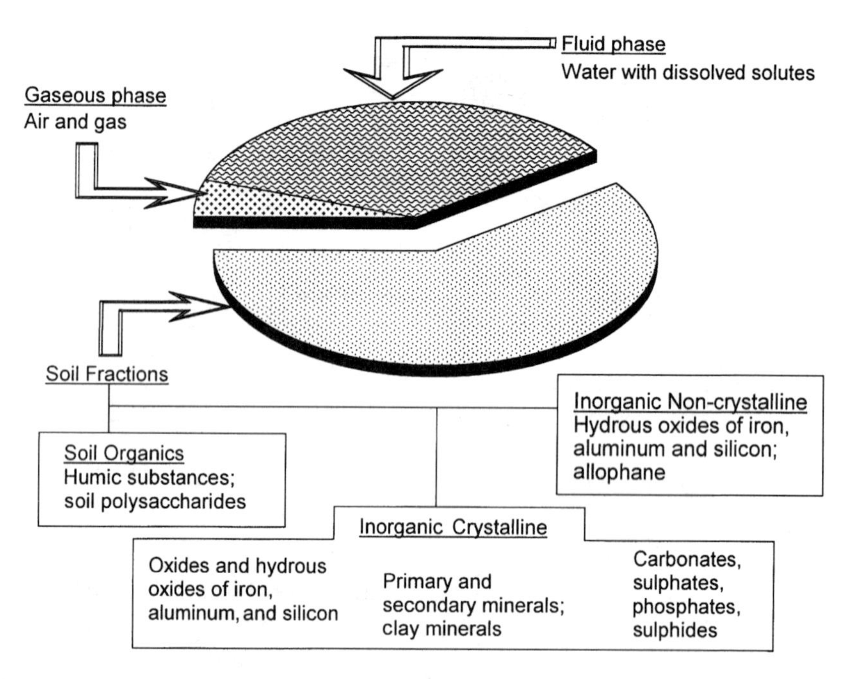

Figure 2.6 Soil fractions in substrate soil material.

To be more precise, one should consider the various soil components in a given soil mass to include the three separate phases: fluid, solid, and gaseous. Within each of these phases are also various components, as shown in Figure 2.6. The *soil fractions* in soil material consist of at least two broad categories as shown in the figure, i.e., soil organics and inorganics. The inorganic solids consist of crystalline and non-crystalline material. We will be concerned with the fluid phase and the various soil fractions in the assessment of the transport and fate of contaminants. The inorganic non-crystalline material can take the form of minerals as well as quasi-crystalline and non-crystalline materials. Soil-organic components primarily include the partly decomposed humic substances and soil polysaccharides.

Insofar as contaminant interaction and attenuation processes are concerned, the inorganic clay-sized fraction, the amorphous materials, the oxides/hydrous oxides, and the usually small yet significant soil-organic content play the most important roles. It is the surface features and the characteristics and properties of the surfaces of the soil fractions that are important in interactions with contaminants. Since many of the bonding relationships between contaminants and the soil surfaces involve *sorption forces*, it is easy to see that the greater the availability of soil sorption forces, the greater is the ability of the soil to retain contaminants. This is accomplished by having *sorption sites* (i.e., sites where the sorption forces reside) and a large number of such sites, generally having a large specific surface area. For a more detailed treatment of soil surface properties and soil behaviour, the reader should consult the specialized texts dealing with this subject, e.g., Yong and Warkentin (1975), Yong et al. (1993), Sposito (1984), and Greenland and Hayes (1985). For this chapter, we are concerned with those physical properties of soil that are important in controlling pollutant transport. The description of the surface properties with direct impact on the interactions between the soil fractions and contaminants will be discussed in Chapter 3.

2.3 SOIL FRACTIONS

It is important to understand that the nature of the surfaces of the soil fractions controls the kinds of reactions established, as mentioned previously. The soil fractions considered here include the clay minerals, amorphous materials, various oxides, and soil organics. Together, these constitute the major solid components of a soil — other than the primary minerals such as quartz, feldspar, micas, amphiboles, etc. These primary minerals are those minerals that are derived in unaltered form from their parent rocks through physical weathering processes, and compose the major portions of sand and silt fractions in soils.

In this chapter, we will be primarily concerned with the physical characteristics and properties of the soil fractions insofar as they relate directly to the various aspects of soil-contaminant interaction. Other considerations pertaining to soil mechanical properties and behaviour are better treated in specialized textbooks dealing with soil properties and behaviour (e.g., Yong and Warkentin, 1975) and with the many books on soil mechanics and geotechnical engineering. The surface and chemical properties of the soil fractions will be considered in detail in Chapter 3 when we discuss the interaction between soil fractions and water, i.e., soil-water relations.

2.3.1 Clay Minerals

Clay minerals are generally considered to fall in the class of secondary minerals (Figure 2.6) and are derived as altered products of physical, chemical, and/or biological weathering processes. Because of their very small particle size, they exhibit large specific surface areas. They are primarily layer silicates (phyllosilicates) and constitute the major portion of the clay-sized fraction of soils. We can group the various layer silicates into six mineral-structure groups based on the basic crystal structural units forming the elemental unit layer, the stacking of the unit layers and the nature of the occupants in the interlayers, i.e., layers separating the unit layers. The basic crystal structural units forming the tetrahedral and octahedral sheets are shown in Figure 2.7. The formation of the unit layers from the basic unit cells and sheets, together with the stacking of these sheets into unit layers is shown in Figure 2.8. The example shown in the figure depicts the arrangement for a typical kaolinite particle. The terminology of *sheet* and *layer* used in this book tries to be consistent with the development of the unit structures shown in Figures 2.7 and 2.8. Depending on the source of information, the literature will sometimes use these terms interchangeably.

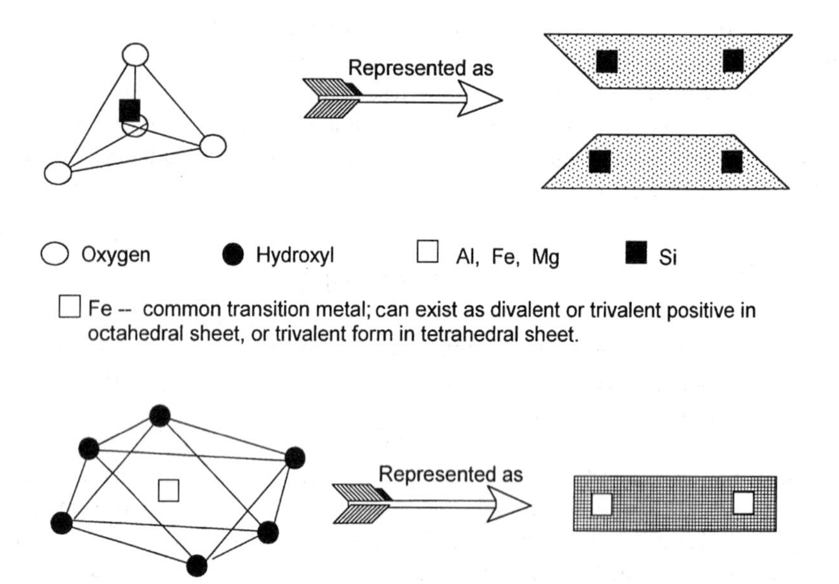

Figure 2.7 Tetrahedral and octahedral structures as basic building blocks for clay minerals.

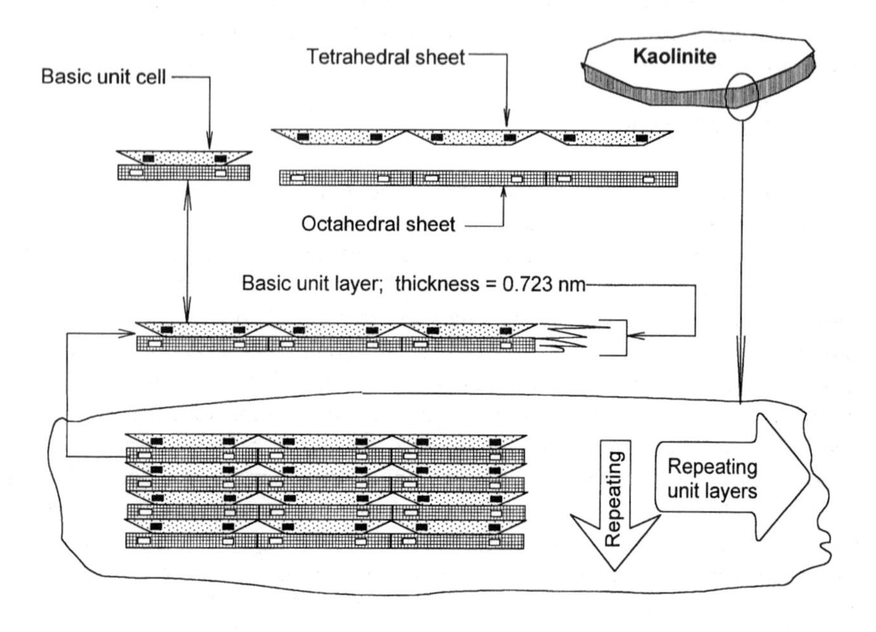

Figure 2.8 Basic unit cell and unit sheets forming the unit layer of kaolinite mineral.

The kaolinites are in the group known as the kaolinite-serpentine mineral-structure group. The typical structure is composed of uncharged 1:1 sheets (tetrahedral and octahedral) forming the basic unit layer. Repeating layers are separated by 0.713 nm. This separation spacing is the thickness of each layer, and is often referred to as the *repeat spacing*. Kaolinite is the only principal group of clay minerals that has a 1:1 sheet structure, i.e., one tetrahedral sheet and one octahedral sheet as seen in Figure 2.8. All the other groups are basically 2:1 sheet arrangements (two octahedral sheets and one tetrahedral sheet), with differences based upon the charged or uncharged nature of the layers and occupancy of the interlayers. The schematic representation of the kaolinite mineral shown in Figure 2.8 indicates that the basic unit cell consists of a stacking of a tetrahedron on top of the octahedral unit. In general, the tetrahedral positions are occupied by Si ions as shown in Figure 2.7, and two thirds of the octahedral positions are occupied by Al ions. The octahedral sheet with Al ions filling two thirds of the available positions is known as the *gibbsite structure*, with chemical formula $Al_2(OH)_6$. When magnesium (Mg) is in the octahedral sheet, all the positions are filled because of the need to balance the structure, and the chemical formula is $Mg_3(OH)_6$.

The total structural unit (tetrahedral unit cell on top of the octahedral) is generally called a *triclinic unit cell*, and has a thickness of 0.713 nm. It is sometimes argued that the structure composed of the lateral combination of these triclinic unit cells (tetrahedral sheet on top of the octahedral sheet) is quite often disordered, and that the term *kaolinite* should be used only for fully ordered minerals which show triclinic symmetry. The term *kandite* is sometimes used in place of kaolinite when doubt exists. This term is a combination of the species of minerals which classify as kaolinite homopolytypes — *ka*olinite, *n*acrite and *d*ick*ite*. The term homopolytype refers to the situation when all the layers involved in translation are similar in composition and structure (Warshaw and Roy, 1961). When such is not the case, as in mixed-layer structures, the term heteropolytype is used to refer to such structures.

For this book, we will use the term *kaolinite* to refer to the mineral and *kandite* to refer to the mineral group comprising of the kaolinite homopolytypes which includes kaolinite, nacrite, and dickite. As noted previously, when the unit cells are joined laterally to form a stacking of a tetrahedral sheet on top of an octahedral sheet, we obtain the basic stack (of the two sheets) identified as the unit layer (Figure 2.8), and the repeat stacking of these unit layers will establish the spatial dimensions of a typical kaolinite (particle) crystal.

Serpentines belong to the same mineral-structure group and thus have similar (to kaolinites) structures except that the octahedral positions may be occupied by magnesium, aluminum, iron, and other ions. In consequence, the mineralogical and chemical properties of such minerals tend to be more complex than the kaolinites even though they both have 1:1 sheet structures. They are not common constituents of soils since the mineral forms of serpentine are generally unstable in weathering conditions, and tend to transform to other minerals. One can obtain, for example, iron-rich smectite (Wildman et al., 1968), and a variety of lateritic material ranging from goethite and gibbsite to chlorite and smectite under accelerated weathering conditions.

Source of charge	Isomorphous substitution	Charge characteristics	Clay mineral	C.E.C meq/ 100g	Surface area sq m/g
edges, broken bonds, (hydroxylated edges)	Dioctahedral 2/3 of positions filled with Al	variable and fixed charges	kaolinites 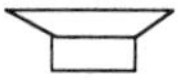	5–15	15
isomorphous substitution, some broken bonds at edges	Usually octahedral Subst. Al for Si	mostly fixed charges	illites 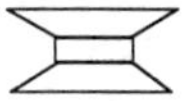	25	80
isomorphous substitution	Dioctahedral; Al for Si Trioctahedral or mixed Al for Mg	mostly fixed charges	chlorites 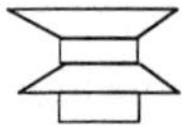	10–40	80
isomorphous substitution	Usually trioctahedral Al for Si	mostly fixed charges	vermiculites 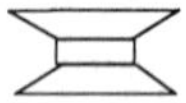	100–150	700
isomorphous substitution, some broken bonds at edges	Dioctahedral Mg for Al	mostly fixed charges	montmorillonites 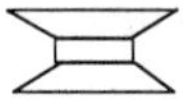	80–100	800

silica layer
alumina layer
2:1 = two silica layers to one alumina layer

Figure 2.9 Some typical clay minerals and sources of charge.

The illites (second row of the table shown in Figure 2.9) have charged 2:1 sheets and potassium as the interlayer occupants. Illites belong to the mica mineral-structure group. In its strictest use, *illite* refers to the family of mica-like clay minerals. This classification term is generally used to refer to hydrous clay micas that do not expand from a 1.0 nm basal spacing (Grim et al., 1937). The difference between these and macroscopic micas can be found in the lesser potassium content and greater structural hydroxyls.

Chlorites (Figures 2.9 and 2.10) also have charged 2:1 sheets forming the basic unit layers. However, they belong to another mineral-structure group because of the octahedral interlayer which joins the trioctahedral layers, as seen in row three of the table in Figure 2.9 and the sketch in Figure 2.10. This octahedral sheet which forms the interlayer has also been called a brucite layer, a gibbsite layer, or an interlayer hydroxide sheet. This hydroxide interlayer differs from the regular octahedral sheet in that it does not have a plane of atoms which are shared with the adjacent tetrahedral sheet. Whilst cations such as Fe, Mn, Cr, and Cu are sometimes found as part of the hydroxy sheets, the more common hydroxy sheets are $Al(OH)_3$ or $Mg(OH)_2$. The typical repeat spacing for the unit layer which consists of the unit shown in the figures is 1.4 nm. As might be anticipated, with this repeat spacing of 1.4 nm, they can be difficult to recognize when any of the minerals such as kaolinite, vermiculite, and smectite are present in the soil.

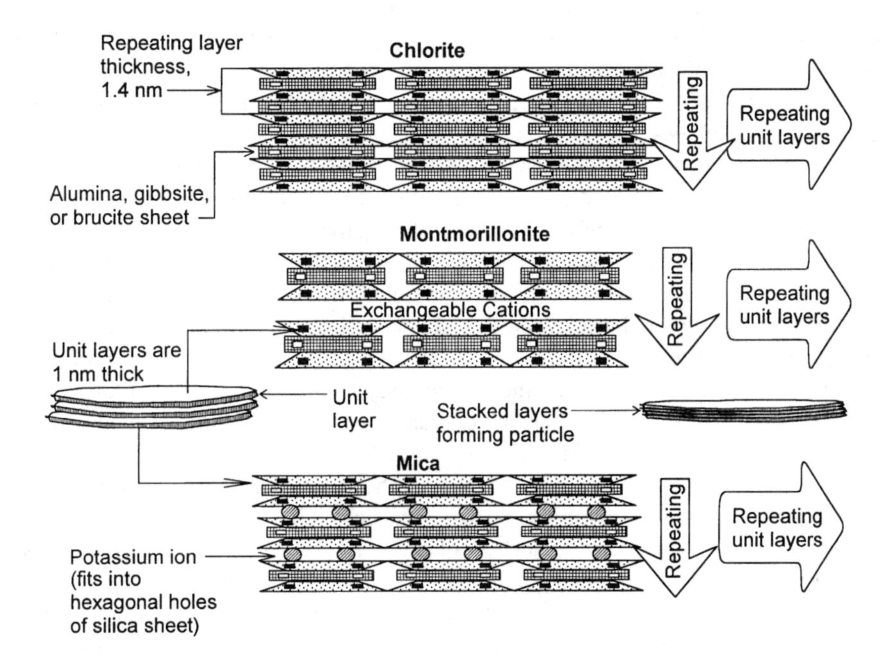

Figure 2.10 Basic unit cells, sheets, and layers for chlorines, montmorillonite, and mica.

Vermiculites fall into the smectite-vermiculite group. The minerals in this group consist of charged 2:1 layers with interlayer cations of variable hydration characteristics upon exposure to moisture. This interlayer water can be easily removed by desiccation to produce the typical dehydrated vermiculite with basal spacing of 1.0 nm. In the fully hydrated state, the basal spacing expands to 1.4 nm. — which corresponds to two molecular layers of water.

The smectites that constitute the other part of the smectite-vermiculite group are well known for the mineral montmorillonite (Figures 2.9 and 2.10) which is quite often confused with the parent term of *smectites*. In the strictest sense, *smectites* represent the group of hydrous aluminium silicate clays containing magnesium and calcium. Included in this group are the dioctahedral minerals represented by montmorillonite, beidellite, and nontronite, and the trioctahedral minerals represented by saponite, sauconite, and hectorite. The dioctahedral smectites are generally obtained as the result of weathering processes, whereas the trioctahedral smectites that appear to be inherited from the parent material are not commonly found as soil fractions.

It is quite common to find the term montmorillonite used to represent bentonite, particularly in more recent engineering practice dealing with clay liners and barriers. Bentonites are derived from alteration of volcanic ash and consist primarily of both montmorillonite and beidellite. Depending on the source of the bentonites, one can find proportions of montmorillonite in bentonite ranging from 90% down to 50% or even less. The hydration characteristics of the interlayer cations will determine the hydrated basal spacing. This aspect of the montmorillonites will be discussed in greater detail in the next chapter when we deal with the phenomenon of *swelling clays* as part of the study of clay-water interactions.

The source of the electric charge imbalance arising because of the formational characteristics of these minerals can be seen in the table shown in Figure 2.9. As we can see, substitution of one ion for another in the clay lattice and imperfections at the surface (especially at the edges) occurring during crystallization or formation of the mineral results in the development of negative electric charges on the clay particles. If the substituting ion has a lower positive valence than the substituted ion, then the lattice is left with a net negative charge. The main substitutions found are aluminum for silicon in the silica sheet, and ions such as magnesium, iron, or lithium substituting for aluminum in the alumina sheet. These substitutions account for most of the charge in the 2:1 and 2:2 minerals, but only a minor part in the 1:1 kaolinites. They produce a characteristic negative charge, which is generally called a *fixed charge*. These are discussed in greater detail in Chapter 3.

Isomorphous substitution, imperfections at the surfaces of the clay particles, and unsatisfied valence charges on the edges of particles all combine to provide a net negative electric charge on the surface of the clay particles. This feature is most important. Because heavy metal pollutants are positively charged, these metal cations will be electrostatically attracted to the negative surface charges of the clay particles. The combination of high specific surface area and significant surface charge make the clay minerals important participants in the contaminant-soil interaction process. The greater the amount of negative surface charges available, the greater will be the potential for attracting positively charged contaminants, i.e., cationic contaminants. This means that if a soil contains more exposed surface areas, i.e., higher specific surface area, the greater will be the capability of the soil to sorb contaminants via "plus-minus" bonding mechanisms (ionic bonding), everything else being equal.

Isomorphous substitution during formation generally results in development of fixed charges for the particles. Clay particle surfaces that provide the fixed charges are called *fixed charge surfaces*. In contrast to the fixed charges that are characteristic of isomorphous-substituted layer lattices, variable charges exist in certain soil particles and constituents, i.e., the sign of the charge being dependent on the ambient pH — the hydrogen ion concentration of the aqueous environment. The particle surfaces associated with variable charges are called *amphoteric surfaces* or *variable charge surfaces,* and the soil fractions that are generally considered as possessing variable charge surfaces include the oxides/hydrous oxide minerals, and a large number of non-crystalline inorganics and soil organics. Thus, for example, the charge characteristics and the CEC for the clay minerals listed in Figure 2.9 show that kaolinites are classified as having variable and fixed charges, i.e., the edges are considered to be variable-charged whereas the surfaces are considered as fixed-charged. This is discussed in greater detail in the next chapter.

2.3.2 Soil Organics

Soil organic material originates from vegetation and animal sources, and occurs in mineral surface soils in proportions as small as 0.5 to 5% by weight. (We consider peat material to be the exception; the proportions can be as high as 100%.) Organic matter is generally categorized along states of degradation (Greenland and Hayes, 1985) or into humic and non-humic material (Schnitzer and Khan, 1978). The formation of humic materials has been the subject of much speculation. Schnitzer and Khan (1972) cite four hypotheses for their formation, i.e.:

1. Formation through plant alteration;
2. Formation through chemical polymerization of degraded animal and plant material;
3. Products of autolysis of cellular materials; and
4. Synthesis by microbes.

If we categorize the soil organics in terms of states of degradation, we obtain:

1. Unaltered organics (fresh and old non-transformed organics); and
2. Transformed organics that bear no morphological resemblance to the original source. These decayed organics are further classified into:
   a. amorphous materials (e.g., humic substances),
   b. decayed materials (compounds that belong to recognizable classes, e.g., polysaccharides, lignins, polypeptides, etc.).

The term *amorphous* used in relation to organics refers to transformed organic materials that do not exhibit properties and characteristics of the parent material, and which cannot be traced back to origin or specific parent material. In that sense, the meaning of the term differs from the term "amorphous" used in relation to soil inorganic constituents. Typical *amorphous organics* in soils are humic substances (humic acids, fulvic acids, and humins). These are highly aromatic polymers and are recognized for their high content of functional groups. A more detailed description of the functional groups will be found in Chapter 3. At that time, we will address the description of functional groups for the various soil constituents (fractions) and evaluate these in respect to soil-water relations.

The most popular method for classifying soil organics is to classify them in relation to humic and non-humic material (i.e., 2a above). The *humic* substances are defined as organics resulting from the chemical and biological degradation of non-humic materials, whereas *non-humic* compounds are organics which remain undecomposed or are only partially degraded. The method of classification is based on the extraction procedure, i.e., procedure for extraction of the material from the parent organic material. The classification of humic substances into humic acids, fulvic acids, and humins is based on their solubility to acid and base as follows:

Humic acids — soluble in base, but precipitate in acid;
Fulvic acids — soluble in both base and acid;
Humins — insoluble in acid and base.

A typical organic extraction technique involves an alkali treatment that separates the fulvic and humic acids from the rest of the soil, and from the humin fraction. The organics dissolved in the supernatant are decanted and the humic acid fraction is precipitated from solution with acid, leaving the soluble fulvic acid fraction in

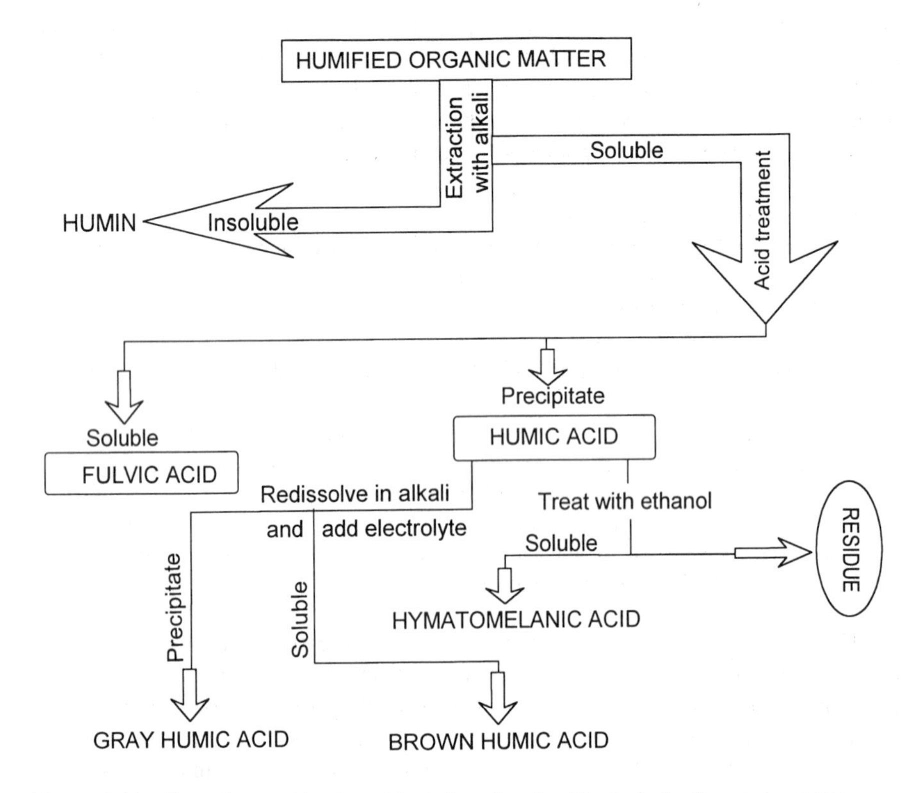

Figure 2.11 Extraction and treatment technique for classification of soil organic matter.

the supernatant (see Figure 2.11). A further refinement of the standard technique can be introduced to determine the presence of polysaccharides, as performed by Yong and Mourato (1988). Their technique, shown in Figure 2.12, uses sequential acid and alkali treatment procedures in combination with the standard method shown in Figure 2.11. With this procedure, Yong and Mourato (1988) were able to extract four distinct organic fractions: humic acids, fulvic acids, humins, and non-humic fractions. The non-humic fractions contained polysaccharides of microbial origin.

2.3.3 Oxides and Hydrous Oxides

The oxides and hydrous oxides are very important soil fractions insofar as contaminant-soil interaction is concerned. *Hydrous oxides* can refer either to the non-crystalline form or to the crystalline form. It is important to distinguish between the two since distribution of the hydrous oxides in the soil is to a very large extent dependent upon whether it is in the mineral form or the amorphous form. Thus, for example, we can obtain coatings of amorphous forms of the oxides onto mineral particle surfaces because of the nature of the net electric charges on the surfaces of both the mineral particles and amorphous materials. Since the nature of the charges on these amorphous materials are pH dependent, evaluation of the interactions obtained between soil fractions in the presence of water requires attention to the

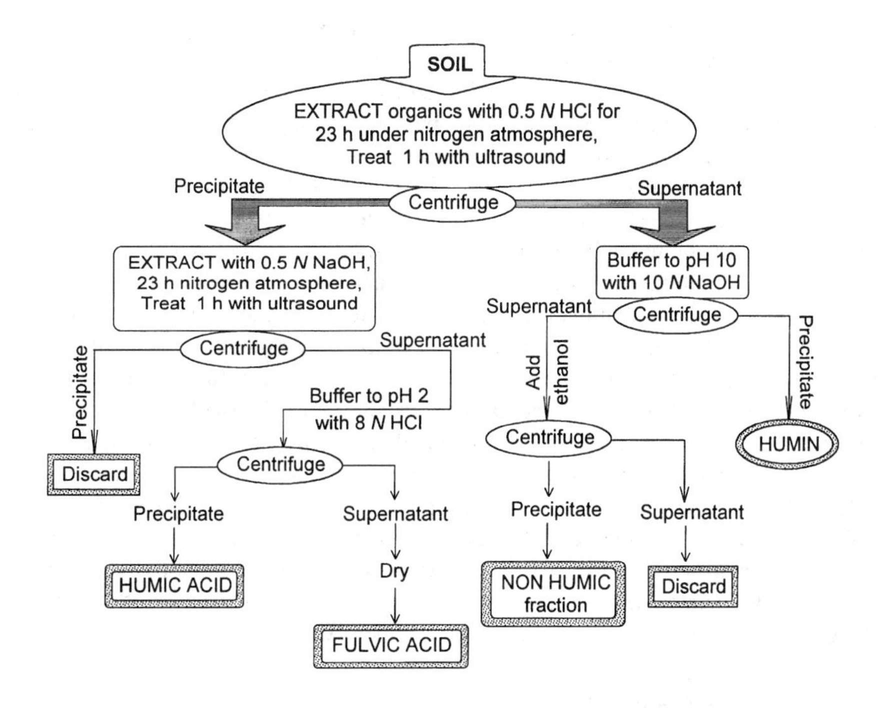

Figure 2.12 Extraction technique for soil polysaccharides, humic acids, fulvic acids, and humins. (Adapted from Yong and Mourato, 1988.)

amphoteric nature of these fractions. These interactions will be discussed in greater detail in Chapter 3.

The *oxides*, as a term, generally refers to the crystalline (mineral) form of the material. Included in the various oxides and hydrous oxide minerals are the principal constituents of highly weathered tropical soils such as laterites and bauxites, e.g., haematite, goethite, gibbsite, boehmite, anatase, and quartz. By and large, these are oxides, hydroxides and oxyhydroxides of iron, aluminum, manganese, titanium, and silicon, and they differ from layer silicate minerals in that their surfaces essentially consist of broken bonds. The low solubilities of the oxides of aluminum, iron, and manganese in the general pH range found in most natural soils mean that their oxide forms are more dominant than the oxides of titanium and silicon. The terminology used to describe the oxides may take different forms, depending upon the source of the information. For example:

$$Al(OH)_3 = \text{aluminum trihydroxide} = \text{aluminum hydroxide}$$

$$AlOOH = \text{aluminum oxide-hydroxide} = \text{aluminum oxyhydroxide}$$

$$\gamma\text{-}Al(OH)_3 = \text{alumina trihydrate} = \gamma\text{-aluminum hydroxide} = \text{gibbsite}$$

$$\gamma\text{-}AlOOH = \text{alumina monohydrate} = \gamma\text{-aluminum oxyhydroxide} = \text{boehmite}$$

The primary structural configuration of the oxides is octahedral. The oxides of aluminum, for example, show octahedral sheets containing OH^- ions with two thirds of the positions occupied by Al^{3+} ions. The OH^- ions in each of the octahedral sheets stacked on top of each other can either be directly opposite to each other (i.e., between the stacked sheets) or in the space formed by the OH^- ions in the opposing stacked sheet. In the first instance where the OH^- ions lie on top of each other in the stacked sheets, the sheets are essentially bonded by hydrogen bonds. This is the structure of gibbsite. In the latter case, the arrangement is termed as a closely packed OH^- ion configuration, representative of the structure of bayerites.

Iron oxides are the dominant form of oxides found in most soils, with goethite being the most common type found. Because of their low solubility and their redox-reversibility behaviour, their presence in soil needs to be carefully considered. As with most of the other oxides, the amorphous shapes of the oxides form coatings surrounding particles which show net negatively charged surfaces (see Chapter 3 for reactive surfaces). This not only changes the charge characteristics of the soil particles, but also changes the characteristic physical and chemical properties of the soil. As we will see in the next chapter, when the oxide surfaces are immersed in an aqueous environment, the broken (i.e., unsatisfied) bonds are satisfied by hydroxyl groups of dissociated water molecules. These oxides and hydrous oxides exhibit charge characteristics that are pH dependent, i.e., they exhibit variable charge properties.

2.3.4 Carbonates and Sulphates

Carbonates and sulphates (i.e., carbonate and sulphate minerals) are generally considered to be relatively soluble in comparison to silica minerals. The most common of the carbonate minerals found in soils is calcite ($CaCO_3$), derived primarily from calcareous parent material and generally obtained in semi-arid regions of the world. Other types of carbonate minerals include magnesite ($MgCO_3$), dolomite ($CaMg(CO_3)_2$), trona ($Na_3CO_3HCO_3 \cdot H_2O$), nahcolite ($NaHCO_3$), and soda ($Na_2CO_3 \cdot 10H_2O$). Carbonates can also be obtained from sedimentary rocks. The most common are calcite and dolomite. Under reducing conditions, siderite ($FeCO_3$) with Fe^{2+} can be formed. Carbonate minerals are thought to be good adsorbers of heavy metals and phosphates.

Although one can list a few sulphate minerals such as gypsum, hemihydrate, thenadite, and mirabilite, gypsum (i.e., $CaSO_4 \cdot 2H_2O$) is the most common of the sulphate minerals found in soils — primarily in arid and semi-arid region soils. Existence of the crystalline form of $MgSO_4$, Na_2SO_4, and other sulphate minerals is generally confined to the soil surface because of their high solubilities. By comparison, gypsum is at least 100 times less soluble than $MgSO_4$ and Na_2SO_4. The relatively high solubilities of carbonates and sulphates compared to the layer silicates and the aluminum/iron oxides, hydroxides, and oxyhydroxides mean that their presence in large amounts is mainly confined to regions where limited leaching and high evaporation occur, typically in arid and semi-arid regions.

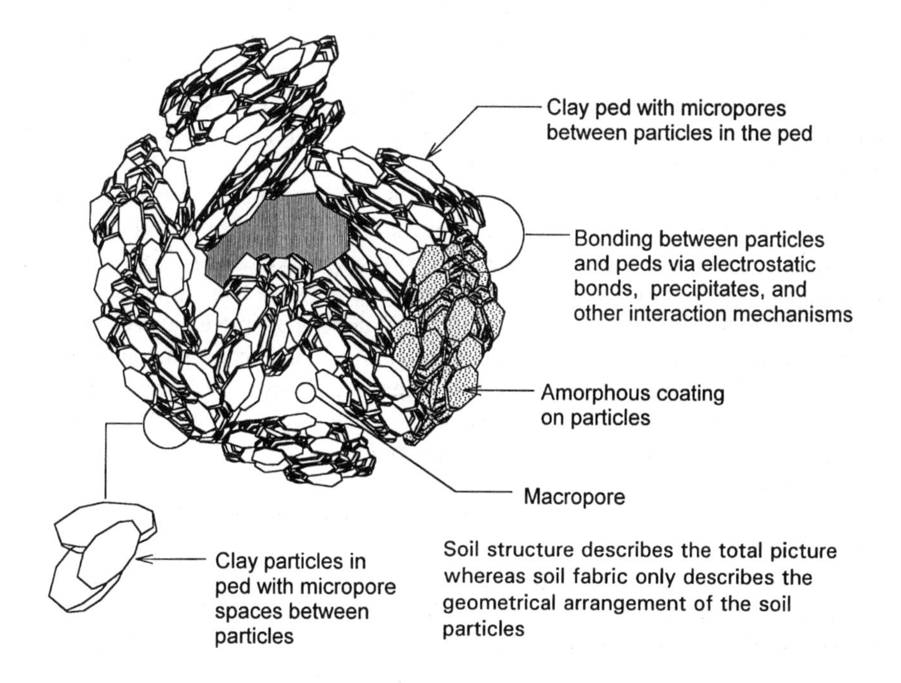

Figure 2.13 Representative elemental volume (REV) of soil composed of clay particles forming peds. Intraped pores are micropores and interped pores are macropores.

2.4 SOIL STRUCTURE

The arrangement and distribution of soil fractions in a soil "mass" will directly influence not only the physical properties of the soil, e.g., density and porosity, but also those performance characteristics that depend upon those properties, such as shear strength, consolidation, and hydraulic conductivity. The geometrical arrangement of soil particles is defined as the *soil fabric*. As such, soil fabric refers solely to particle arrangements and pays no attention to the other factors and items that contribute to the stability of the arrangement of particles. The term *soil structure* refers to that property of a soil derived from the nature of the soil fractions and their distribution and arrangement. Soil structure includes soil fabric and the various other soil characteristics derived from the types and proportions of soil fractions which constitute the soil, and is the term that is more commonly used when one describes the features of a soil.

The schematic picture shown in Figure 2.13 is meant not only to provide the overall view of soil structure and soil fabric, but also to show the existence of macro- and micropores in the representative soil volume. In the later chapters, we will pay more attention to the role of the various soil fractions in development of soil structure.

For this present discussion, the schematic diagram shows various soil (clay) peds arranged in an arbitrary fashion. We can consider the total picture shown in Figure 2.13 to represent a typical soil volume element which we can denote as a REV (representative elemental volume). The REV picture shows the macrostructure and the picture of the peds, with the arrangement of the particles in the peds which will determine the microstructure.

The role of the various other soil fractions (other than clay mineral), such as oxides, carbonates, and soil organic matter, in the control of both soil fabric and especially soil structure depend on the proportions of such fractions. To a very large extent, their distribution amongst the other fractions, and their influence on the physical and chemical properties of the total soil depend on interactions established in the presence of water. In particular, formation of bonds between soil particles, and coating of soil particles are features that can significantly alter the properties of the soils. These will be evident when we address the subject of soil-water relations in the next chapter. For the present, we will confine ourselves to the general fabric and structure of soils.

The stability or strength of the REV is determined by the soil structure. This means that the integrity of the REV depends on the strength of the individual units, the arrangement of these units, the bonds between the units, and how well the units act to support each other, i.e., the interaction mechanisms established. Because of the nature of the soil fractions, and because of the manner in which the soils are obtained or developed, it is possible to obtain particle and ped arrangements that favour a horizontal orientation. This gives rise to anisotropic behaviour of the soil, and is particularly important in considerations of transport processes in soils. The factors that need proper consideration include not only orientation of fabric elements, but also orientation of the macro- and micropores.

Figure 2.14 shows various modes of fabric orientation of both ped units and total REV elements. The relationships between soil structure and the mechanical properties of soil have been well documented in many research studies. Since the attention in this book is directed toward interactions between pollutants and soil fractions, with the intent to determine the fate of pollutants, these relationships lie outside our scope. The reader is advised to consult textbooks on soil behaviour for such information (e.g., Yong and Warkentin, 1975). For the present, the role of soil structure will be demonstrated more in terms of the distribution of soil fractions, and the kinds of reactive surfaces presented to the pollutants. In addition, the presence and distribution of the micro- and macropores will be seen to be significant when we need to consider the transport of pollutants through soil. Since different types of soils, densities, particle arrangements, etc. present different fabric arrangements and structure, transport of pollutants will be through macropores and micropores. Depending on the continuity established between macropores, and depending on the type and density of micropores, diffusion transport of pollutants will be severely impacted. A combination of both diffusion and advection transport through soil is not uncommon because of the presence and distribution of the macropores and micropores.

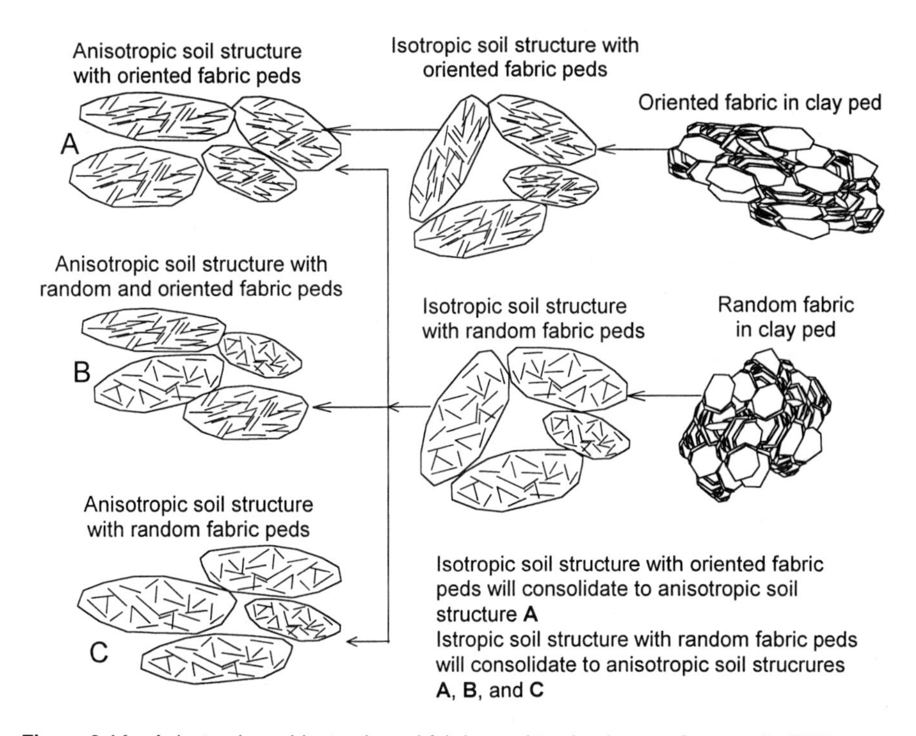

Figure 2.14 Anisotropic and isotropic ped fabrics and total anisotropy/isotropy in REV.

2.5 PHYSICAL PROPERTIES

The main physical soil properties of relevance to soil contamination problems can be broken down into two principal groups: (a) those dealing with contaminant transport processes, e.g., permeability, porosity, density, soil structure, and water saturation; and (b) those dealing with the strength and compactibility of the soils, i.e., those physical/mechanical properties pertinent to establishment of stability of soil materials used as engineered barriers and those properties involved in the "dig and dump" methods of contaminated site remediation. These physical properties and the behaviour of soils have been studied in great detail and can be found in textbooks specifically written to deal with these subjects (e.g., Yong and Warkentin, 1975). For the purpose of this book, we are interested in those physical properties that impact directly on the transport of pollutants in the soil substrate, and the influence of contaminants on the integrity of compacted or natural soil. Since the transport and fate of pollutants in the soil and in situ remediation techniques utilizing pump and treat procedures involve the movement of fluids and contaminants in the soil, the specific property of interest is the permeability or hydraulic conductivity of the soil.

2.5.1 Hydraulic Conductivity

In general, the transmission property of a soil in respect to the movement of a fluid through the soil is measured in terms of the rate of flow of the fluid through a representative volume of the soil. The reason why attention is paid to the hydraulic conductivity of the soil in pollutant transport considerations is because we need to be aware of whether propagation of the pollutant plume in the soil is due to the advective performance of the carrying fluid and/or the diffusion of the pollutants in the carrying fluid. In the fully water-saturated state and with zero turbulence in the flow of water through the soil, the rate of macroscopic flow of water through a representative volume of the soil can be related to the hydraulic gradient via an impedance factor. This impedance is identified as the *hydraulic conductivity* of the soil. Thus if v represents the velocity of flow of the permeating fluid (water), and i = hydraulic gradient, the following relationship is obtained:

$$v = ki = k\frac{\Delta h}{\Delta x} \tag{2.1}$$

where k is a constant defined as the hydraulic conductivity (also known as the *coefficient of permeability*), h = hydraulic head, x = spatial distance, and $i = \frac{\Delta h}{\Delta x}$ = hydraulic gradient. This relationship, commonly referred to as the *Darcy Law,* is valid at low flow velocities of water in soils, especially for smooth granular-sized particles. The coefficient of permeability k is quite often referred to as the Darcy coefficient k and is a measure of the resistance of the soil to the flow of water (through the soil). Since the hydraulic gradient i is dimensionless, k will have the same dimensional units as the velocity term.

The properties of the water are not taken into account in Equation 2.1. This becomes important when contaminants are present in the porewater since these will likely change the properties of the porewater. We need to establish a new relationship as follows:

$$v = k^*\frac{\gamma g}{\eta}i \tag{2.2}$$

where k^* = intrinsic permeability coefficient, γ = density of permeating fluid, g = gravity, and η = viscosity of the permeating fluid.

Equation 2.2 provides a deterministic evaluation of the hydraulic conductivity of a soil where some attention is paid to the physical properties of the permeating fluid. Since contaminants and pollutants in the porewater will alter the viscosity of the porewater, this relationship is a more realistic portrayal of the assessment of hydraulic conductivity in the soil. No specific attention is paid to the physical surface properties of the soil and to the nature of the pore spaces through which the fluid must flow. Permeation of fluid into various types of soils involves physical interaction between the fluid and the surfaces of the many different kinds of soil fractions. Incorporation of the surface area of a soil and its respective porosity requires an analysis such as the one described by the Kozeny-Carman relationship. The conceptual model used for development of this relationship relies on the Poiseuille equation for viscous flow of fluids through narrow tubes.

$$v^* = \frac{\gamma r^2 \Delta\psi}{8\eta\Delta x} \tag{2.3}$$

where v^* = mean effective flow velocity through a tube of radius r, and $\Delta\psi$ = potential difference between front and end of tube of length Δx. To make the narrow tube model applicable to soils, we recognize that v^* represents the flow through the soil pores, and that the effective length of flow is greater than the straight-line distance because of the presence of pore spaces which are not interconnected in a straight line. Hence the effective velocity, i.e., the porewater velocity, should now be written as:

$$\begin{aligned} v^* &= \frac{v\Delta x_e}{n\,\Delta x} = \frac{v}{n}T \\ &= C_s\frac{\gamma r^2}{\eta}\frac{\Delta\psi}{\Delta x} \end{aligned} \tag{2.4}$$

The shape factor C_s is introduced in Equation 2.4 in place of the fraction ⅛ to account for the fact that the pore spaces in the soil are not circular. In general, this factor is assumed to be 0.4 unless otherwise specified. The tortuosity term T in the equation is the ratio of the effective path length Δx_e to the direct path length Δx, i.e., $T = \Delta x_e/\Delta x$. The effective path length Δx_e is the flow path length which depends on the arrangement of the void spaces through which continuous hydraulic flow is established. If is often argued that because the micropores in the peds (Figure 2.13) are considerably smaller than the macropores that exist between the ped, advective flow occurs mostly in the channels defined by the macropores. Continuity of macropores is fundamental to the establishment of the effective path length. The role of soil structure in the control of contaminant transport is exercised not only in the arrangement of the peds, but also in the nature of the micropores in the peds. The various ped fabrics shown in Figure 2.14 will control the diffusion of the contaminants.

Since surface area is an important property that needs to be considered when water flow across the surface occurs, and since not all surfaces are exposed in the soil mass, the relationship between the narrow tube radius r and surface area S_a can be described as follows:

$$r = \frac{n}{S_w} = \frac{n}{S_a(1-n)} \tag{2.5}$$

where S_w represents the wetted surface area. The changes in the original conceptual model of viscous flow of fluids through narrow tubes — to account for non-circular pores, non-connected and non-regular pores, flow tortuosity, and wetted surfaces — now permits us to obtain the working relationship as follows:

$$k^{**} = \frac{C_s \gamma}{\eta T^2 S_a^2} \frac{n^3}{(1-n)} \tag{2.6}$$

where k^{**} denotes the hydraulic conductivity. The superscript ** is used deliberately to distinguish this from the Darcy k coefficient.

Up to now, we have only focused on the simple physical aspects of hydraulic conductivity, where the factors and parameters determining soil permeability fall into two groups: (a) those physical factors associated with the permeant, e.g., viscosity and density, and (b) physical factors associated with the soil, e.g., porosity, tortuosity, and surface area. Other factors need to be considered. These fall into the grouping of chemical properties and parameters. However, for a simple assessment procedure, it is possible to use the values obtained using only the physical considerations discussed.

2.5.2 Soil Fractions and Physical Properties

The influence of soil type on its physical properties is obtained from the types and distribution of its soil fractions. The key features are the surface properties of the soil solids such as specific surface area (SSA), cation exchange capacity (CEC), and the surface functional groups of the various soil fractions. The first surface feature (SSA) falls into the category of physical properties whereas the other two (CEC and functionality) are the chemical properties of the soil and will be discussed in the next two chapters.

The SSA of the soil refers to the total surface area per unit weight of dry soil, generally expressed in terms of m^2/g of soil. Referring to Figures 2.13 and 2.14, this means that individual particles in each of the soil peds contribute *all* of their surface areas to the total surface area of the soil. The problem of determination of the SSA where all of the particles are "sampled" is always challenging. It is clear, however, that the greater the percentage of fines in the soil mass, the greater the SSA. This physical soil property is a key one. As we can see from the schematic portrayal of the soil particles in the peds and the arrangement of the peds, accurate measurements of the SSA of soils can be quite difficult. Most laboratory techniques rely on coating *all* particle surfaces with some form of adsorbate to provide the basis for SSA calculations. Errors arise when: (a) not all the surfaces of individual particles are coated; (b) coatings of particles are not uniform (i.e., coatings should be one molecular layer of the adsorbate); and (c) reactions caused between the adsorbate and the soil fractions result in creating artifacts. In the final analysis, we need to consider the measured SSA as an operationally defined property of the soil. This is discussed further in Chapter 4.

2.5.3 Utilization of Information on Soil Properties

To demonstrate the use of information about the physical properties of a soil for examination of the soil's potential as a contaminant buffering material, we cite the studies conducted on soil samples (Table 2.1) obtained from six different site locations

Table 2.1 Site Location and Samples Obtained

Samples	Description and Location
MR1–MR4	Weathered mudrock from Bryn Pica Landfill, Aberdare.
GT1–GT5	Glacial till from Bryn Pica Borrow Pit near Aberdare.
NEA1–NEA5	Estuarine alluvium from landfill site in Neath.
PEA1–PEA5	Estuarine alluvium from landfill site in Newport.
CEA1–CEA5	Estuarine alluvium from Cardiff Landfill site.
SGT1–SGT5	Glacial till from Swansea (coal mine open pit).

in Wales, U.K. (Yong et al., 1998a). These include weathered mudrocks (MR series), glacial till material from Aberdare (GT series) and from Swansea (SGT series), and estuarine alluvium from Neath (NEA series), Newport (PEA series), and Cardiff (CEA series).

The estuarine alluvium and glacial till materials contain a higher proportion of illite, in contrast to the mudrock soils which contain almost equal proportions of illites and kaolinites. We can list the recognized minerals in decreasing order of abundance as follows:

Estuarine alluvium from Cardiff: *CEA series*: illite > kaolinite > chlorite > quartz

Glacial Till Soil: *GT series*: illite > kaolinite > quartz > feldspar

Mudrock soils: *MR series*: illite = kaolinite > quartz; traces of chlorite and feldspar.

Table 2.2 shows many of the physical properties of two sets of samples from each series given in Table 2.1. The variations in particle sizes and distribution, organic content, and types of minerals in the three kinds of soil series with respect to compactibility, permeability (as determined by the Darcy coefficient of permeability k), consistency limits, and specific surface area (SSA) can be seen in Table 2.2.

The consistency limits, i.e., plastic and liquid limits, that have been generally considered as useful indicators of the soil's plasticity, reflect the soil's water-holding capacity. To a very large extent, this characteristic is dependent on the specific surface area of the soil and the nature of the surface forces associated with the soil fractions. We will discuss surface forces in relation to the surface functional groups in the next few chapters. Everything else being equal, we can see from Figure 2.15 the influence of SSA, organic content, and percentage of fines on the development of the plasticity characteristics of the soils. All the individual samples identified in Table 2.1 are shown in this figure. In a sense, the SSA is a direct reflection of the percentage of fines and organic content, and as might be expected, Table 2.2 shows a correlation between them.

The Casagrande "A" line (Casagrande, 1947) shown in the chart (Figure 2.15) separates the plasticity characteristics into six categories. By and large, clays occupy the regions above the "A" line and silts occupy the regions below the "A" line. The categories are as follows:

Table 2.2 Physical Properties of Samples from Table 2.1

Sample No.	ω_o %	Gs	LL %	PI %	G %	S %	M %	C %	$\gamma_d$max Mg/m^3	Wopt %	k $\times 10^{-10}$ m/s	Orgs. %	SSA m^2/g
MR1	14.4	2.57	31.8	13.1	49	21	9	21	1.93	10.0	3.9	2.32	46.38
MR4	14.9	2.50	32.2	13.2	43	19	23	15	1.99	10.1	2.4	1.95	39.53
GT1	21.8	2.64	35.5	13.5	23	30	17	30	1.84	13.0	3.5	2.77	69.87
GT4	13.9	2.52	27.7	10.4	23.3	33.1	29.9	13.7	1.93	12.5		0.7	30.231
NEA1	45.7	2.59	59.7	28.6	0	8	47	45	1.47	23.0	2.5	5.53	71.12
NEA4	51.8	2.49	65.8	30.0	0	8	42	50	1.36	30.0	2.2	5.11	73.34
PEA1	96.8	2.58	75.2	38.1	0	2	46	52	1.38	27.0	2.6	5.97	84.61
PEA4	34.1	2.63	50.8	25.9	0	3	39	58	1.62	20.8	1.9	2.74	80.69
CEA1	31.1	2.53	46	23.3	0	1	49.6	49.4	1.69	21.6	2.22	3.75	83.5
CEA4	30.5	2.48	47.1	24.2	0	0	47	53	1.66	20	1.8	5.78	73.59
SGT1	20	2.58	38	19.6	9.2	4.4	52.9	33.5	1.82	14.5		1	51.2
SGT4	26	2.62	24.9	11.1	23.7	35	25.1	16.2	2	10.9		1.1	43.763

ω_o = water content; Gs = specific gravity; LL = liquid limit; PI = plasticity index; G = gravel; S = sand; M = silt; C = clay; $\gamma_d$max = maximum dry density;

Wopt = optimum water content; *k* = Darcy coefficient of permeability; Orgs. % = percent organics; SSA = specific surface area.

Adapted from Yong et al., 1998a.

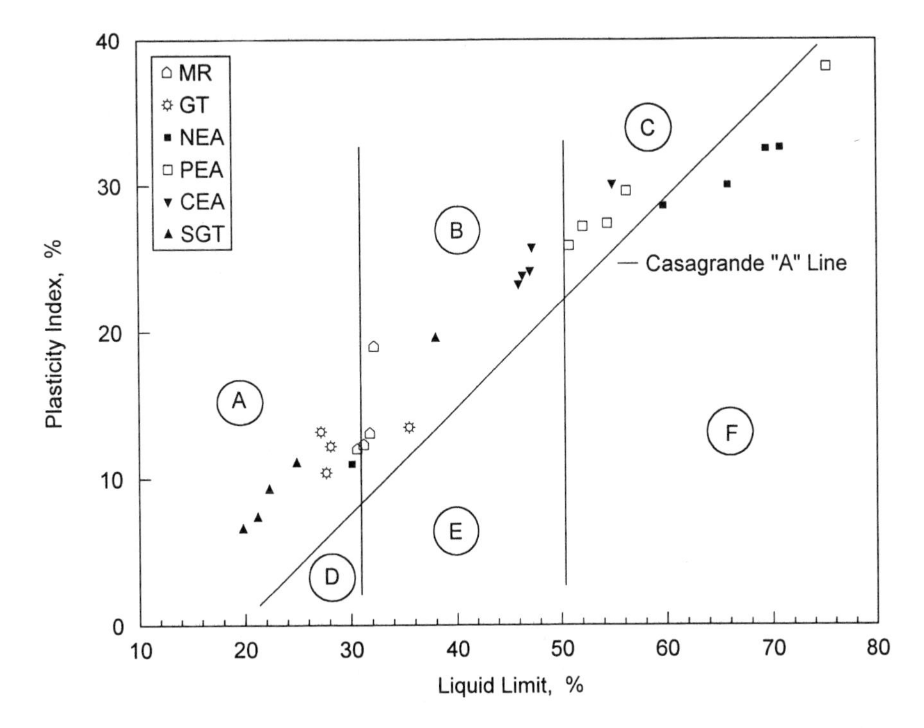

Figure 2.15 Plasticity chart showing plasticity characteristics of soil samples in Table 2.1. Soil types characterized by the Casagrande "A" line are classified according to soil plasticity.

- **Region A** — Inorganic clays with low plasticity and soils with relatively low cohesion occupy this region. Most of the glacial till soils (SGT and GT series) fall into this class.
- **Region B** — This region contains inorganic clays of medium plasticity. The weathered mudrocks (MR series) are shown at the lower portion of this region, and some of the estuarine alluvium soils (CEA series) occupy the higher end of this class. Note that not all the estuarine alluvium soils are contained in this class.
- **Region C** — This plasticity region is characteristic of inorganic clays of high plasticity. The PEA series of estuarine alluvium fit into this class.
- **Region D** — Soils with very low compressibility and cohesion occupy this "plasticity" region, e.g., inorganic silts.
- **Region E** — Inorganic silts with some cohesion and organic silts fall within this class.
- **Region F** — Organic clays and highly compressible inorganic silts are contained in this class, along with the NEA series of estuarine alluvium.

The compactibility of soils is also related directly to the same factors that control soil consistency. The compaction curves shown in Figure 2.16 demonstrate the significant differences between the soils identified in Tables 2.1 and 2.2. The samples tested refer to MR1, GT1, and CEA1. The higher organic content and higher clay

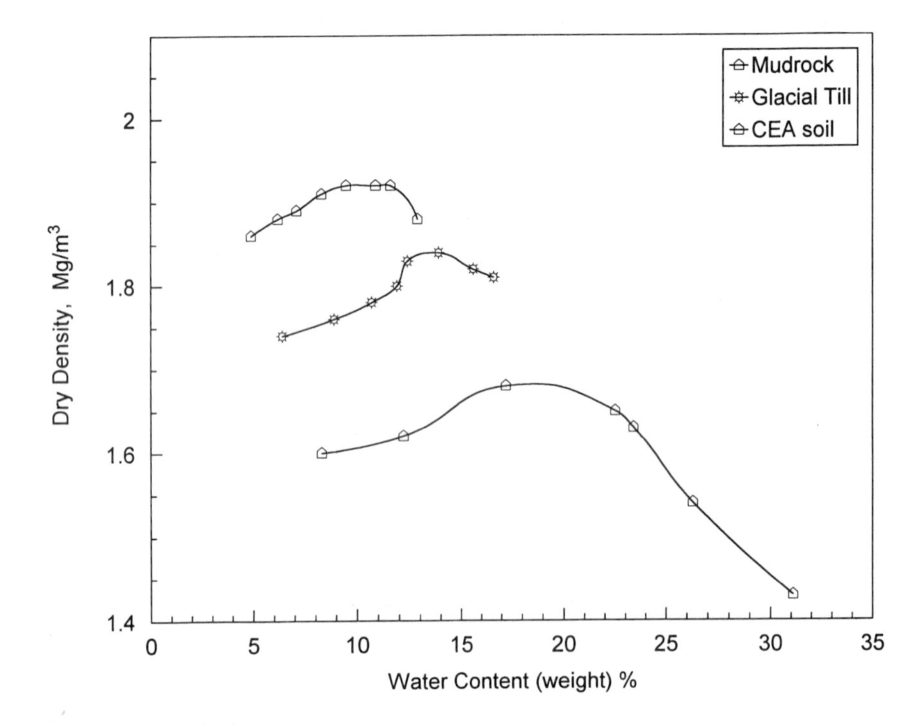

Figure 2.16 Proctor compaction results for three soils, showing difference due to soil type.

content in the glacial till soil GT1 in comparison to the weathered mudrock MR1 means that more water is needed for the soil to reach its optimum compaction density. The specific surface area comparison between the two confirms that the SSA of the GT1 soil is almost 50% higher than the MR1 soil. These considerations provide some insight into soil compaction characteristics. Similar comparisons can be made to include the estuarine alluvium CEA1.

2.6 CONCLUDING REMARKS

Historically, we have been concerned primarily with the physical characteristics and properties of subgrade soils because most of the problems and/or projects concerned with ground engineering have focused on activities associated with construction. In problems and projects dealing with contamination of the ground, we need to recognize that the interactions between contaminants determine the fate of the contaminants (pollutants). In that regard, we need to have a better appreciation of what it is in the soil that provides for interaction characteristics and properties of direct impact in control of pollutant fate.

A proper study of the nature of soil can encompass considerably more scope and detail than that which has been given in this chapter. The same can be said for the physical properties of soils. We have been concerned primarily in demonstrating the simple points that require attention. These refer to the kinds of soil fractions such as coarse and fine-grained soils, soil organics, amorphous materials, etc. This is particularly important because all these different soil fractions demonstrate significantly different capabilities in sorbing (retaining) contaminants, i.e., they have different sorption capacities. To better understand this, we need to examine their surface properties. This will be dealt with in the next two chapters. For the present concern, we have tried to provide the essential elements of what constitutes the basic soil fractions, and how these demonstrate themselves in control of physical properties.

We have directed our attention only to those properties that impact significantly on problems associated with the movement of contaminants and certain ground remediation procedures. The fluid transmission properties and the structure of soils are characteristic of the type of soil and the regional controls. These are significant considerations in the transport of contaminants in the substrate. Soil plasticity is a very simple first guess tool for assessment of the likely status of the soil in respect to many surface and physical properties.

CHAPTER 3

Soil-Water Systems

3.1 SURFACE RELATIONSHIPS

The processes involved in pollutant transport and fate determination in contaminated soils involve the properties of both the soil material and the pollutants. To a very large extent, the properties of their surfaces are of paramount significance since the mechanisms of interaction between pollutants and soil fractions are via the various sets of physico-chemical forces associated with their respective surfaces. Depending upon the level of detail and perspective required, we can study the mechanisms of interaction by trying to quantify these interactions in terms of intermolecular forces and/or energy relationships. These relationships can be viewed in thermodynamic terms as, for example, by studying the thermodynamics of soil water, or in physico-chemical terms through considerations of chemical bonds and electrostatic energy relationships.

When we are confronted with the existence of pollutants in the ground, i.e., a contaminated site, one of the first requirements is to determine the degree of risk posed by these pollutants in respect to human health and the environment. There is one very pressing question that needs to be addressed in assessing the problem:

> How are the pollutants retained in the ground? i.e., What controls the *fate* and *persistence* of each of the pollutants?

The *fate and persistence of pollutants* will be discussed in detail in the next four chapters. For this chapter, we are concerned with: (a) gaining the basic information that would permit us to understand the interactions between the soil fractions and water, and (b) obtaining a better understanding of how the results of these interactions define the soil-water system. Here again, because we are dealing with the soil-water system, we will use more general term *contaminants* (i.e., substances foreign to the natural soil system) in this chapter to include both pollutants and non-pollutants. We consider a *soil-water system* to mean a soil mass that includes the soil fractions and the aqueous phase contained within the soil mass. While the general term *soil* is most often used in place of the term soil-water system, it is sometimes necessary

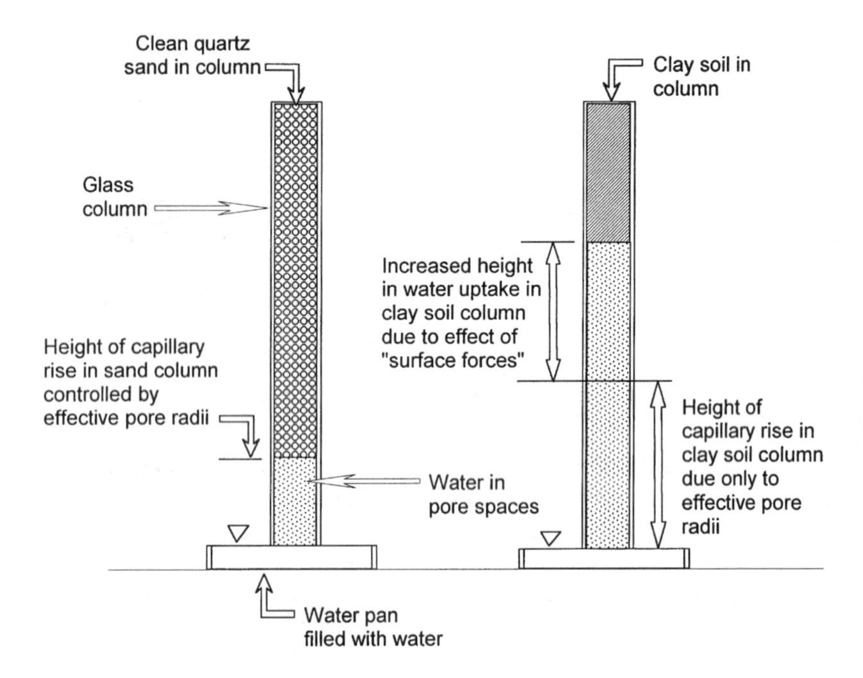

Figure 3.1 Capillary rise experiments. Capillary rise in sand column is shown on the left, and water uptake experiment is shown on the right.

to emphasize the term soil-water system when interactions between the soil solids and the aqueous phase are considered. Some of the questions we want to address include those concerned with determination of:

- Why some soils retain (*hold*) more contaminants in comparison to others;
- Why some contaminants are retained in the soil-water system longer than others; and
- How contaminants are partitioned and distributed within the soil-water system.

To address these concerns, it is necessary for us to obtain a better appreciation of the nature of the surfaces of the various soil fractions and how the types of forces associated with theses surfaces are described. A very good example of how the effect of surface forces is demonstrated can be found by conducting a thought experiment. For this, we consider two open-ended glass cylinders standing vertically with their bottom ends resting in dishes (pans) containing water as shown in Figure 3.1. One glass cylinder contains a clean dry quartz sand sample whilst the other contains a dry clay soil sample.

The height of capillary rise in the left-hand tube which contains the clean sand will be given by the following relationship:

$$h = \frac{2\,\sigma\cos\alpha}{r\,\gamma_w} \tag{3.1}$$

where h = height of capillary rise of the water, σ = surface tension of the water, α = angle of contact of the water with the surfaces of the soil solids, r = effective radius of the average pore size in the sand column, and γ_w = density of water. Similarly, if only the effective radius of the average pore sizes in the clay soil column in the right-hand tube was responsible for the water uptake in the column, we would show the height at somewhere close to the middle of the tube. However, the illustrative sketch (Figure 3.1) shows that the height of water uptake rises up close to the top of the tube, with the extra height being due to the effect of forces associated with the surfaces of the clay soil particles. The uptake of water in the glass tube containing the clay soil to heights above that of an equivalent capillary rise is commonly credited to the matric suction of the soil (see Section 3.6).

3.2 SURFACES OF SOIL FRACTIONS

The soil fractions with reactive surfaces include layer silicates (clay minerals), soil organics, hydrous oxides, carbonates, and sulphates. This section examines the relationship between the structure of these fractions and their respective surface properties.

3.2.1 Reactive Surfaces

We define *reactive surfaces* as those surfaces of soil fractions that react chemically with dissolved solutes in water contained in the pore spaces of the soil-water system. This is the aqueous phase of the overall soil-water system. For simplicity in terminology, we will define this aqueous phase as the *porewater*, i.e., water in the pores of the soil mass. Reactions between the surfaces of the soil fractions and dissolved solutes are generally evaluated in both chemical and physico-chemical terms.

A very good example of this can be found in the surfaces of 1:1 and/or 2:1 layer-lattice structures of clay minerals. The disruption of these layers during formation and other processes will result in broken bonds at the layer surfaces and edges. These broken bonds are crystal atoms for which valences are not completely satisfied or compensated. Cations and anions in the porewater compensate or satisfy these broken bonds through mechanisms of interaction which are classified either as chemical adsorption (chemisorption) or physical adsorption (non-specific adsorption). The chemisorption process is sometimes also defined as specific adsorption. In this process, the ions in the porewater penetrate the coordination shell of the structural atom and are bonded by covalent bonds through the O and OH groups to the structural cations. The non-specific adsorption process (physical adsorption), on the other hand, involves compensation of the broken bonds by electrostatic attraction.

Chemically reactive groups (molecular units) associated with the surfaces of substances under consideration are defined as *surface functional groups*. These render the surfaces of the soil fractions reactive. Whilst surface functional groups are most often used in conjunction with the description of soil organic material, inorganic soil solids also possess such groups. The surface hydroxyls (OH group) is the most common surface functional group in inorganic soil fractions (soil solids), such as clay minerals with disrupted layers (e.g., broken crystallites), hydrous oxides, and amorphous silicate minerals. It is quite common to refer to the surfaces of these inorganic soil fractions as *hydroxylated surfaces* to reflect the presence of these surface hydroxyl groups.

3.2.2 Surface Functional Groups — Soil Organic Matter

A greater variety of surface functional groups exists in the case of soil organic matter. The results obtained in the study reported by Yong and Mourato (1988), using IR spectra information on the soil organics obtained with the extraction method shown in Figure 2.12, are shown in Table 3.1. The schematic in Figure 3.2 shows the common functional groups associated with soil organic matter. Even though these surface functional groups are classed as organic molecular units, they cannot be diluted since they are part of the organic matter itself. The literature shows wide ranges and values for the proportions of each kind of functional group. To a very large extent, this is due to differences in soil organic matter composition, i.e., source material, degradation, and various other processes. Extraction and testing procedures are also prominent factors which contribute to the wide range of values reported. The values shown in Table 3.2 are representative values (Schnitzer and Khan, 1978; Griffith and Schnitzer, 1975; Schnitzer et al., 1973; and Hatcher et al., 1981).

The basic structure of all soil organics is formed by carbon bonds that are combined in saturated or non-saturated rings (salycyclic or aromatic rings, respectively) or as chains. As shown in Figure 3.2, carbon and nitrogen combine with oxygen and/or hydrogen to form the various types of surface functional groups that control most of the properties of organic molecules and their reactions with other materials in the soil-water system. The most common functional groups are hydroxyls,

Table 3.1 IR Spectra of Organic Fractions Extracted from Soil Material Obtained from a Site (Adapted from Yong and Mourato, 1988)

Adsorption Band (cm^{-1})	Organic Fraction	Description
3400	All fractions	OH stretching of free hydroxyls and hydration molecules
3000–2800	Fulvic acids, humins	Aliphatic C–H bonds
2200–2100	Humic acids, humins	COOH vibrations
1725	Humic acids, humins	C=O stretching of fulvic acids' COOH and ketones
1600 (large)	Humic acids, fulvic acids, humins	Aromatic bonds and some overlapping of strongly H-bonded C=O groups
1600 (small)	Non-humic fractions	C–H deformations of aliphatic groups
1400 (large)	Non-humic fractions	C–H deformations of aliphatic groups
1400 (small)	Humic acids, fulvic acids, humins	C–H deformations
1240	All fractions	C–H stretching and OH deformation of COOH
1140	All fractions	OH deformation of phenolic and alcoholic functional groups
1100–1000	Non-humic fractions	Polymeric carbohydrates
950–450	Humins	Vibration of aluminium and silicon elements

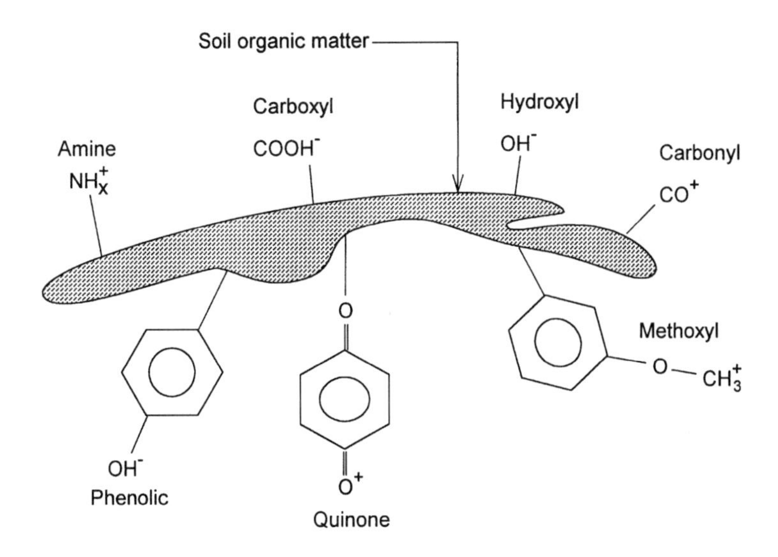

Figure 3.2 Sketch of a soil organic matter (SOM) showing the various kinds of functional groups associated with such a soil material.

Table 3.2 Composition and Functional Groups for Fulvic Acid, Humic Acid, and Humin

	Fulvic Acid	Humic Acid	Humin
% Carbon content	40–50	50–60	50–60
% Oxygen	40–50	30–40	30–35
% Hydrogen	4–7	3–6	NA
Carbonyl, %	Up to 5	Up to about 4	NA
Carboxyl, %	1–6	3–10	NA
Quinone, %	2±	1–2	NA
Ketones, %	2±	1–4	NA
Alcoholic *OH*, %	2.5–4	Up to 2	NA
Phenolic *OH*, %	2–6	Up to about 4	NA

NA = Non-applicable.

carboxyls, phenolic, and amines. They can protonate or deprotonate depending on the aqueous environment pH, i.e., they will develop positive or negative charges depending on the pH of the soil and their respective pK_a or pK_b (i.e., their respective log acidity and log basicity constants). The carboxyl group is the major contributor to the acidic properties of the soil organics.

In addition to the oxygen, hydrogen, and nitrogen contents shown in Table 3.2, these soil organic materials contain some sulphur in the range of 0 to 2% for humic acids and up to about 4% for fulvic acids. Because the hydrogen in the oxygen-containing functional groups can be dissociated, the presence of these surface functional groups will endow the material with acidic properties. The extent of dissociation depends on the pH of the soil-water system and the concentration (and species) of the cations in the fluid phase of the system. Ion exchange will be possible. By and large, carboxyls and phenolic *OH* groups contribute significantly to the cation exchange capacity (CEC) of the soil organic material, and are considered to be the most important functional groups. They contribute significantly to the source of negative charge, which has been reported to range from 2 to 4 meq/g (Greenland and Hayes, 1985). This compares with the charge range of 0.01 to 2 meq/g for clay minerals.

Non-humic materials of soil organics obtained using the procedure given in Figure 2.12 are generally composed of large numbers of aliphatic rings, typical of polysaccharides. These have been shown in Table 3.1. The decrease in aromaticity between humic acids, fulvic acids, and humins reflects the biodegradation sequence of humins, beginning with degradation of non-amorphous organics into humic acids, continuing on to fulvic acids and finally humins.

3.2.3 Surface Functional Groups — Inorganic Soil Fractions

The inorganic soil fractions are composed of clay minerals, oxides and hydrous oxides, carbonates and sulphates, feldspars, zeolites, and amorphous silicate minerals. We mentioned in the first part of this section that surface hydroxyls (*OH* group) constitute the most common surface functional group for clay minerals, oxides and hydrous oxides, and allophanes (amorphous silicate minerals). The nature of the reactive surfaces of the inorganic fractions can be examined in relation to the structure of the layer lattice structures shown previously in Figure 2.3 and the typical structures shown in Table 2.1.

The group of clay minerals with the 1:1 structure (kaolinites and chlorites, Figures 2.8 and 2.10) show siloxane and gibbsite surfaces on opposite basal surfaces of the particles — by nature of the 1:1 structural arrangement. The siloxane surface is defined by the basal plane of oxygen atoms which bound the tetrahedral silica sheet as shown in Figure 2.8. These basal planar surfaces are typical of minerals whose structures have bounding tetrahedral sheets. Thus, whilst kaolinites and chlorites possess one siloxane surface by virtue of the 1:1 structure, the illites, montmorillonites, and vermiculites portrayed in Figures 2.9 and 2.10 show siloxane-type surfaces on both bounding surfaces.

Siloxane-type surfaces are reactive surfaces by virtue of the structural arrangement of the silica tetrahedra and the nature of the substitutions in the layers. The regular structural arrangement of interlinked SiO_4 tetrahedra, with the silicon ions underlying the surface oxygen ions, provides for the development of cavities bounded by six oxygen ions in ditrigonal formation. Where no substitution of the silica in the tetrahedral layer and lower valence ions in the octahedral layers occurs, the surface may be considered free of any resultant charge. However, as we have seen in Figure 2.9, replacement of the ions in the tetrahedral and octahedral layers by lower valence ions occurs through isomorphous substitution. This results in the development of resultant charges on the siloxane surface, thus rendering it as reactive.

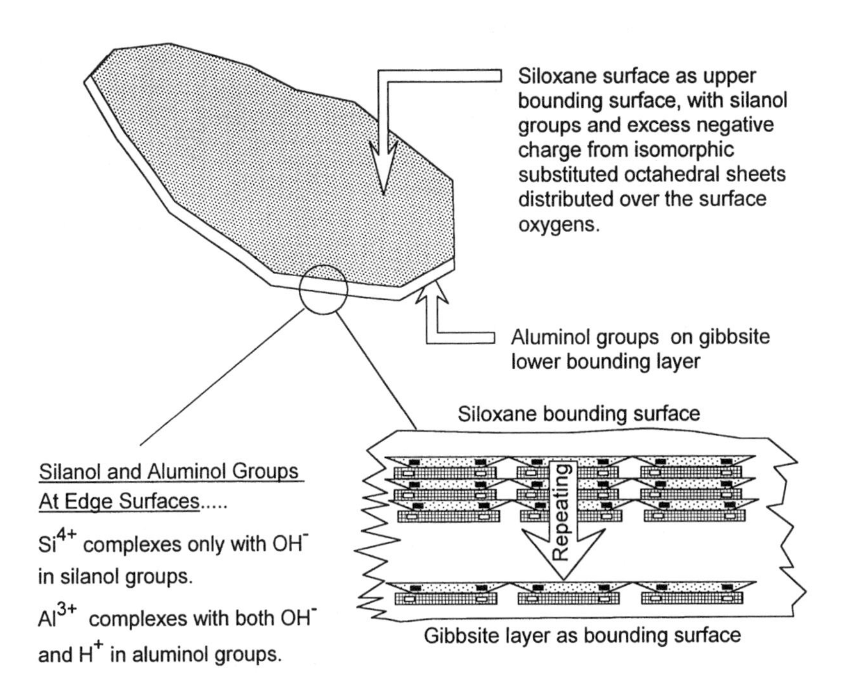

Figure 3.3 Silanol and aluminol on bounding and edge surfaces of kaolinite mineral particle.

When the edges of the layer lattice minerals (phyllosilicates) are broken — as is the case for most crystallites — we have hydrous oxide-types of edge surfaces. In the example sketch shown of a typical kaolinite particle in Figure 3.3, the 1:1 structure gives us a siloxane upper bounding surface and a gibbsite layer at the opposite bounding surface. The amount of silanol groups on the siloxane bounding surface depends upon the crystallinity of the interlinked SiO_4 tetrahedra. Isomorphous substitution in the octahedral sheets with lower valence ions will result in excess negative charges which will be distributed over the surface of the particle.

The surface of the edges of the particle of kaolinite shown in Figure 3.3 contain both silanol and aluminol groups. Whilst the Al^{3+} in the exposed edges of the octahedral sheets complex with both H^+ and OH^- in the coordinated OH groups, the Si^{4+} will complex only with OH^-. Association of the surface hydroxyls with a proton occurs below the point of zero charge (zpc), thereby endowing the surface with a positive charge. Conversely, the donation or loss of a proton by the surface hydroxyls above the zpc will bestow the surface with a negatively charged surface. The gibbsite sheet that acts as the bounding surface will also have aluminol groups. However, whilst these surface aluminol groups will show some of the characteristics as the edge aluminol groups, they do not appear to affect the net negative charge distributed on the bounding surface (Greenland and Mott, 1985).

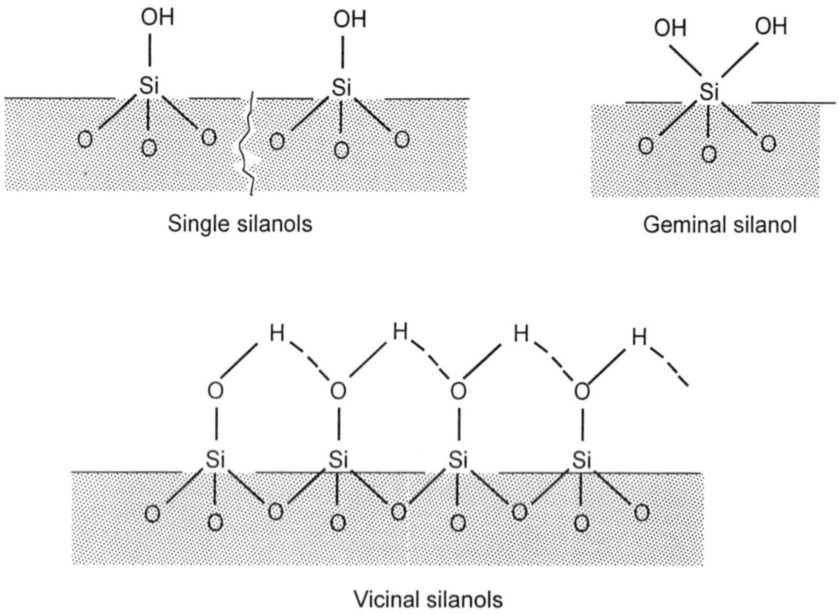

Figure 3.4 Silanol groups on surfaces of silica colloids. (Adapted from Bergna, 1994.)

The surfaces of the hydrous oxides (of iron and aluminium, for example) show coordination to hydroxyl groups which will protonate or deprotonate in accordance with the pH of the surrounding medium. Exposure of the Fe^{3+} and Al^{3+} on the surfaces provides development of Lewis acid sites when single coordination occurs between the Fe^{3+} with the associated H_2O, i.e., $Fe(III) \cdot H_2O$ acts as a Lewis acid site.

By and large, silanol [–SiOH] and siloxane [–Si–O–Si–] functional groups can exist together on the surfaces of the silica tetrahedra. Of the various types of silanol groups, isolated (i.e., single silanols), geminal (silanediol), and vicinal groups are more common (Figure 3.4). Silanol groups may also be found within the structure of the particles. *Siloxane bridges* are formed from the condensation of combined surface and internal silanol groups. The hydrophobicity of the siloxane surface is due to the presence of siloxane groups. Siloxanes tend to be unreactive because of the strong bonds established between the *Si* and *O* atoms and the partial π interactions.

When surface silanol groups dominate, the surface will be hydrophilic. The surface silanol groups are weak acids. However, if strong *H*-bonding is established between silanol groups and neighbouring siloxane groups, the acidity will be decreased. In silanol surfaces, the *OH* groups on the silica surface become the centres of adsorption of the water molecules. If internal silanol groups are present, hydrogen bonding (with water) could exist between these internal groups, in addition to the bonding established by the external silanol groups.

Specific indicators, such as the Hammett and arylmethanol indicators, can be used to distinguish between Lewis and Brønsted acidity (Johnson, 1955). Since Hammett indicators are proton and electron acceptors, the protons from the SiOH and AlOH groups associated with the layer-lattice structures of the clay minerals, together with the electrons from the structural Al, Fe, and exchangeable Fe, will be accepted. Total acidity will be registered as being a combination of both Lewis and Brønsted acidities. By using arylmethonal indicators that will only react with Brønsted acid sites, the Lewis acidity will be obtained as a simple product of the subtraction between measurements obtained from arylmethonal and Hammett indicators. With such techniques, we obtain an appreciation of the contributions made from each of these acid sites.

3.2.4 Electric Charges on Surfaces

The various functional groups at the basal and edge surfaces of the inorganic soil fractions, together with the results of substitution in the lattices of the phyllosilicates, are physically expressed as negative and positive charges distributed on the surfaces of the soil fractions — as is the case for the surface functional groups associated with the soil organics. The example shown in Figure 3.3 demonstrates the above, i.e., the broken octahedral sheets provide for Lewis acid sites $[Al(III)\cdot H_2O]$ that can bind *OH* groups in single coordination. Table 3.3 shows the surface charge densities for some of the more common clay minerals. The primary sources of the charges distributed on the basal surfaces and also on the edges of the mineral particles are shown. Since the minerals have basal planes characterized as

Table 3.3 Charge Characteristics for Some Clay Minerals

Soil Fraction	Surface Area, m^2/g	Range of Charge meq/100 g	Reciprocal of Charge Density nm^2/charge	Source of Charges
Kaolinite	10–15	5–15	0.25	Surface silanol and edge silanol and aluminol groups (ionization of hydroxyls and broken bonds)
Clay micas and chlorite	70–90	20–40	0.5	Silanol groups, plus isomorphous substitution and some broken bonds at edges
Illite	80–120	20–40	0.5	Isomorphous substitution, silanol groups and some edge contribution
Montmorillonite[1]	800	80–100	1	Primarily from isomorphous substitution, with very little edge contribution
Vermiculite[2]	700	100–150	1	Primarily from isomorphous substitution, with very little edge contribution

Note that ratios of external:internal surface areas are highly approximate since surface area measurements are operationally defined, i.e., they depend on the technique used to determine the measurement.

[1] Surface area includes both external and intra-layer surfaces. Ratio of external particle surface area to internal (intra-layer) surface area is approximately 5:80.

[2] Surface area includes both external and internal surfaces. Ratio of external to internal surface area is approximately 1:120.

siloxane and/or hydrous oxides types of surfaces, it is not surprising that the resultant charges distributed on the particle surfaces reflect those charges associated with such surfaces.

The hydrous oxides such as goethite [α-FeOOH] and gibbsite [γ-$Al(OH)_3$] are not shown in the table because the nature and magnitude of their charges are dependent upon: (a) their structure; (b) the specifically adsorbed potential-determining ions; and (c) the pH of the porewater. Charge reversal occurs when we change the pH of the system from pH ranges below the zpc (point of zero charge) to values above the zpc.

3.3 SURFACE CHARGES AND ELECTRIFIED INTERFACE

Probably one of the most significant properties of soil particles is the electrified interface that is manifested when these particles are brought into contact with water. Clustering of counterions around the particles results because of electroneutrality requirements for the reactive surfaces of the particles. The reactions at the particle solid-liquid interfaces are intermolecular interactions. These are both chemical reactions with surfaces and electrical interactions at the surfaces. The chemical interactions are all short-range, and can be evaluated by considering interactions between the surface charges arising from ionization of surface functional groups, complex formations, and proton transfer. The electrical interactions at the surfaces are longer range. These include electrostatic (Coulombic) and polarization interactions.

The combination of chemical and electrical interactions provides the basic forces which, together with the surface properties of contaminants, will control the fate of contaminants.

3.3.1 Net Surface Charges

The charge density for any soil particle is the sum of all the charges acting on the total surface of the particle. Because of the possibility of differences in the signs of the charges between surface and edge charges (minus and plus), strictly speaking, we should refer to the charge densities as the *net surface charge densities*. However, since the term *charge density* is now in common usage — in place of the net surface charge density — this term will be used in this book. Charge reversal occurring in the face of changing pH values is a significant characteristic of several kinds of soil fractions, e.g., hydrous oxides and kaolinites. Hydroxylation produces unequal amounts of H^+ and OH^- on the surfaces, and the specific adsorption of H^+ and OH^- and other cations and anions can be considered as surface coordination reactions at the interface. Charge reversal at the surfaces of the soil particles because of pH changes is the result of proton transfers at the surfaces.

In the absence of *potential-determining ions* (pdis) such as those cations and anions involved in surface coordination reactions, the total surface charge density of a soil particle σ_{ts} can be considered to consist of σ_s the permanent charge due to the structural characteristics of the soil particle (isomorphous substitution), and σ_h

the resultant surface charge density due to hydroxylation and ionization (net proton surface charge density), i.e.,

$$\sigma_{ts} = \sigma_s + \sigma_h \tag{3.2}$$

The net proton surface charge density σ_h can be written as

$$\sigma_h = F\,(\Gamma_H - \Gamma_{OH})$$

where F = Faraday constant, and Γ refers to the surface excess concentration, i.e., surface concentration in excess of the bulk concentration. Since these surface excess concentrations are the adsorption densities, Γ_H and Γ_{OH} refer to adsorption densities of H^+ and OH^- ions and their complexes. At some particular soil solution pH, we will obtain the condition that $|\Gamma_H| = |\Gamma_{OH}|$. At that time, we will reach the *point of zero net proton charge* (pznpc) and the pH associated with this is designated as the pH_{pznpc}. Thus, from Equations 3.2 and 3.3, we obtain

$$\sigma_{ts} = \sigma_s + F(\Gamma_H - \Gamma_{OH}) \tag{3.4}$$

The point of zero net proton charge (pznpc) is not to be confused with the point of zero charge (zpc) or the isoelectric point (iep). Several slightly differing definitions exist in the literature for the zpc and iep. These appear to be related to methods of determination of these particular charge density relationships, and the role of counterions in the inner and outer Helmholtz planes. There is agreement that the pH_{pzc} and pH_{iep} are unique pH conditions that are defined operationally. The *point of zero charge* (zpc), refers to the pH at which titration curves intersect (Figure 3.5), whereas the *isoelectric point* (iep) refers to the pH at which the zeta potential ς is zero (Figure 3.6). The zeta potential ς refers to the electric potential developed at the solid-liquid interface as a result of movement of colloidal particles in one direction and counterions in the opposite direction. It is a measure of the colloidal stability obtained by the balance between positive and repulsive energies. In terms of a soil-water system, the Stern-layer water that generally moves with the soil particle is considered to be part of the colloidal particle. Accordingly, the shearing plane that defines solid-liquid interface for the slipping movement between the clay particles and the counterions contained between the surface of the particles is the outer Helmholtz plane, and the associated potential is generally assumed to be the ς potential. As will be evident in the discussion of diffuse double-layer (DDL) models, the ψ potential describes the potential in the diffuse ion layer, with the maximum value at the outer Helmholtz plane assumed to be the ς potential. The minimum value ψ occurs at a point equidistant between adjacent clay particles.

The two graphs shown as Figures 3.5 and 3.6 indicate that the zpc and iep for the same kaolinite soil are almost identical. This is generally not always the case because the methods for determination of the influence of the charge densities and (influence) of the ions in the inner and outer Helmholtz planes are not the same. The zeta potential ς is computed from experimentally derived measurements made

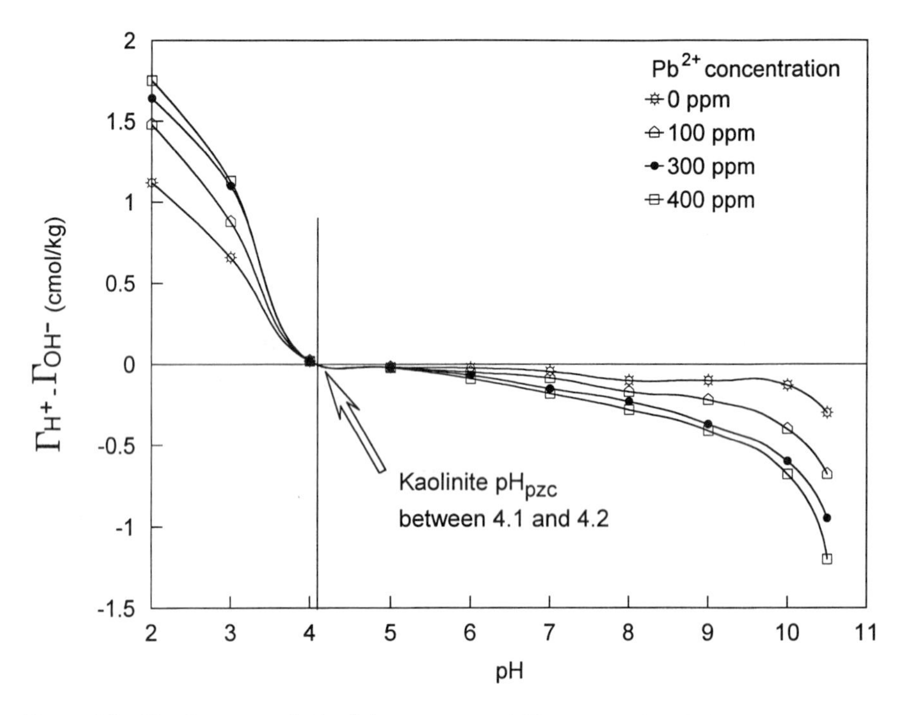

Figure 3.5 Titration curves for kaolinite at constant KCl concentration. Pb^{2+} added in the form of $Pb(NO)_3$.

with a zetameter using the Helmholtz-Smoluchowski relationship. Thus comparisons of iep and zpc made with the different techniques are best viewed as information obtained to describe the operationally defined nature of these unique pH points.

The zpc and iep can also be distinguished according to whether specific adsorption of cations or anions is considered. Defining specifically adsorbed ions as potential-determining ions (pdis), we see that when H^+ and OH^- ions constitute the only potential-determining ions, the pH condition at which the adsorption densities of H^+ and OH^- ions and their complexes are equally balanced is characterized as the pH_{iep}. Denoting Γ as the surface excess concentration, this means that the pH_{iep} is obtained when $|\Gamma_H| = |\Gamma_{OH}|$. This distinction is necessary to distinguish it from the situation where contributions to the particle surface charge from adsorbed ions (from solution) cause changes in the potential of the particle. The presence of specifically adsorbed cations will decrease the pznpc, whereas specifically adsorbed anions will increase the pznpc. This is shown by the net proton surface charge density relationship σ_h in Figure 3.7. The middle pznpc in the figure represents the proton balance condition where H^+ and OH^- ions are the only pdi's. With specifically adsorbed cations (wc), a lower pznpc is registered (shown by $pznpc_{wc}$) and with specifically adsorbed anions (wa), a higher pznpc is obtained ($pznpc_{wa}$).

When Γ_H and Γ_{OH} include contributions to the surface charges from pdi's other than H^+ and OH^- ions, as is the case for many earth alkaline and heavy metals, the pH condition where the sum of all the surface charges is zero is identified as the

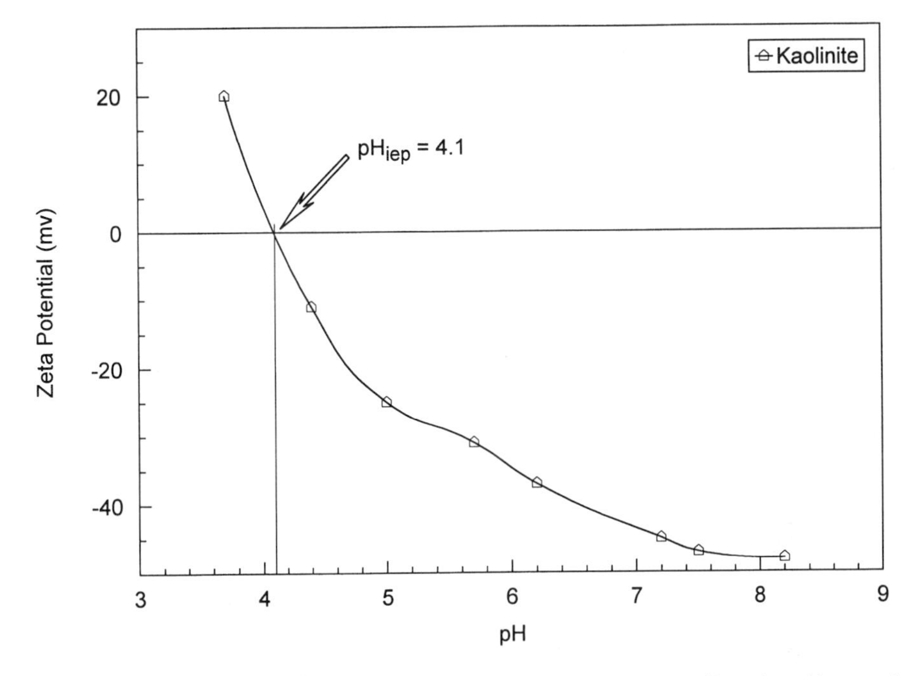

Figure 3.6 Zeta potential for the same kaolinite shown in Figure 3.5. (Data from Yong and Ohtsubo, 1987.)

point of zero charge (pzc). The associated pH value is designated as pH_{pzc}. Strictly speaking, since the point of zero charge is really the point of zero net charge, i.e., $\sigma_{ts} = 0$ (Equation 3.2), the condition pH_{pznc} should be used in place of pH_{pzc}. The condition where the pH_{pznc} is equal to the pH_{iep} has been defined as the *pristine point of zero charge* pH_{ppzc} (Bowden et al., 1980).

3.3.2 Electric Double Layer

When soil particles are brought into contact with an aqueous solution, the reactive surfaces of the soil particles will interact with the ions and molecules in the solution. Because of the charged nature of the surfaces of the particles, the counter-charged ions (i.e., counterions) in solution will interact with the surface charges. Since the net charge on clay particle surfaces is negative, this means that the cations in solution will tend to accumulate near the surface of the particles. The interaction between a negatively charged soil particle surface and the cations in the soil-water will generate an electric double layer (EDL). Together with the diffuse swarm of ions beyond the EDL, the schematic shown in Figure 3.8 provides the basic elements of the diffuse double-layer (DDL) interaction model which permits one to calculate interaction energy between ions and particles.

Diffuse double-layer and other similar types of models have been the subject of considerable study in electrochemistry and colloid chemistry, where the structure of electrified interfaces is a major concern. The assumptions or constraints invoked in

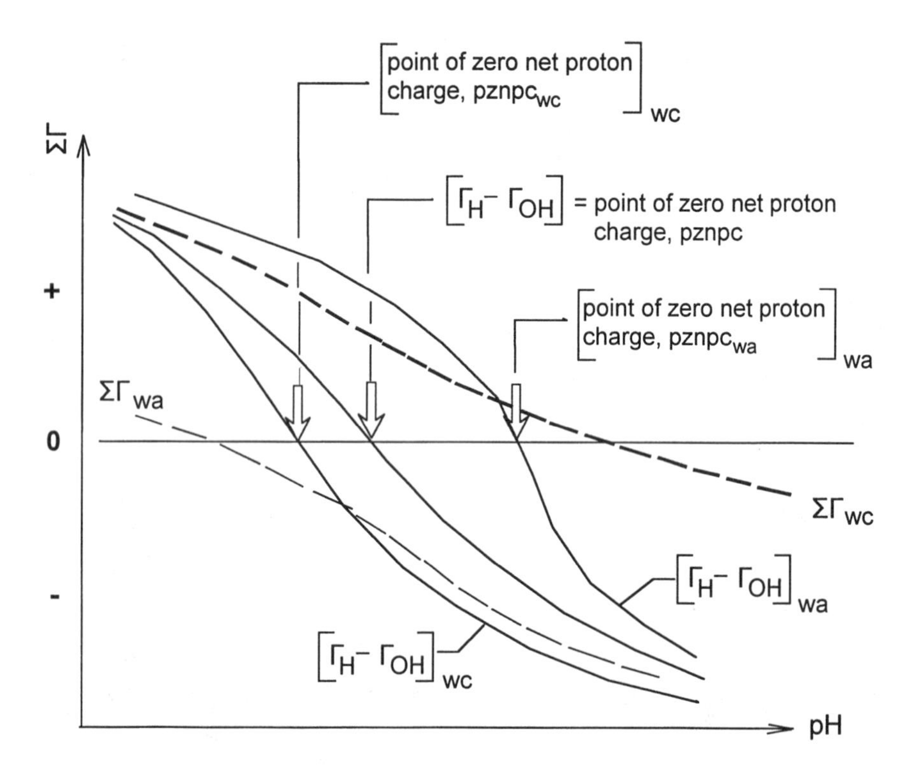

Figure 3.7 Net charge σ_h as determined by proton balance. The subscripts "wc" and "wa" signify *with specifically adsorbed cations* and *with specifically adsorbed anions*, respectively. Note that the middle "pznpc" refers to the condition with only H^+ and OH^- ions as pdis.

respect to the electrified interface impact directly on the calculations and determination of the mechanisms of interactions between solutes in the aqueous solution and the charged surfaces. Macroscopic overviews of soil-water and contaminant-soil interactions can be obtained with the aid of these DDL models. For example, the effect of a change in valency of the counterions in solution can be determined through calculations that will show differences in the equilibrium concentrations of these ions, as illustrated in the bottom diagram shown in Figure 3.8. Whilst the mechanisms of interaction in the Stern layer are somewhat more complex and deserve detailed study and attention, the interactions developed in the diffuse layer portion of the DDL can be well represented by calculations using the model. We will deal with the methods for calculation in the diffuse layer before embarking on a discussion of the layers immediately adjacent to the charged particle surfaces (i.e., Stern layer).

3.4 DIFFUSE DOUBLE-LAYER (DDL) MODELS

The schematic model shown in the top diagram in Figure 3.8, which is highly simplified, represents the basic elements of a charged surface and a swarm of

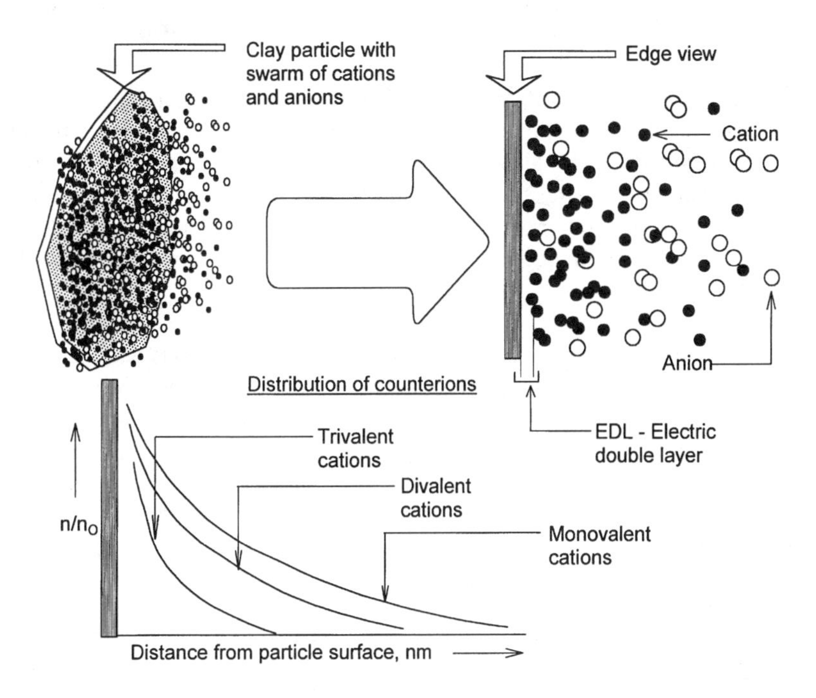

Figure 3.8 Schematic diagram of clay mineral particle with cations and anions in the diffuse ion layer.

counterions in the aqueous solution. The electric double layer (EDL), shown in the top right-hand drawing of Figure 3.8, is contained within the Stern layer. The EDL consists of the layer of surface charge from the soil particle and a single layer of counterions immediately adjacent to the charged particle surface. Whilst convenience in representation shows the negative charges and positive charges (ions) arranged in the fashion shown on the diagram, it would be a mistake to imagine that these ions are stationary and arranged in an orderly fashion. The ions in solution are free to move. Accordingly, the parallel-plate condenser double-layer model first proposed by Helmholtz and Perrin does not strictly apply. In that model, the double layer is represented by two sheets of negatively and positively charged sheets of equal magnitude, and the positive charges are assumed to be stationary. The Gouy-Chapman model of the DDL overcomes these restrictions and allows for the mobility of ions in solution, and permits one to compute the average electrical potential ψ as a function of distance from the charged particle surface. The calculations can provide for a reasonable set of answers for ψ in the diffuse layers for totally parallel particle systems. The Gouy-Chapman model does not, however, provide for an accurate description of ψ immediately adjacent to the charged particle surface because of other mechanisms associated with chemical bonding and complexation. Hence, the simple electrostatic interaction calculations do not apply in the region within the vicinity of the charged particle surfaces.

A correction to the Gouy-Chapman model considering the behaviour of the ions at the interfaces was first proposed by Stern (1924). However, because of its apparent complexity, it was not given much significance. The theory was revised by Kruyt (1952) and Verwey and Overbeek (1948) to account for deflocculation of particles, and by Schofield (1946) to provide the means for calculation of negative adsorption and osmotic pressures. Detailed developments of the basic relationships for the Gouy-Chapman DDL model can be found in Yong and Warkentin (1975) and Yong et al. (1992a). These will not be repeated here. Instead, we will provide the elements of the analytical model in order to highlight the usefulness, applicability, and limitations of DDL models.

By and large, the basis for determination of interaction between the charged ions and solutes (cations and anions) in solution and the charged (soil) particle surfaces derives from the assumption that these interactions are Coulombic in nature. Furthermore, the ions in solution are considered to be point-like in nature, i.e., zero-volume condition. The mathematical description of these interactions obeys two specific conditions:

1. The Coulombic interactions given in terms of the potential ψ are described by the Poisson relationship in respect to variation of ψ with distance x away from the particle surface; and
2. The density of the charges ρ due to the assumed point-like ions that contribute to the interactions (i.e., space charge density) can be described by the Boltzmann distribution.

The resultant basic relationship obtained for ψ is commonly identified as the Poisson-Boltzmann equation, i.e.,

$$\frac{d^2\psi}{dx^2} = -\frac{4\pi}{\epsilon}\sum_i n_i z_i e \exp\left(\frac{-z_i e\psi}{\kappa T}\right) \tag{3.5}$$

where n_i and z_i represent the concentration and valency of the ith species of ion in the bulk solution, and e, κ, ϵ, and T are the electronic charge, Boltzmann constant, dielectric constant, and temperature, respectively.

The solution of Equation 3.5 can be found in many reference textbooks, e.g., Kruyt (1952), van Olphen (1977), Bockris and Reddy (1973), Singh and Uehara (1986), Yong and Warkentin (1975), and Yong et al. (1992c). For the boundary conditions of $y = 0$, and $dy/dx = 0$, where y is a dimensionless potential $= -(e\psi/\kappa T)$, Yong and Warkentin (1975) show that as $x \to \infty$, the following is obtained:

$$\frac{dy}{dx} = 2\sqrt{\frac{8\pi z_i^2 e^2 n_i}{\epsilon\kappa T}}\sinh\left(\frac{y}{2}\right) \tag{3.6}$$

By imposing the boundary condition where $y \to \infty$ at $x = 0$, we will obtain the solution for y. Replacing the dimensional potential y with $-(e\psi/\kappa T)$ provides the solution for the distribution of the potential ψ in respect to distance x from the soil particle surface.

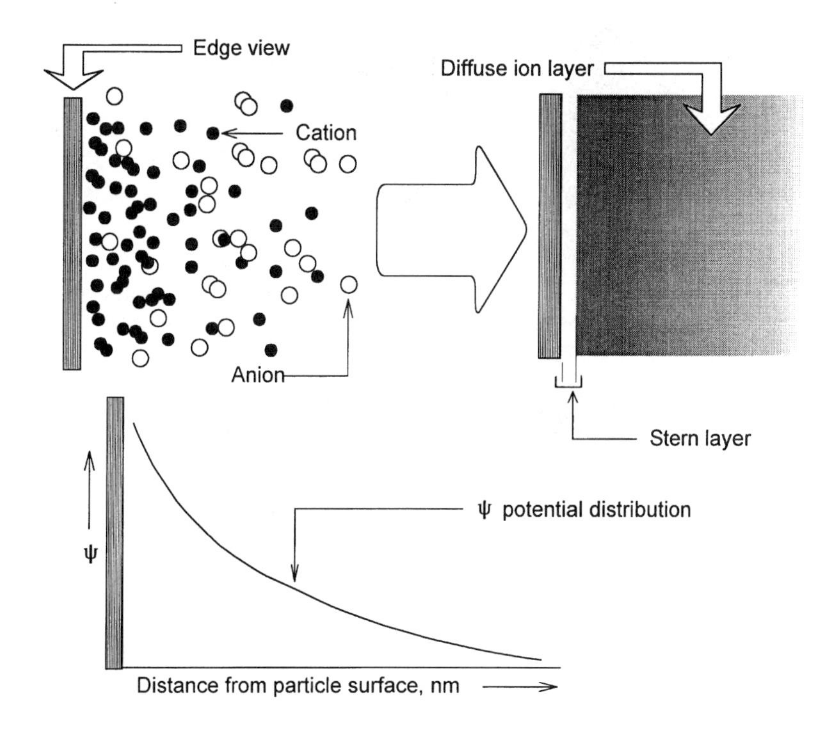

Figure 3.9 ψ potential distribution.

$$\psi = -\frac{2\kappa T}{e}\ln\coth\left(\frac{x}{2}\sqrt{\frac{8\pi z_i^2 e^2 n_i}{\epsilon\kappa T}}\right) \tag{3.7}$$

The negative sign on the right-hand side of Equation 3.7 indicates that the potential ψ decreases as one proceeds further away from the particle surface, as can be seen in the bottom diagram in Figure 3.9 and in the diagram given in Figure 3.10 which describes the resultant ψ obtained from the interaction between two adjacent parallel particles. The diagrams show the potential ψ beginning from the outer boundary of the Stern layer. The distribution of ψ within the Stern layer will be discussed in the next subsection. The distribution of ions in relation to distance x from the particle surface can be obtained from the Boltzmann relationship and Equation 3.7. Thus for example, the distribution of cations n_+ shown in the bottom diagram in Figure 3.8 can be obtained from:

$$n_+ = n_i\left(\coth\frac{x}{2}\sqrt{\frac{8\pi z_i^2 e^2 n_i}{\epsilon\kappa T}}\right)^2 \tag{3.8}$$

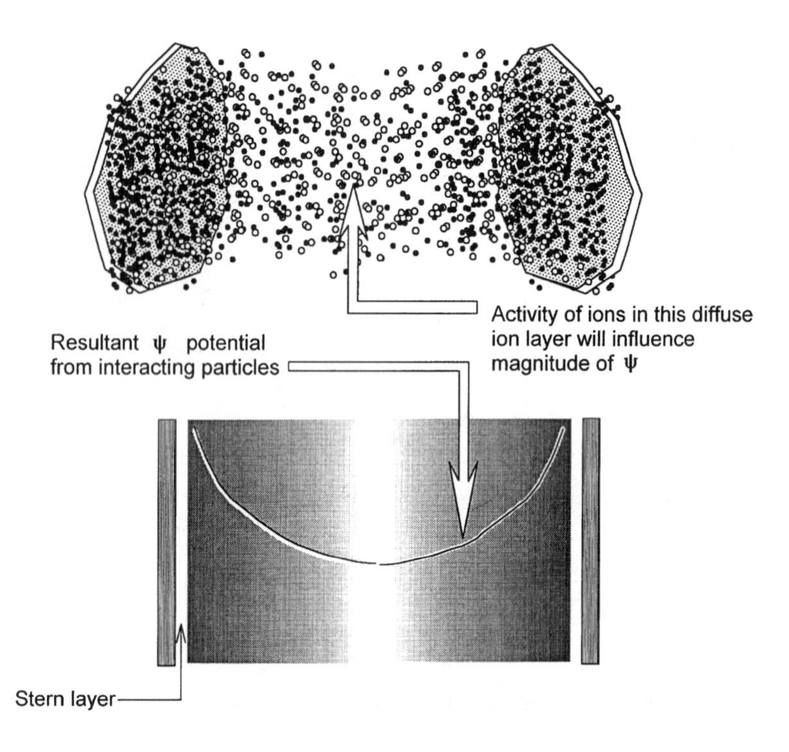

Figure 3.10 Interaction between two adjacent particles and resultant ψ.

The example given by Yong et al. (1992a) shows that for the situation of a 0.001 *M* salt with monovalent ions, the value of n_+ at a distance of $x = 50$ *nm* is calculated to be 0.016 *M*, which is 16 times greater than the bulk solution concentration. In the case of divalent ions, the calculated concentration is 0.004 *M*, which is one-fourth the value of the monovalent cations.

3.4.1 Stern and Grahame Models

The top right-hand diagram in Figure 3.8 shows the electric double layer (EDL). This layer is included within the Stern layer. The Stern model (1924) considers that the total counterions needed to balance the net negative charges from the reactive particle surface consists of two distinct groups (of ions). The first group is arrayed immediately next to the particle surface at a separation distance of δ from that surface. The counterions that constitute this group are considered to be adsorbed to the surface. This group of counterions and the surface charges on the clay particle surface essentially constitute the EDL. Note that the assumption of negative charges in the soil particle being concentrated at the surface of the particle (i.e., surface charges) is made to simplify calculations in most DDL models. In a sense, the combination of the negative charges on the particle surface and this first layer of adsorbed ions is a molecular condenser. The other group of counterions is considered to be diffused in a cloud surrounding the particle, i.e., similar to the diffuse layer of the Gouy-Chapman model, and can be described by the Boltzmann distribution. The surface charge σ_s is balanced by the Stern layer charge σ_δ and the diffuse-layer charge σ_{ddl}, i.e., $\sigma_s = \sigma_\delta + \sigma_{dl}$.

The Stern model addresses the problem of high capacity calculated from the Gouy theory by limiting the capacity to the properties of the molecular condenser. Although the Stern model initially considered the capacity of the molecular condenser to be a constant value, present interpretation considers the capacity to be property dependent. This point is particularly significant since these properties are a function of specifically adsorbed ions in the immediate array adjacent to the particle surface. The surface potential ψ_s varies in accord with the electrolyte concentration and charge characterization of the soil particle, i.e., whether the surface of the particle is a constant charge surface or a pH-dependent charge surface. It drops from ψ_s to a Stern layer potential ψ_δ as one progresses from the surface of the reactive particle to the outer boundary of the Stern layer. Beyond this boundary, the potential ψ is governed by the relationship given in Equation 3.7. We should point out that in electrokinetics, it is commonly argued that ψ_δ can be considered as equal (or almost equal) to the zeta potential ς.

The relationship between the surface charge density σ_s and surface potential ψ_s is

$$\sigma_s = \left(\frac{2n_i \epsilon \kappa T}{\pi}\right)^{1/2} \sinh \frac{z_i e}{2\kappa T} \psi_s \tag{3.9}$$

Since the surface charge potential ψ_s is constant for a specific pH value for variable charge minerals, the Nernst relationship (Equation 3.10) for surface potential can be used to account for the dependence of the surface potential ψ_s on the presence of potential determining ions (pdis).

$$\psi_s = \frac{2.303 \kappa T}{\epsilon}(pH_o - pH) \tag{3.10}$$

where pH_o = pH at which the surface potential $\psi_s = 0$.

The Grahame modification considers this Stern layer to consist of pdis as the immediate array of sorbed ions. This array of ions constitutes the *inner Helmholtz plane* (ihp). These pdis are *specifically adsorbed ions*. This means that the ions are adsorbed by forces other than the electric potential and thus have the ability to influence the sign of the potential. Since ions that are chemisorbed will affect the sign and magnitude of the surface charge, we can conclude that pdis are chemically adsorbed ions. Accordingly, they will determine the interfacial potential difference (galvanic potential) between the solid and liquid phases.

The hydrated layer of ions next to the specifically adsorbed ions reside in the *outer Helmholtz plane* (ohp). The ohp is considered to be the outer boundary of the

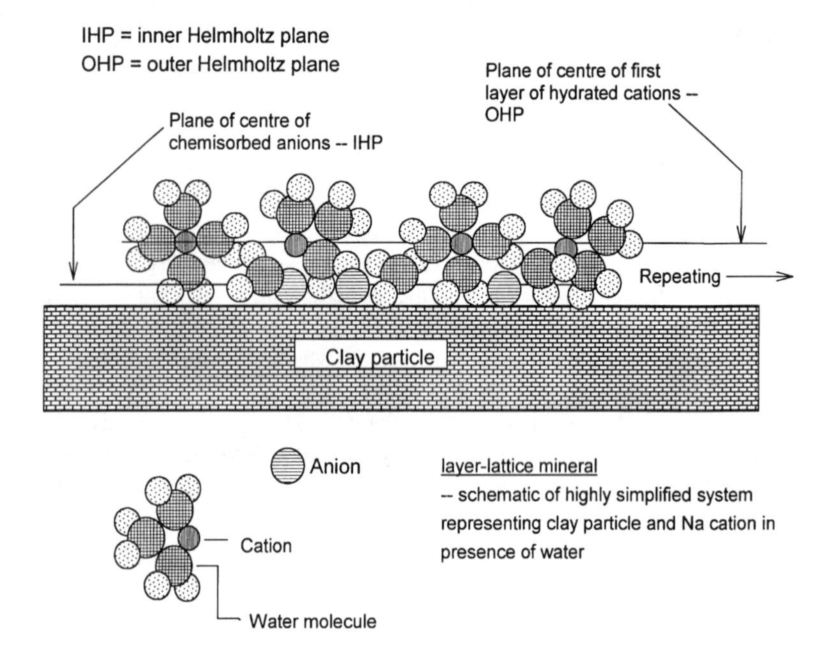

Figure 3.11 Schematic diagram showing distribution of cations and anions immediately adjacent to reactive particle surface.

Stern layer, and strictly speaking, this Grahame modification of the Stern model should be identified as the *Stern double layer* model. Accordingly, the Stern layer potential ψ_δ is assigned to the ohp. Figure 3.11 shows a simplified schematic of the hydrated anions and cations immediately adjacent to a reactive particle surface. The location of both the ihp and the ohp are also shown in the diagram. The relationship between the various potentials can be seen in Figure 3.12. Designating the potential at the ihp and ohp as ψ_{ihp} and ψ_{ohp}, respectively, and ψ_s as the surface potential, the schematic diagram shows their relationship to distance x away from the particle surface. At the ohp, ψ_{ohp} is taken to be equal to ψ_δ, and, from that point onward, ψ can be determined from the relationship given as Equation 3.7.

3.4.2 Validity of the DDL Models

The basic DDL model (i.e., Gouy-Chapman model) is most often used to provide for scoping calculations in contaminant-soil interactions. By this, we mean that rough calculations can be made regarding contaminant interaction forces and energies with soil particles, for purposes of qualitative comparisons. Accordingly, it is important to recognize its limitations and applicability. First and foremost, the model assumes ideal conditions and behaviour of the ions in the double layer. Contaminant-soil

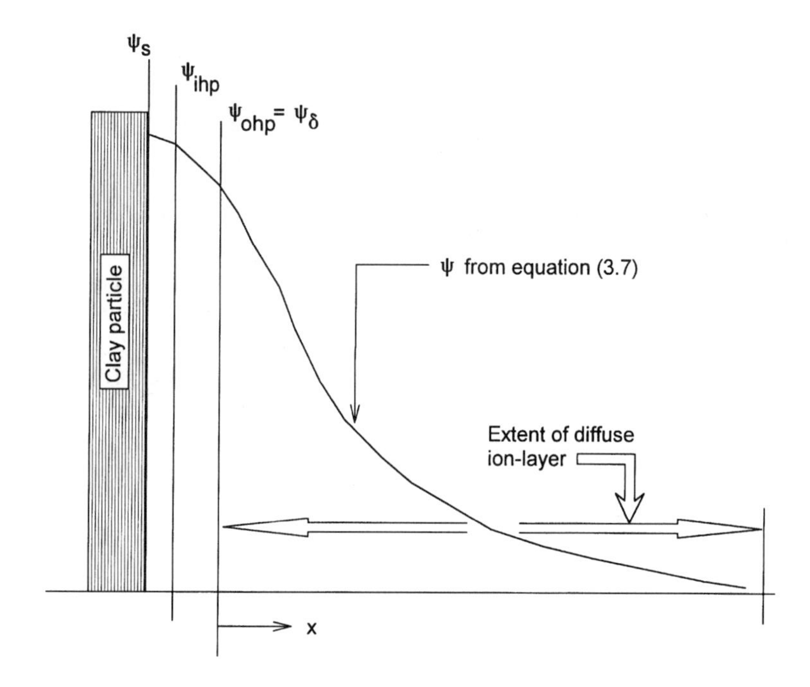

Figure 3.12 Location of the various potentials in relation to particle surface.

interactions in field situations are most often furthest away from ideal conditions and behaviour. The basic Boltzmann relationship leading to Equation 3.5 equates the potential energy of the ion in the region of the charged soil particle to the work involved in bringing that same ion from the bulk solution — to the exclusion of all other energy components. The energies of interaction between the ion and surrounding ions, particle, and water, together with the polarization energy of the ion in the electric field and the influence of dielectric saturation in the diffuse ion layer constitute the energy components that could significantly affect the calculations. These have been studied by Bolt (1955) who developed approximate relationships to account for the effects of these other energy components on the electric double layer. These were incorporated into the generalized Poisson-Boltzmann equation described in Equation 3.5, and the results obtained showed that when the concentrations of the co-ions were small, the corrections required were correspondingly small — particularly if the surface charge density does not exceed 2×10^{-7} meq/cm^2, and when the surface charge density is constant.

3.4.3 Interaction Energies

Recognizing the limitations imposed by assumptions of idealized conditions and behaviour of ions, we will determine the interaction energies between the ions, molecules and the charged surface sites using the Stern double-layer model. The concentration of ions of species i in the *ihp* can be obtained from the Boltzmann relationship as follows:

$$n_{i,\,ihp} = n_i \exp\left(\frac{-z_i e \psi_{ohp}}{\kappa T}\right) \exp\left(\frac{E_{ohp} - E_{ihp}}{\kappa T}\right) \tag{3.11}$$

where E_{ihp} and E_{ohp} refer to the interaction energies in the *ihp* and *ohp*, respectively. The interaction energy E_{ihp} in the *ihp* results from Coulombic interaction between the ions and the forces associated with the negative charge sites on the particle. The interaction energy E_{ohp} in the *ohp*, which is also the interaction energy of the Stern layer E_δ, consists of four energy components: (a) Coulombic interaction energy E_c; (b) ion-dipole interaction E_{id}; (c) dipole-dipole interaction E_{dd}; and (d) dipole-site interaction E_{ds}. Yong et al. (1992a) gives these interaction energies as follows:

$$E_{ihp} = \frac{z_i e^2}{\epsilon R} \tag{3.12}$$

$$E_{ohp} = \frac{z_i e^2}{\epsilon R} + \frac{\mu e}{\epsilon r^2}\left(1 - \frac{D_n \mu^2}{\epsilon r^3}\right) + \frac{\mu e}{\epsilon r_1^2}$$

where the dipole moment of the water molecule $\mu = 1.8 \cdot 10^{-10}$ esu cm, r = sum of the radii of the ion and water molecule, R = distance between the centre of ion i and the negative charge site on the particle, r_1 = distance between centre of the dipole and corresponding negative charge site on the particle, and D_n = geometrical factor = 0.334 for 3 water molecules (1.188 for 6 water molecules).

The relationships can be used to calculate the distribution of ions in the Stern layer and also in the diffuse ion layer. This has been shown by Alammawi (1988) in Figure 3.13 for a montmorillonite particle interacting with 10^{-3} *M* NaCl. The abscissa in the figure is given in terms of the number of layers of water distant from the particle surface. In this instance, the space occupied by the *ihp* is presumed to be equivalent to a single molecular layer of water. The same assumption also applies to the *ohp* which also serves as the boundary of the Stern layer. The distances represented as equivalent layers of water molecules from layer 3 onward are the diffuse ion layer. As expected, the calculations show high concentrations of sodium ions in the *ihp* and *ohp*. These concentrations drop by one order of magnitude as one enters the diffuse ion layer.

Whilst the calculations obtained using these relationships account for many of the other energy components identified in the previous section, care must be taken to ensure that the assumptions involved in idealizing the system are closely matched by the actual systems. The presence of other interacting particles could complicate the calculations. However, the energies of interactions for all of these can be accommodated in the final sets of calculations, e.g., using the Van't Hoff relationship. The use of these kinds of calculations give us an insight into particle-solute interaction and water-holding capacities. These are important phenomena in considerations dealing with the transport and fate of pollutants.

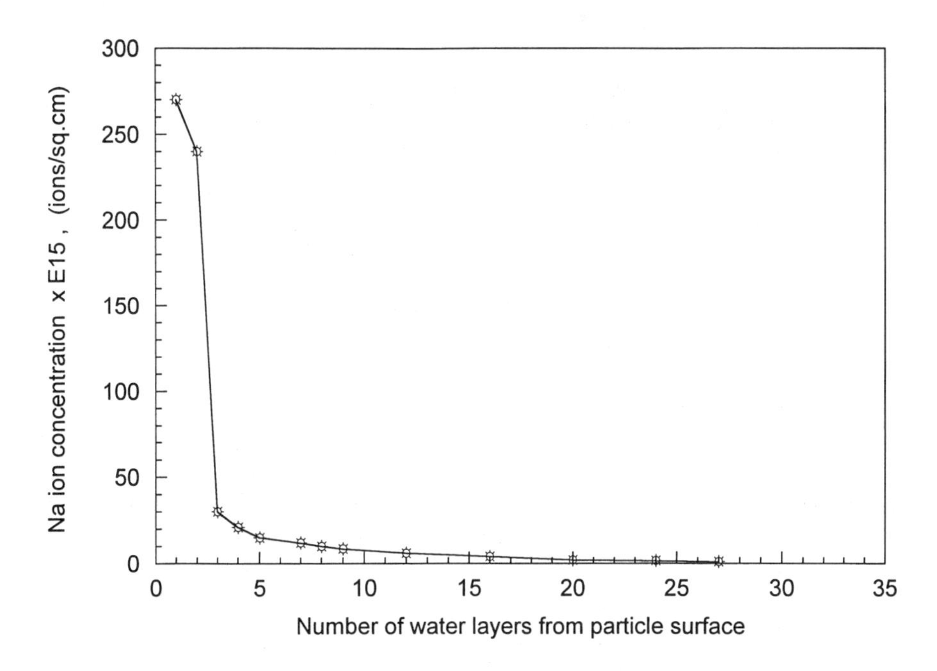

Figure 3.13 Distribution of sodium ions at distances away from the surface of a montmorillonite particle in interaction with 10^{-3} *M* NaCl. (Data from Alammawi, 1988.)

3.4.4 DLVO Model and Interaction Energies

The DLVO model (named for Derjaguin, Landau, Verwey, and Overbeek) is essentially an interaction energy model. It takes into account the nature of the charged soil particle surfaces, the chemical composition of the soil-water system, and the soil fabric (particle arrangement and particle separation distances) in its calculation of the interparticle or interaggregate forces. The calculations consider van der Waals' attraction and the DDL repulsion developed in the Gouy (diffuse ion) layer as the primary factors in the development of the energies of interaction between the particles. The particle interaction models reported by Flegmann et al. (1969) are used as the basis for calculation of the maximum energies of interparticle action for similar charged surfaces in face-to-face and edge-to-edge particle arrangement. In the case of dissimilar charged surfaces (edge-to-face), the relationship given by Hogg et al. (1966) is used.

Assuming (a) constant surface potential surfaces and (b) potential determining ion influence on surface potentials, the energy of repulsion between interacting parallel-faced particles can be obtained as:

$$E_2^{ff} = \frac{4n_i\kappa T(yz_i)^2\exp\left(\frac{-D_H d}{2}\right)}{D_H[1+\exp(-D_H d)]} \tag{3.13}$$

where:

E_r^{ff}	= long-range repulsion energy,
superscript ff	= parallel-faced particle interaction,
z_i, n_i	= valence and bulk concentration of counterions, respectively,
κ	= Boltzmann constant,
T	= absolute temperature,
y	= the dimensionless potential previously used in Equation 3.6 = $-e\psi/\kappa T$
D_H	= Debye-Hueckle reciprocal length, and
d	= distance between particles.

The energy of repulsion between particle faces and particle edges is modelled on the basis of interaction between a large sphere with radius a_f with a corresponding potential ψ_f, and a small sphere of radius a_e and its corresponding potential ψ_e. Using the superscipts fe to refer to particle surface-particle edge, and denoting the dielectric constant as ϵ, the repulsion energy E_r^{fe} can be obtained as:

$$E_r^{fe} = \frac{\epsilon a_f a_e(\psi_f^2 + \psi_e^2)}{4(a_f + a_e)}\left[\frac{2\psi_f\psi_e}{\psi_f^2 + \psi_e^2}\ln\frac{1 + \exp(-D_H d)}{1 - \exp(-D_H d)} + \ln[1 - \exp(-2D_H d)]\right] \tag{3.14}$$

From the account given in Yong et al. (1992a) the repulsion energy for particle edge-particle edge interaction to edge interaction, E_r^{ee}, and the attraction energies for particle surface-particle edge interaction, E_a^{fe}, and particle edge-particle edge interaction are obtained as follows:

$$E_r^{ee} = \frac{\epsilon a_e \psi_e^2}{2}\ln[1 + \exp(-D_H d)] \tag{3.15}$$

$$E_r^{ee} = -\frac{A}{12}\left(\frac{1}{d_a^2 + 2d_a} + \frac{1}{d_a^2 + 2d_a + 1} + 2\ln\frac{d_a^2 + 2d_a}{d_a^2 + 2d_a + 1}\right) \tag{3.16}$$

$$E_a^{fe} = -\frac{A}{12}\left(\frac{r_a}{d_a^2 + d_a r_a + d_a} + \frac{r_a}{d_a^2 + d_a r_a + d_a + r_a} + 2\ln\frac{d_a^2 + d_a r_a + d_a}{d_a^2 + d_a r_a + d_a + r_a}\right) \tag{3.17}$$

The attraction energies for interaction of parallel-faced particles, E_a^{ff}, uses the London-van der Waals attraction energy relationship as follows:

$$E_a^{ff} = \frac{A}{12\pi d^2} \tag{3.18}$$

where A represents the Hamaker constant, and is generally assumed to be about $4.4 \cdot 10^{-13}$ ergs.

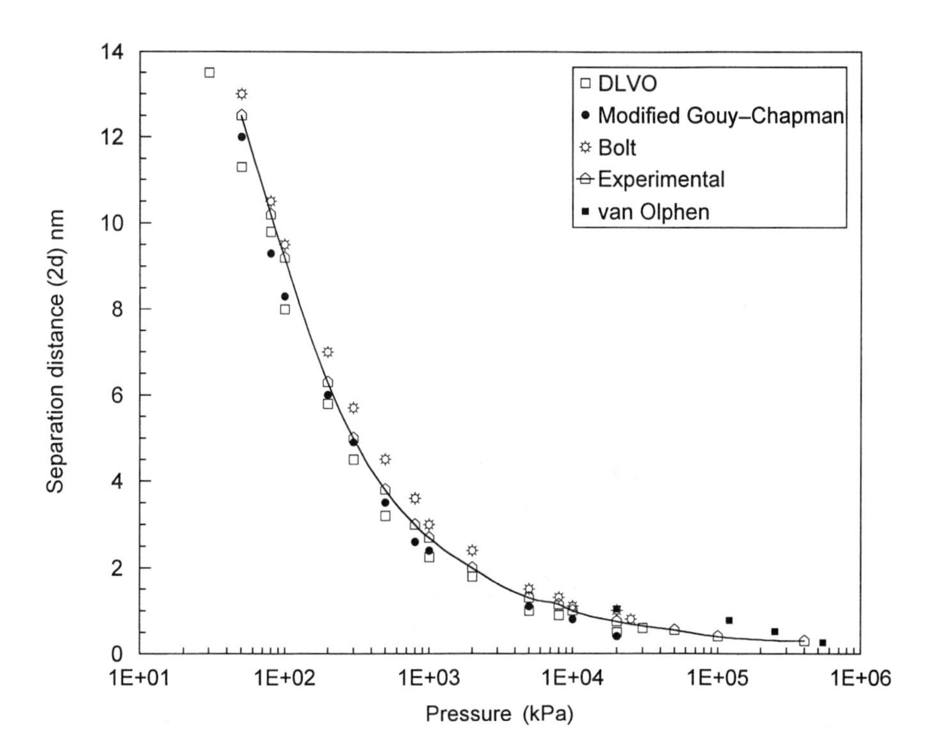

Figure 3.14 Comparison of calculated swelling pressures with measured values. Experimental curve is for tests reported by Alammawi (1988).

Calculations of swelling pressure (i.e., repulsion energies converted to swelling pressure equivalents) in relation to interparticle spacings, using the DLVO model, can be obtained for ideal systems where particle separation distance is above 3 nm. So long as the interparticle separation distance is above 3 nm, close matching between calculated and measured values can be obtained. At lower particle separation distances, the repulsion energies tend to be overwhelmed by van der Waals attractive forces. The results shown in Figure 3.14 are from Yong (1999b). The reported data points are calculated values of swelling pressure at various interparticle separation distances for a 10^{-3} M NaCl montmorillonite, using the DLVO and modified Gouy Chapman models. The data points attributed to Bolt are from the results reported by Bolt (1956), and the data points for van Olphen were interpreted from calculations reported by van Olphen (1977) for pressures required to remove the first four layers of water immediately next to a montmorillonite soil. The results from the DDL models and the DLVO model show close agreement with the experimental values reported by Bolt (1955, 1956). The measured values obtained from experiments conducted by Alammawi (1988) are identified as "experimental" in the figure.

3.5 INTERACTIONS AND SOIL STRUCTURE

We discussed soil structure and soil fabric in general terms in the last chapter without paying particular attention to interactions of the fractions with water, i.e.,

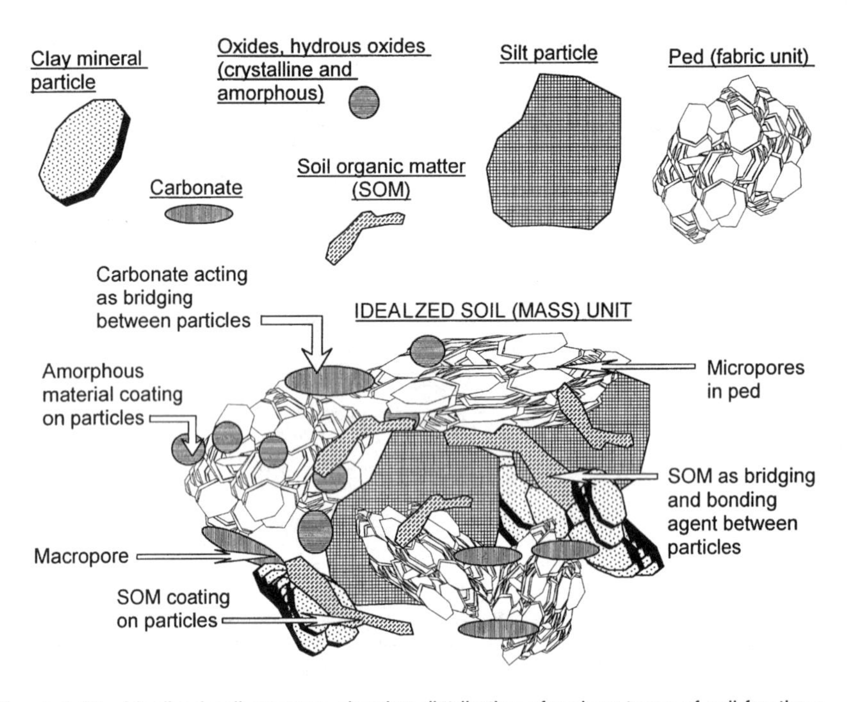

Figure 3.15 Idealized soil structure showing distribution of various types of soil fractions.

the soil-water system. The immediate structure of a soil is a direct function of the type and distribution of reactive surfaces and functional groups available, and the processes of hydration of the various fractions. The principal features of soil structure that impact directly on the study of contaminated soils are those related directly to the surface and transport properties of the soils — with the essential ones being the cation exchange capacity (CEC) of the soils, specific surface area (SSA), hydraulic conductivity, and partition coefficient. These properties are all seen to be dependent on the macro and micro structure of the soil, and on the manner in which the various fractions are distributed in the soil. In particular, the coating of particles by amorphous and soil organics, and the bonding between particles and peds are processes that directly affect the surface properties of the soil. Figure 3.15 shows a schematic picture of the major features of an idealized soil structure.

Carbonates in soils function either as individual particles that connect with other particles, or as coatings on soil particles. In any event, they will alter the SSA of the soil because of aggregation of particles resulting from carbonate connections. In all likelihood, the SSA will be decreased if carbonates are present in the soil-water system, in comparison to a carbonate-free soil. The results from Quigley et al. (1985), shown in Figure 3.16 regarding the influence of carbonate content on the activity and SSA of a freshwater varved clay, highlight the aggregating effect of carbonate presence in the clay. The relative activity shown in the ordinate is the ratio of the plasticity index (i.e., liquid limit ω_l, minus plastic limit ω_p) divided by the SSA of the soil. This is a useful index since it combines the influence of the

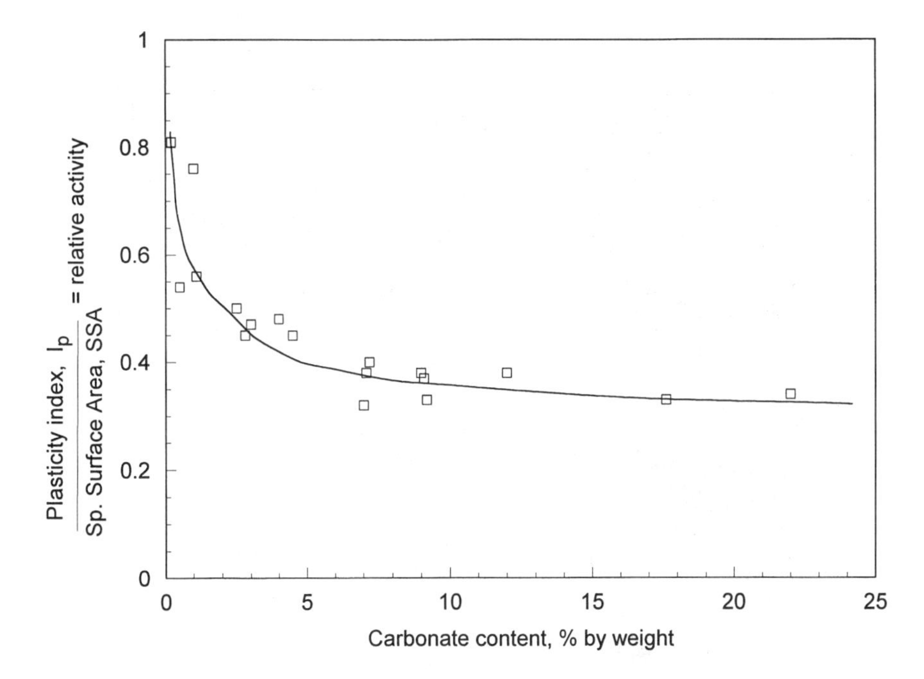

Figure 3.16 Influence of carbonate content on relative activity of a freshwater varved clay. (Data from Quigley et al., 1985.)

carbonates on the activity of the clay (I_p) with the aggregating effect (by the carbonates) through the use of the SSA. The CEC may or may not be changed since the CEC of carbonates is relatively small, and its contribution to the overall CEC of the soil may not be effectively felt.

Calcite ($CaCO_3$ (s)) mineral functions well as a buffer for pH changes in the soil-water. Its precipitation and dissolution characteristics contribute to the chemical regime in the porewater. The surface charges on the minerals are influenced by Ca^{2+} and CO_3^{2-}. The surface charges are predominantly positive at normal groundwater pH values above its zpc of around pH 4.5. Below the zpc, the surface charges become increasingly negative in character.

Amorphous materials (e.g., amorphous iron, alumina, and silica hydrous oxides) form coatings on soil particles and bonds between particles. In doing so, they not only create aggregation of particles, but also change the surface charge characteristics of the particles that are coated. The schematic model shown in Figure 3.17 is derived from analyses of performance of a sensitive marine clay (Yong et al., 1979). The structural organization of the various elements of the amorphous material model is similar to the Cloos et al. (1969) model for amorphous silico-aluminas. The highly simplified sketch shown in Figure 3.17 illustrates the coating and bridging effect. The amorphous material consists of a core unit and an outer layer. The core of the amorphous material consists primarily of silicon in tetrahedral coordination with some isomorphic substitution of Si with Fe or Al, and is partially coated with Fe or Al in octahedral coordination. The outer layer of the amorphous material consists

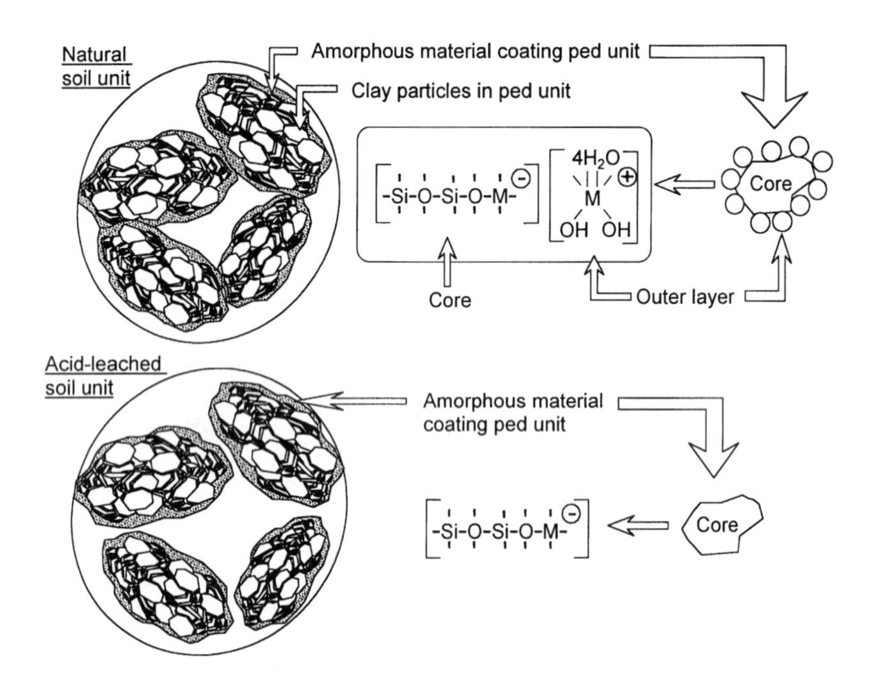

Figure 3.17 Schematic of a natural sensitive marine clay showing amorphous material coating ped units. Top diagram shows natural soil structure and bottom diagram shows structure after acid leaching. (Adapted from Yong et al., 1979.)

of Fe and Al, and is destroyed when exposed to acid leaching. The central core that remains is negatively charged, as seen in the bottom diagram in Figure 3.17.

The manner or sequence by which amorphous materials are exposed to the various soil fractions can also influence the development of surface charges on the resultant reactive surfaces. Because of the amphoteric nature of the surface of amorphous oxide materials, reversal of the sign of the surface charges occurs as one progresses from below the pH_{iep} to pH levels above the pH_{iep}. Figure 3.18 shows what happens when these materials are brought into contact with a variable charge clay mineral such as kaolinite. The ferrihydrite (amorphous iron oxide) prepared initially at a pH of 3.0 shows a dominant positively charged surface. At this pH, the edge charges of the kaolinite mineral particles are also positive. The surface charges of the kaolinite that are primarily due to isomorphous substitution effects remain negative. The top left-hand drawing in Figure 3.18 shows the kaolinite mineral particle well coated by the ferrihydrite, similar to the previous model shown in Figure 3.17. Increasing the pH of the mixture of ferrihydrite and kaolinite (top of Figure 3.18) serves only to change the sign of the surface charge of the coated kaolinite particles — going from predominantly positive (due to the positive charges of the amorphous coating) to predominantly negative. At pH levels above 8, the ferrihydrite begins to become predominantly negative in surface charge.

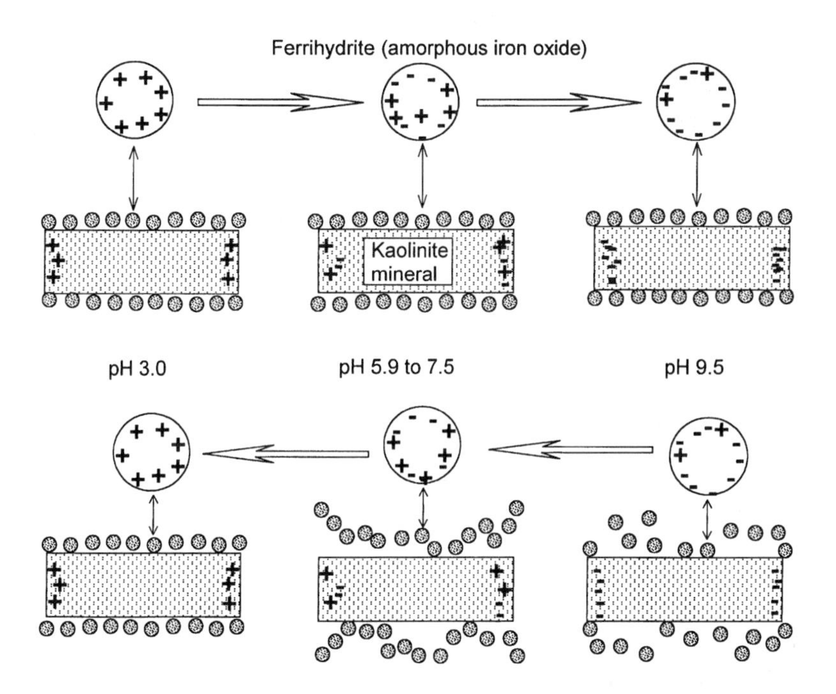

Figure 3.18 Model of mixtures of ferrihydrite (amorphous iron oxide) and kaolinite prepared at different pHs. Top row shows mixture prepared at pH 3.0 whilst bottom row shows mixture prepared at pH 9.5. (Adapted from Yong and Ohtsubo, 1987.)

The bottom "pictures" in Figure 3.18 show that when the mixture is formed at high pH levels (pH 9), the negative surface charges on the ferrihydrite will not be accommodated by the negatively charged surfaces (and edges) of the kaolinite particles. As the pH level is reduced, both the ferrihydrite and the edges of the kaolinite begin to become positively charged. At pH 3, we observe the same effect as when the mixture was formed at pH 3. What is interesting about the entire picture is apparent in the middle and right-hand portion of Figure 3.18. The arrangement (distribution) of the amorphous material (ferrihydrite or amorphous iron oxide) remains well-sorbed to the kaolinite particles when the mixtures are formed below the iep of both the kaolinite and the ferrihydrite. The same is not true for the mixtures formed at pH 9. The Bingham yield stress measurements shown in Figure 3.19 for the mixtures and the kaolinite and ferrihydrite confirm the arrangement (distribution) and structure of the amorphous materials, as shown in the models depicted in Figures 3.17 and 3.18. The yield stress of the iron oxide by itself reaches a peak value at around a pH of 9. The differences in the yield stress curves for mixtures formed at the high pH and low pH confirm that the iron oxide remains unattached to the kaolinite particles when formed at high pH. When the mixtures are formed at low pH, interactions producing the resultant yield stress are controlled by the oxide-coated particles.

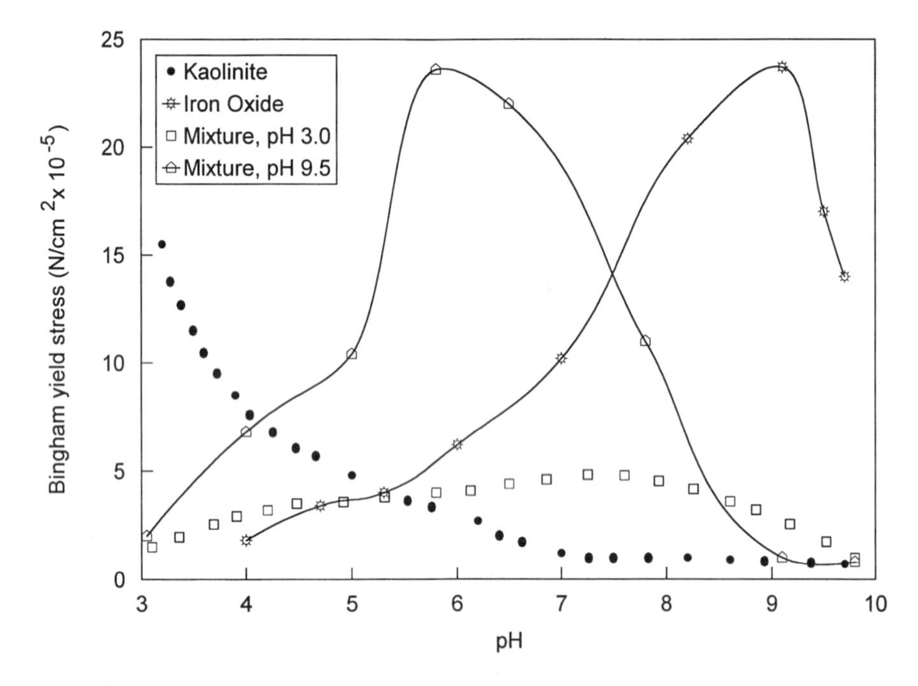

Figure 3.19 Variation of Bingham yield stress for kaolinite, ferrihydrite, and mixtures in relation to pH. (Adapted from Yong and Ohtsubo, 1987.)

Depending on the nature of the surfaces of soil organics (i.e., charge and surface functional groups) they can significantly influence the surface properties of soils through aggregating and coating effects. Studies on interactions and soil (kaolinite) structure development with anionic carboxylic and nonionic hydroxylic *non humic organics* (NHO) produced by soil bacteria (Yong and Mourato, 1990) show:

- Development of edge attraction between kaolinite protonated edge hydroxyls and negative functional groups of the NHO. Of the free hydroxyls on the edge surfaces, IR studies showed that the higher energy hydroxyls participated more in complexation than the less energetic ones. The characteristic high energy band –OH (3950 cm^{-1}) and the kaolinite (3280 cm^{-1}) band disappeared, whilst the less energetic energy band –OH (3900 cm^{-1}) remained visible.
- Direct clay lattice interactions with NHO are most probably due to chemisorption, water bridging (H bonding), and ion-dipole interactions.

3.5.1 Swelling Clays

Swelling clays are commonly used liner materials in the management of leachate transport through specially designed engineered barriers. The unit layer structure of smectites and the various features that render some of them as swelling clays (e.g., montmorillonites, beidellites, and nontronites) have been discussed in Chapter 2. Greater details can be found in many textbooks and research studies, e.g., Sposito

(1984), Pusch and Karnland (1996), and Yong and Warkentin (1975). A significant feature in swelling clays can be seen in the water uptake characteristics of the 2:1 unit layer structure of montmorillonites saturated with different exchangeable cations (see Figure 2.10). The basal spacing *d*(001) of 0.95 nm to 1.0 nm for the anhydrous smectite will expand from 1.25 nm to 1.92 nm, depending on the amount of water intake (hydration). Except for Li and Na as exchangeable cations in the interlayer, basal spacing expansion for montmorillonites containing other exchangeable cations appear to reach a maximum value of about 1.92 nm, which is about the thickness of 4 layers of water.

The size of a unit particle of montmorillonite varies according to the exchangeable cation in the interlayer. Sposito (1984) provides estimates of the number of unit layers in face-to-face orientation for montmorillonite, ranging from a single unit layer for Li and Na as exchangeable cations, up to 6 to 16 for Ca as the exchangeable cation, depending upon the technique used for determination of unit-layer stacking. The sequence of water uptake in a swelling soil depends on the nature of the exchangeable cations in the interlayer spaces and the initial water content of the partly saturated soil. Upon first exposure to water (or water vapour), hydration processes dominate and water sorption is an interlayer phenomenon. For swelling soils containing Li or Na as the exchangeable cation, continued uptake beyond hydration water status occurs due to double-layer forces. The solvation shell surrounding small monovalent cations consists of about 6 water molecules if the solution is dilute (Sposito, 1984), reducing to about 3 for concentrated solutions. No secondary solvation shells are associated with added water intake. Water uptake beyond interlayer separation distances of about 1.2 nm occurs because of double-layer swelling forces, resulting in the formation of a solution containing dispersed single unit layers. In the case of divalent cations, both primary and secondary solvation shells are obtainable. These move together as a solvation complex. The primary shell is seen to be composed of about 6 to 8 molecules and the secondary shell contains about 15 water molecules. Sposito maintains that in dilute suspensions, the homoionic forms of Na-montmorillonite, which are unit layer particles are different from Ca-montmorillonite particles, which consist of 6 to 8 unit layers stacked in face-to-face array. This is in accord with the observations reported by Farmer (1978) who indicated that most Li and Na-smectites swell in dilute solution or in water into a gel-like state with average interlayer separation in proportion to $1/\sqrt{c}$ — where c is electrolyte concentration in the liquid phase. Initial water uptake by nearly anhydrous clays is strongly exothermic, with the water being firmly held in the coordination sphere of the cation and in contact with the surface oxygens.

The relative proportions of water between interlayer (water) and interparticle water, together with the energy status of each is a function of the nature of the soil and factors associated with either interlayer or interparticle phenomena, e.g.,

- Interlayer or interlamellar expansion due to sorption of water (hydration) is determined by the layer charge, interlayer cations, properties of adsorbed liquid, and particle size.
- Expansion of interlayer distance beyond expansion due to hydration can be ascribed to mechanisms represented by the diffuse double-layer models.

The mechanism for sorption from very low water contents or from the anhydrous state is due to not only the presence of charge sites and exchangeable ions in the interlayers, but also to the attractions between water molecules and the polar surface groups. From a basal spacing *d*(001) of 0.95 nm to 1.0 nm for the anhydrous state, sorption or hydration will increase the spacings to between 1.25 to 1.9 nm. As stated previously, this is a function of the nature of the exchangeable cations. Thus, for example, with Li and Na as the exchangeable ions, the basal spacings reach 1.24 nm at a relative humidity of 0.5, whilst with Mg and Ca exchangeable ions, the basal spacings reach 1.43 and 1.5 nm, respectively, for the same relative humidity. The basal spacings at 100% humidity level for montmorillonite have been reported by Suquet et al. (1975), and with Quirk (1968), who show the Li and Na montmorillonites swelling as >4.0 nm. Mooney et al. (1952) indicates that the 1.24–1.25 nm spacing corresponds to one monolayer of water in the interlayer region, and that basal spacings of 1.5, 1.9, and 2.2 nm correspond to 2, 3, and 4 layers of water between each alumino-silicate layer. We must note that the volume change swelling in the interlayer between 1.0 nm and 2.2 nm should only be identified as *crystalline swelling* if there exists a definite hydration structure to the water. Sposito (1984) reports the following relationship for a sodium montmorillonite suspended in NaCl:

$$V_{ex} = 0.5524 + 0.3046c^{-1/2}$$

where the exclusion volume V_{ex} is a measure of the volume change per unit mass of soil due to the osmotic activity of the ions of concentration c in the solution. The apparent interlayer separation relationship with c and the fact that the interlayer space contains no discernable hydration states allows us to separate crystalline swelling and double-layer swelling phenomena. This means to say that expansion beyond 4 layers of water is most likely due to osmotic forces, resulting in dilution of the ionic concentration in the interlayers.

We can conclude that the hydration structure of the water provides the distinct difference between crystalline swelling, i.e., interlayer separation, and subsequent swelling due to osmotic activities of the cations. Water movement into the various pore sizes (see Figures 2.13 and 4.19) is in response to the specific water-uptake mechanisms operative within the pore space. These include:

- Crystalline swelling forces associated with the matric potential (see next section) which will result in interlayer swelling up to about 1.0 nm.
- Subsequent swelling (if any) which will be due to forces associated with double-layer swelling.

A distinction must be made between completely dry and moist soil conditions in the analysis of volume change and swelling, particularly if the soil is a highly swelling soil. For an initial dry swelling soil, the first sequence of volume change is due to wetting by hydration forces. Volume changes as high as 100% of the original volume of the dry clay can be obtained when 4 monolayers of water enter between the layers of a montmorillonite clay. Volume change is related to the hydratability of the cations. The degree of separation of the unit layers depends on the affinity of the cations for water relative to their affinity for the interlayer surfaces. As stated above, volume change beyond the hydration volume change is most likely due to swelling associated with double-layer forces.

Yong and Warkentin (1959, 1975) show that when the concentrations of Na become greater, tactoids or domains consisting of more than 2 or 3 stacked unit layers are obtained. Calculations of swelling pressure using DDL models become less than accurate unless account is given to the modification of the total surface area of interaction. Similarly, tactoids or domains are also formed when divalent cations and/or mixed cations are present as exchangeable cations. Yong and Warkentin (1959) have shown that for a calcium-saturated montmorillonite, if one uses an average 4-unit layer stacking arrangement with 2 nm interlayer spacings for the particles (tactoids), good agreement between measured and calculated swelling pressures can be obtained.

3.6 SOIL-WATER CHARACTERISTICS

Interactions between soil fractions and water are characterized in terms of energy relationships. The soil-water characteristics provide a useful means to assess the capability of the soil to retain water. It is a measure of the water-holding capability of the soil under the various conditions defined by its water content and the forces within the soil responsible for water uptake. Insight into the transmission properties of the soil and the interaction capability of the soil fractions with solutes in the porewater can also be gained by determination of the various thermodynamic potentials which are characteristic of the soil. As might be deduced, these characteristics are important in the control of the fate of pollutants.

3.6.1 Soil-Water Potentials

In Figure 3.1 in Section 3.1, we showed the example of the height of rise of water in two columns — one containing clean sand and the other containing a clay soil. The height of capillary rise in the sand column was given in Equation 3.1 as:

$$h = \frac{2\,\sigma \cos\alpha}{r\,\gamma_w} \qquad \text{(repeated, 3.1)}$$

The contact angle of the water with the soil particles α is a function of these two elements, whereas the effective radius of the average pore size r is a direct function of soil type and soil density. Since the surface tension of the water σ and the density of water γ_w are properties of the water, it is clear that the height of capillary rise h will be determined by the type of soil and the packing of the soil solids (density). The smaller the value of r, the greater is the height of capillary rise h. For the clean sand in the left column in Figure 3.1, where the height of capillary rise is represented by Equation 3.1, we can define a capillary potential ψ_c which is responsible for establishing the height of capillary rise h. In essence, this *capillary potential* ψ_c is a measure of the energy by which water is held by the soil particles by capillary forces. Buckingham (1907) defined it as the potential due to capillary forces at the air-water interfaces in the soil pores holding water in the soil. It is a measure of the work required per unit weight of water to pull the water away from the mass of soil.

Instead of considering the potential in respect to the soil, it is more convenient to consider the work required to move water into and out of a soil mass. Accordingly, we need to define the potential in terms of the water in the soil. This defines the total work required to move the water in the soil as the *soil-water potential* ψ. When we do so, we change the algebraic sign. This point is particularly significant since confusion can arise when the wrong frame of reference is used. A very good example of this is the concept of *soil suction*. This suction is said to be responsible for the water-holding capability of a soil. This is a useful concept since it is not difficult to imagine water being "sucked up" by a soil, and associate the "sucking" phenomenon with the internal "suction" property of the soil. In cohesionless soils, for example, the soil suction is considered to be due to mechanisms associated with air-water interfaces (*capillary suction*). In terms of potentials, the capillary suction is defined as the capillary potential ψ_c. If we define the capillary potential in respect to the soil solids — instead of the soil-water — and call this the *soil capillary potential* (scp) (to avoid confusion with ψ_c) we will see that the scp decreases as the soil water content increases. This means that the soil is less capable of "pulling" water upward in the column shown in Figure 3.1 as its water content increases. In soil mechanics terminology, we will say that the capillary suction decreases as the water content in the soil increases. In that sense, the scp, which is defined in respect to the soil solids, will bear the same concept as the capillary suction. Whilst this similarity in perspectives might be useful, the requirements of analyses of the soil-water system as a whole are better satisfied if ψ_c is considered in respect to the soil-water. If we define the potential ψ_c in respect to the water in the soil, we see that ψ_c increases as the soil water content increases, since the capillary potential is a negative quantity (in terms of the soil water). What this means is that the capillary potential ψ_c is *less negative* in value as the water content increases.

The soil-water potential ψ describes the water-holding capability of soils, i.e., it describes the energy by which water is held to (or attracted) the soil fractions in the soil. As seen from the preceding, it is particularly useful in providing a simple picture of the kinds of internal forces that will contribute to water movement and retention in soils. Terms such as *soil-water tension* and *soil suction* have been used as simple descriptive terms. The former (soil-water tension) refers to the water in the soil in equilibrium with a pressure less than atmospheric, and the latter (soil suction) refers to the physical action of "sucking of water" by the soil solids in the soil itself. This is consistent with the previous explanations concerning the thermodynamic description of the status of water, i.e., *soil-water tension* is the thermodynamic description of the status of the water, and the mechanics description of the actions of the soil solids in the soil (*soil suction*).

Because potentials are defined with a reference base in mind, this base is normally considered to be a reference pool of free water at the same elevation and temperature of the soil and under one atmospheric pressure. The definitions given by Yong (1999b) are used herein to describe the total potential ψ and its components.

All the potential terms, i.e., total potential ψ and its components, are considered in respect to the reference pool of water.

- ψ = total potential = total work required to move a unit quantity of water from the reference pool to the point under consideration in the soil. It is a negative number.
- ψ_m = matric potential = property of the soil matrix, pertaining to sorption forces between soil fractions and soil-water. This is often mistakenly assumed to be the capillary potential. For granular materials, this assumption may be quite valid. However, for clay soils, complications surrounding microstructural effects and influences do not permit easy resolution in terms of capillary forces.
- ψ_g = gravitational potential = $-\gamma_w gh$, where γ_w = density of water, height h is the height of the soil above the free water surface, and g = gravitational constant. We need to note that if the point in the soil under consideration is below the surface, h is a negative quantity, and hence the relationship becomes positive.
- ψ_π = osmotic potential = ψ_s = solute potential for non-swelling soils. For such cases, $\psi_\pi = \psi_s = nRTc$, where n = number of molecules per mole of salt, c is the concentration of the salt, R = universal gas constant, and T = absolute temperature. In the case of swelling soil, the assumptions and constraints discussed previously for the DDL model apply.
- ψ_p = pressure potential due to externally applied pressure, and transmitted through the fluid phase of the soil-water. In soil mechanics and geotechnical engineering, this potential is directly related to the positive porewater pressure.
- ψ_a = pneumatic (air) pressure potential arising from pressures in the air phase. This is a consideration in partly saturated soils.

3.6.2 Measurements of Soil-Water Potentials

The three most common types of systems used to obtain a measure of the soil-water potential include: (a) tensiometers using the Haines procedure (Haines, 1930); (b) pressure plates or pressure-membrane systems; and (c) thermocouple psychrometry. The first two systems do not provide direct measurements of the potentials. Instead, they measure the equivalent negative pressure in the porewater. Thus, for example, the tensiometer shown as the left-hand "A" apparatus in Figure 3.20 measures the water tension in the soil by determining the length of the column of water that can be "hung" from a soil sample. Since this "hanging" system can be easily mishandled, a mercury manometer is most often used in place of the hanging column. However, because of the nucleation of dissolved air bubbles in the water and in the vapour phase, this system is only useful for measurements of soil-water tensions where pressure differences are less than one atmosphere.

To overcome the problems of the Haines method, and to obtain higher porewater tension values, the pressure membrane apparatus shown as the right-hand apparatus "B" in Figure 3.20 is generally used. This is similar to the pressure technique first reported by Richards (1949). The high air entry porous disk controls the maximum value of pressure that can be applied to the samples in the cell. By introducing air pressure into the cell, one is attempting to drive the porewater from the soil samples. At equilibrium under the applied air pressure, the water remaining in the soil samples is considered to be held to the soil solids under sets of forces that are at least marginally greater than those applied by the air pressure introduced into the cell.

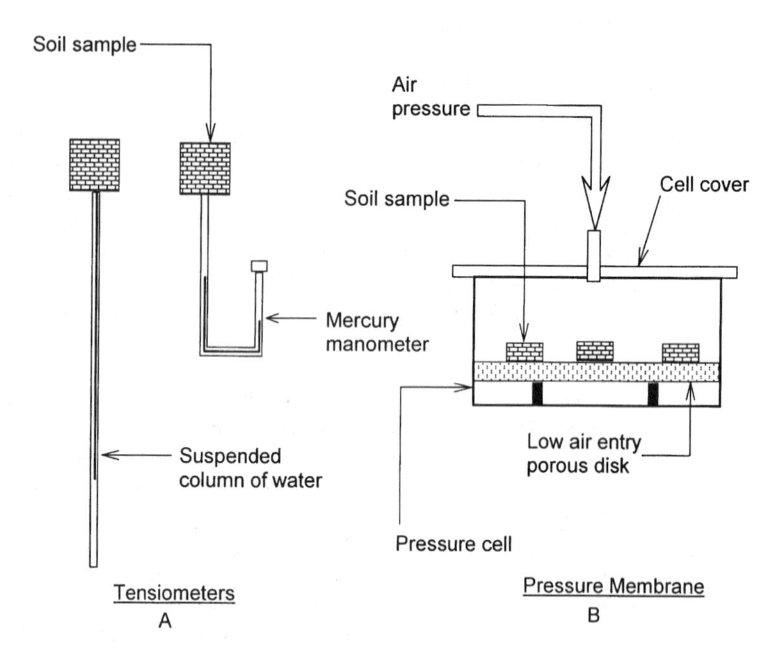

Figure 3.20 Laboratory devices for measurement of water-holding capability of soils.

One assumes that when the applied air pressure is removed prior to sample retrieval for water content determination, the porewater in the sample will go into tension at the same magnitude as the air pressure under which it was equilibrated. The typical soil suction relationships shown in Figure 3.21 illustrate the differences in relationships obtained because of soil type.

The units used in the abscissa are expressed as pF values. These represent a log scale system which expresses the pressure applied to the soil samples in terms of equivalent height of water in centimetres on a logarithmic basis. If h represents the equivalent height of water pressure, then the pF value is given as: $pF = \log_{10}h$. This is a very convenient means of pressure expression that was proposed by Schofield (1935), and can be easily converted into present-day conventional units, e.g., kilopascals (kPa), or into equivalent atmospheric pressures (bars). In the relationships shown in Figure 3.21, it is obvious that major differences in the samples in respect to water-holding capacity can be traced to differences in (a) the specific surface area, and (b) surface active forces of the soil solids.

The thermocouple psychrometer measures the soil-water potential by determining the relative humidity of the immediate microenvironment surrounding the psychrometer. The psychrometer probe essentially consists of a small ceramic bulb within which the thermocouple end (i.e., juncture) is embedded. Cooling of the juncture is obtained by passing an electrical current through it (Peltier effect). By cooling it below the dew point, condensation occurs at the juncture and when the

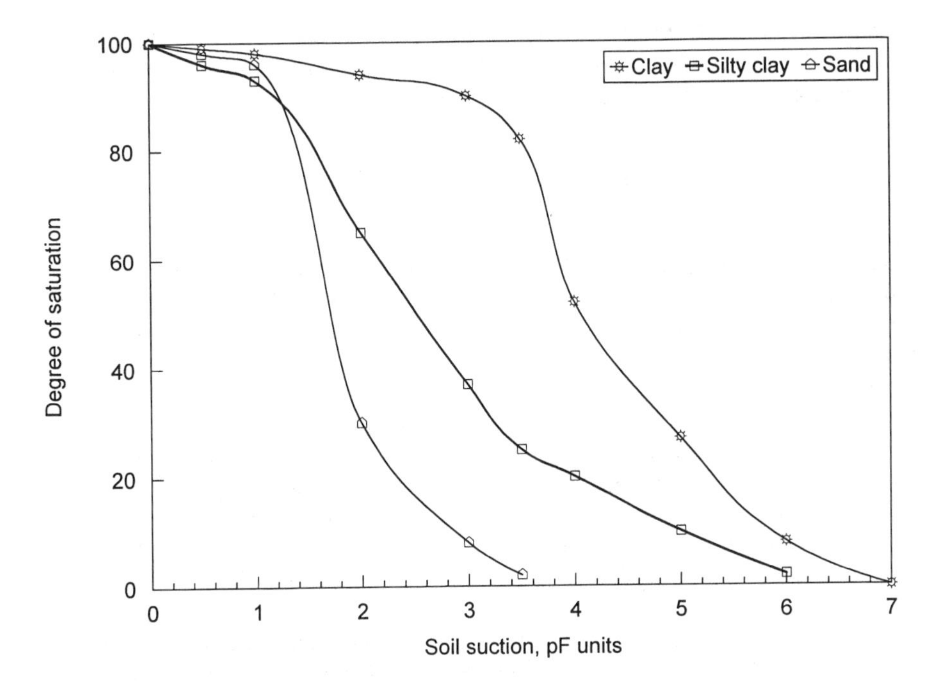

Figure 3.21 Soil suction curves for three types of soil.

electrical current is discontinued, evaporation of the condensed water occurs. The rate of evaporation of the condensed water is inversely related to the vapour pressure in the psychrometer bulb. The evaporation of the juncture water causes a drop in the temperature which is measured as the voltage output of the thermocouple. The magnitude of the temperature drop depends on the relative humidity and temperature of the immediate volume surrounding the psychrometer. By and large, the drier the surroundings, the faster the evaporation rate, and hence the greater the temperature drop. The relationship between ψ and the relative humidity is given as:

$$\psi = \frac{RT}{V_m} \ln \frac{p}{p^o} \tag{3.19}$$

where R = universal gas constant, T = absolute temperature, V_m = molal volume of water, p = vapour pressure of the air in the soil voids, p^o = vapour pressure of saturated air at the same temperature, and the ratio of p/p^o = relative humidity.

3.6.3 Evaluation of Measured Soil-Water Potentials

In addition to T, the absolute temperature, if we consider P and m, the pressure and mass of the soil, to be independently chosen variables, we can express the thermodynamic potential (Gibbs energy) in terms of partial molar free energies using the condition where at equilibrium, G^* the partial molar free energy of water is everywhere the same. By differentiating the Gibbs energy relationship in terms of the partial molar free energy, and considering soil as a three-phase system (air, water, and solids), Yong and Warkentin (1975) have shown that in the absence of significant gravitational and other external force field effects

$$dG_{\omega}^{*} = V_{\omega}^{*}dP - S_{\omega}^{*}dT + \frac{\partial G_{\omega}^{*}}{\partial \theta}d\theta \tag{3.20}$$

where V_{ω}^{*} and S_{ω}^{*} are the partial specific volume and partial specific entropy of the soil-water. The volumetric water content is represented by θ, and the partial molar free energy of the soil-water G_{ω}^{*} is the chemical potential μ_{ω} of the soil water. According to Sposito (1981),

- The partial specific volume of the soil water is by definition equal to zero for a soil which does not shrink or swell as a result of water content changes $V^* = 0$.
- $V^* = 1/\rho_w$ when the soil is fully saturated (ρ_w = mass density of water).
- The soil-water potential $\psi = \mu_{\omega} - \mu_{\omega}^{o}$ (where μ_{ω}^{o} = standard state for μ_{ω}).
- The matric potential ψ_m includes the effects of dissolved components of the soil-water system on the chemical potential μ_{ω}, i.e., $d\psi_m = (\partial\mu_w/\partial\theta)_{T,P,Pa}d\theta$.

Because water and solutes can pass through the membrane that separates the tensiometer soil sample from the water (Figure 3.20), measurements obtained with the tensiometer include the effects of the dissolved solutes. At equilibrium, the condition for the soil-water (porewater) will be given as:

$$\mu_{\omega}(soil\text{-}water) = \mu_{\omega}(tensiometer) = \mu_{\omega}^{o} + \frac{1}{\rho_{\omega}^{o}}(P_t - P^o) \tag{3.21}$$

where the superscript "*o*" refers to the standard state for the respective parameters, and P_t refers to the pressure in the tensiometer. Using τ_{ω} to denote the soil-water tension measured by the tensiometer as a gauge pressure in pascals or atmospheres, the following is obtained (Sposito, 1981):

$$\tau_{\omega} \equiv P^o - P_t = -\rho_{\omega}^{o}[\psi_p(P,\theta) + \psi_m(P^o,\theta)] \tag{3.22}$$

For soils that do not undergo volume change when exposed to water, i.e., rigid porous soils, $\psi_p(P,\theta) = 0$, since $V_{\omega}^{*} = 0$. Accordingly, we see from Equation 3.22 that the soil-water tension τ_{ω} (or soil suction) determined by the tensiometer measures the matric potential ψ_m and the solute (or osmotic) potential ψ_s of the soil. The matric potential ψ_m can be obtained by determining ψ_s separately through a determination of the properties of the extracted solution using, for example, freezing point depression data, or from the general relationship: $\psi_\pi = \psi_s = nRTc$ as stated previously in the discussion on the components of the soil-water potential ψ.

Pressure measurements obtained from the pressure-membrane apparatus (Figure 3.20) can be considered in a similar fashion. The thermodynamic analysis of the processes associated with the procedure shows that if P_ω represents the applied pressure, and if the soil is fully saturated and initial pressure in the pressure-membrane apparatus is zero, i.e., P_ω (initial) = 0, then P_ω provides a direct measure of the matric and solute potentials ψ_m and ψ_s. As in the case of tensiometer measurements, we need to recognize that since the effect of dissolved solutes is included in the measurements obtained, the potential will be designated as ψ_{ms} the matric-solute potential, to distinguish this from the matric potential ψ_m, which does not include the effect of dissolved solutes. There are at least two different concepts of the matric potential ψ_m. These revolve around whether the matric potential includes the effects of solutes. So long as one is careful in differentiating between the various effects, either concept is acceptable.

In the technique used in thermocouple psychrometry, the difference in temperature between a reference temperature and a wet junction temperature is measured, using the air-soil-water interface as a semi-permeable membrane. This provides a measurement of the properties of air that is in equilibrium with the soil-water. The relationship that describes the soil-water potential ψ has been given previously as Equation 3.19. As with the measurements obtained from the tensiometer and pressure-membrane apparatus, the soil-water potential ψ that is computed from the relationship given in Equation 3.19 includes both the matric ψ_m and osmotic ψ_π (or solute ψ_s) components of the soil-water potential.

3.6.4 Matric ψ_m, Osmotic ψ_π Potentials and Swelling Soils

Of the various components of the soil-water potential ψ that are responsible for the water-holding capacity of soils, we can consider the matric ψ_m, and osmotic ψ_π potentials as being the most responsible components. In partly saturated moisture movement, in the absence of externally applied gradients and under isothermal conditions, these two components are most often considered sufficient in describing partly saturated moisture movement. While the contributions from the gravitational ψ_g, pressure ψ_p, and pneumatic ψ_a components are sometimes considered in analysis of the transient energy status of soil-water, these are generally considered as part of the book-keeping exercise.

Determination of the role of the various potentials in water movement in partly saturated swelling soils is sometimes complicated by the phenomena associated with such soils, e.g., swelling pressure and volume expansion. The mechanistic model that describes the situation can benefit from the following simplifying assumptions: (a) no external pressures are imposed on the soil water during soil swelling, i.e., $\psi_p = 0$; (b) ψ_p is not equal to zero when constraints are placed on soil volume expansion, i.e., constraints on volume change are applied; and (c) ψ_g is vanishingly small. We interpret from the discussion given in Section 3.5.1 the following in respect to partly saturated flow and other soil-wetting processes in swelling soils:

- Soil volume expansion will result if the soil is not constrained. The nature of the volume expansion is a function of the water inlet source distribution and soil structure, and the chemistry of the uptake water. Assuming no restrictions on source-water

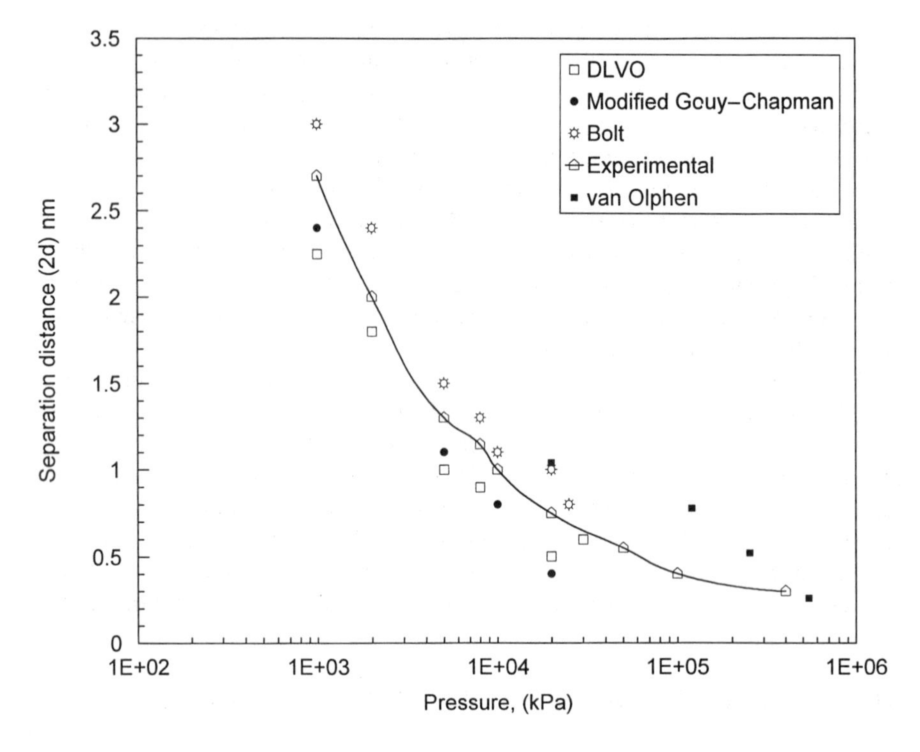

Figure 3.22 Enlarged view of the low particle spacing results of Figure 3.14.

availability and volume expansion, volume change due to swelling terminates upon dissipation of those swelling forces responsible for the volume expansion, i.e., when soil suction becomes zero. At this stage, the soil-water potential ψ is also considered to be zero.

- Volume expansion will be restricted if external pressures are applied to restrict volume change, dependent on confining pressures, and/or availability of water. In any event, full dissipation of those forces responsible for swelling (volume change) does not occur.

Section 3.5.1 has shown that water uptake in the interlayer spaces due to hydration forces will provide for a different form of water structure, and that this volume expansion is defined as crystalline swelling due to ψ_m. The low water content results of Figure 3.14 are highlighted in Figure 3.22. The results presented as "van Olphen" are derived from calculations interpreted from van Olphen (1963). He, along with many other researchers on swelling soils, maintain that at interlayer spacings of up to about 1 nm, the dominant mechanism responsible for interlayer separation (the 2d distance shown in Figures 3.14 and 3.22) during water uptake is the result of the actions due to the adsorption energy of water at the clay particles' surfaces. This, along with the calculations shown in Section 3.5.1, provides the mechanisms that distinguish between:

- Crystalline swelling, i.e., swelling due to sorption of the first 2 to 3 water layers between the unit layers of the 2:1 dioctahedral series of alumino-silicate clays (interlayer or interlamellar sorption of water), and
- Interparticle sorption or water uptake.

The calculations reported by Alammawi (1988) for a Na-montmorillonite with 10^{-3} *N* NaCl can be interpreted and reported as the hydration characteristics in the form shown in Figure 3.13. The calculations show that dilution of the ions occurs rapidly after the first 2 layers of water. The considerable dilution of the concentration of ions beginning with the 3rd water layer suggests that crystalline swelling is confined to the first 2 water layers, and that double-layer swelling occurs from the 3rd water layer onward. This is consistent with the interlayer separation distances reported by Suquet et al. (1975) and Quirk (1968) for Li and Na-montmorillonite.

The preceding results and discussion would indicate that crystalline swelling results from sorption forces associated with the matric potential ψ_m, and that interlayer or interlamellar swelling beyond this point (of crystalline swelling) stems from interactions described by the osmotic potential ψ_π. The following points are noted:

- The presence of air-water interfaces is not a necessary requirement for water uptake by forces associated with the matric potential ψ_m. This suggests that the interlayer spaces during hydration are fully saturated during, and as a result of, crystalline swelling.
- The effect of dissolved solutes is included in the interlayer water structure.
- Interlayer volume expansion beyond crystalline swelling is due to DDL forces that can be determined from the osmotic potential ψ_π.

If engineering terminology of suction is to be used, the corresponding suctions for the preceding (matric and osmotic) are S_m and S_π, the matric and osmotic suctions, respectively. Because measurements of soil suction S using tensiometers and pressure membrane techniques include the effects of dissolved solutes, the matric potential ψ_{ms} is obtained where the subscript *ms* is used for ψ_{ms} to distinguish this from ψ_m, which refers to the matric potential associated with crystalline swelling. The measurement techniques require that constant soil volume be maintained. However, for swelling soils, the transient nature of both water content and volume require that ψ_{ms} must be referred to the particular condition of water content and volume existing at the time of the measurement. In the case of psychrometer measurements, determination of ψ_m is generally obtained by subtracting calculated values of ψ_π determined from extracts of the liquid phase from the total ψ. Hence, the ψ_m obtained from this procedure will not include the effects of solutes.

Measurements of ψ or S using tensiometers, pressure membranes, psychrometers, etc., are conducted as bulk (macrostructural-type) measurements. Because of the technique or the size of the measuring tool, what is measured is the equilibrium status of the soil-water in the macropores (pore spaces separating soil peds). The equilibrium states of interlayer water and the micropore soil-water (water in the pore spaces separating soil particles in the ped) are not measured or determined. We can consider that the equilibrium state of the macropore porewater is defined by the energy states

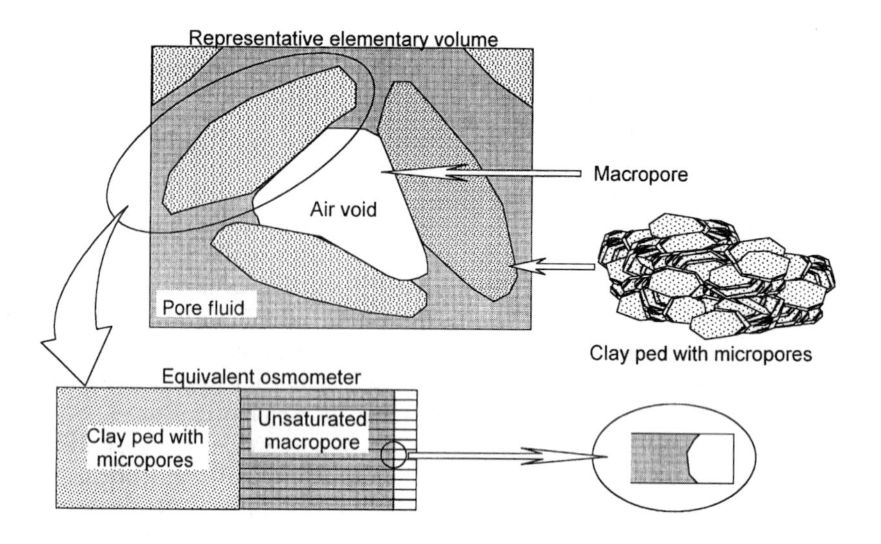

At equilirium, the osmotic potential ψ_Π in peds (due to saturated micropores) is balanced by the matric potential ψ_m in the macropores (due to microcapillary phenomena, van der Waals forces, and weak osmotic activity)

Figure 3.23 Balance between osmotic and matric potentials responsible for partly saturated water movement in a representative elementary volume of soil. At equilibrium, a balance is reached between these two component potentials.

in the micropores and in the interlayers. At that time, the osmotic potential ψ_π in the peds will be balanced by the ψ_m in the macropores, as shown in Figure 3.23. To explain this, we need to consider the different pore sizes in a representative elementary volume (macrofabric). At least three kinds of pore spaces make up a total macrofabric: (a) pore spaces obtained as interlayer separation distances (for convenience, this is identified as interlayer pores); (b) micropores representing pore spaces between particles in clay peds; and (c) macropores as shown in Figures 3.23 and 4.20. For a partly saturated swelling soil, intake of water into the clay peds due to the osmotic potential (in the peds) is restricted because of the matric sorption forces in the macropores (illustrated as microcapillary and other phenomena in Figure 3.23). Water movement in the macrofabric would occur as transport at the boundaries of the particles or peds, clusters, etc., but not necessarily in the large pore spaces.

3.7 CONCLUDING REMARKS

The contributions from the various soil fractions to soil structure cannot be neglected when it comes to evaluation of the surface properties of soils. By and large, because of the pH dependency of the surface properties of such fractions as soil organics, amorphous materials, and even some clay minerals, interactions occurring between various soil fractions will change the characteristic SSA and CEC of the soils — because of coating and agglomeration (aggregation) of particles. Many soil organics act much in the same manner as flocculants, and because of the alteration in soil structure, i.e., alteration in both soil fabric and forces between particles, the reaction of such soils to the influx of contaminants will be correspondingly changed.

The energy characteristics of soils, as demonstrated through the soil-water relationships, are the most relevant soil characteristics in assessments or evaluations of interactions between pollutants (contaminants) and soils. The surface properties of the soil fractions, or soil solids, are the result of the nature of the soil fraction. These are very important: they provide the basis for the interactions between pollutants and the soil solids. We see from the discussion given in the early part of this chapter that the surface properties of the inorganics, such as the mineral particles, are more or less dominated by the hydroxyl surface functional groups, whereas the soil organics possess a greater variety of surface functional groups. We can speculate that the soil organics would perhaps be more capable of retaining pollutants (contaminants) than the mineral particles. This type of speculation must also take into account the specific surface area and cation exchange capacity of the soil.

The energy characteristics defined by the soil-water potentials provide us with an insight into how strongly water is held to the soil solids, i.e., how strongly water is held in soils. The various components of the soil-water potential tell us which component is more or less responsible for the water holding capacity. As we might expect, the matric potential (component) ψ_m is by far the most important soil-water potential component. This has been defined as the potential, which is a property of the soil solids' surfaces. Measurements of the changes in soil-water potential (with psychrometers) have been obtained in conjunction with unsaturated flow of water in soils — as shown, for example, in Figure 3.24 for a swelling clay, and in Figure 3.25 for unsaturated flow into a non-swelling clay. The results shown in Figure 3.25 have been shown in terms of soil suction to demonstrate the engineering aspect of the use of psychrometers. These types of measurements and correlations have provided the basis for development of relationships describing unsaturated flow using the soil-water potential as the driving force, as will be seen in Equation 4.17. Whilst this does provide a sound basis for utilization of soil-water potentials, we must be careful to distinguish between the various component potentials that participate in the retention of water in the soil-water system. This is particularly important since these impact directly on the contaminant-soil interaction characteristics.

The calculations for mid-plane potentials based upon interactions of counterions and soil-particle reactive surfaces in the DDL-type models are by and large in accord with the osmotic potential ψ_π. Much work remains to fully reconcile the matric component ψ_m with the potentials at the Stern layer or at the inner Helmholtz plane. We have information concerning the nature of the reactive surfaces of the soil fractions and the manner in which they react in the presence of water. Water-holding capacities and soil-water potentials, together with DDL-type models, provide us with the basic elements of interactions between water and soil. This gives us a platform from which studies of the effect of contaminants in the form of solutes derived naturally or from anthropogenic sources (pollutants) are introduced into the soil-water system.

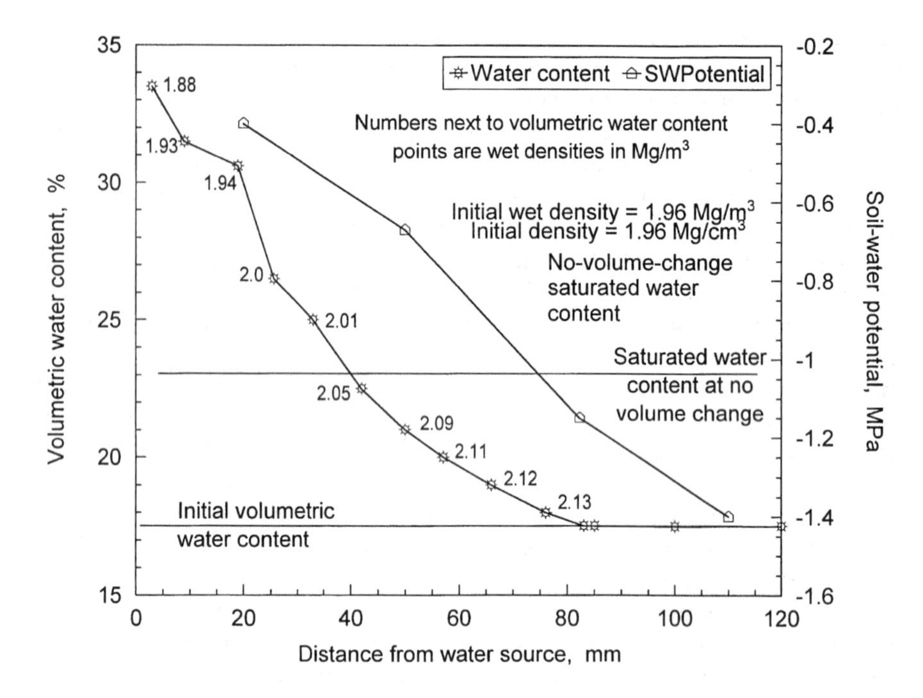

Figure 3.24 Results from unsaturated flow in a soil column, showing volumetric water content development along the soil column and corresponding soil-water potential. Water entry is from the left.

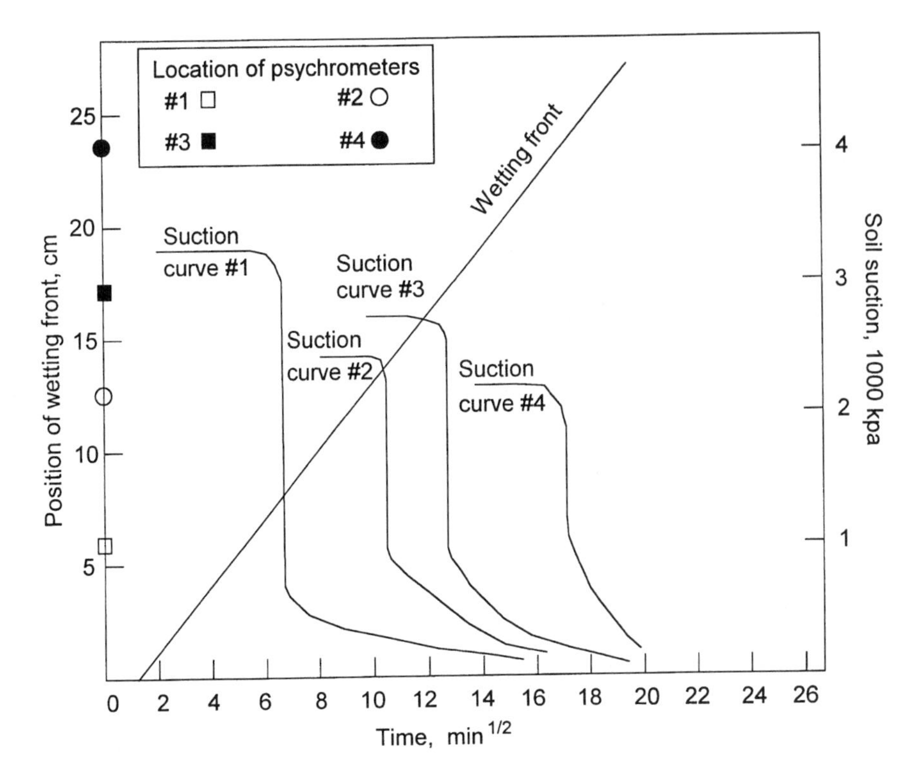

Figure 3.25 Development of soil suction at the wetting front as a result of unsaturated flow into a coarse kaolin soil. The suction curves shown as "suction curve #1," etc. refer to the suctions recorded at the psychrometer positions identified on the ordinate — i.e., distance along sample from inlet water source. (Data from Yong and Sheeran, 1972.)

CHAPTER 4

Interactions and Partitioning of Pollutants

4.1 POLLUTANTS, CONTAMINANTS, AND FATE

We consider, in this chapter, the general mechanisms and processes involved in the interaction between contaminants (pollutants and non-pollutants) and soil fractions, with attention to the general processes involved in the partitioning of pollutants. The details of partitioning inorganic (heavy metals) and organic chemical pollutants will be considered separately in the next two chapters. In Chapter 1, we referred to *pollutants* as contaminants that are considered potential threats to human health and the environment. These pollutants are both naturally occurring substances, e.g., arsenic and Fe, and anthropogenically derived such as the various kinds of chlorinated organics. Most, if not all, of these kinds of substances or compounds can be found on many hazardous and toxic substances lists issued by various governments and regulatory agencies in almost all countries of the world. Amongst these are the *Priority Pollutants* list given in the Clean Water Act, the *Hazardous Substances List* given in the Comprehensive Environmental Response, Compensation, and Liability Act (CERCLA) and the *Appendix IX Chemicals* given in the Resource Conservation and Recovery Act (RCRA).

We do not propose to enter into a debate at this time over the health threats posed by: (a) naturally occurring substances (contaminants) because of high concentrations, e.g., fluoride ion F^-, which can be found in fluorite (CaF_2) and apatite; (b) naturally occurring health-hazard substances, e.g., mercury, which is found as a trace element in many minerals and rocks; and (c) substances such as solvents and heavy metals produced or resulting from anthropogenic activities. Whilst it is tempting to consider *pollutants* as contaminants originating from anthropogenic activities, this simplistic distinction may not serve us well inasmuch as natural pollutants can also be severe health threats. The fundamental premise that governs pollution mitigation (i.e., removal or reduction of pollutant concentration) and remediation of contaminated lands should be protection of health of biotic species and land environment. Accordingly, as in Chapter 1, we will use the term *pollutant* to emphasize the contamination problem under consideration, and also when we mean to address known health-hazard

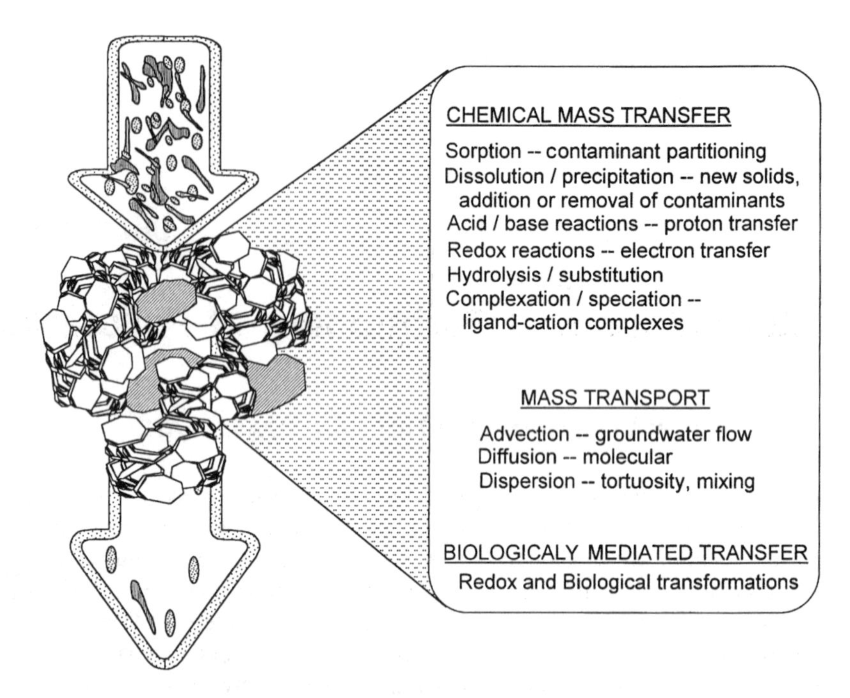

Figure 4.1 Interactions and processes involved in the determination of fate of contaminants and pollutants in soil.

contaminants (specifically or in general). We will continue to use the term *contaminant* when we deal with general theories of contaminant-soil interactions.

The description of the ultimate or long-term nature and distribution of pollutants introduced into the substrate is generally described as the *fate* of pollutants. The fate of pollutants depends on the various interaction mechanisms established between pollutants and soil fractions, and also between pollutants and other dissolved solutes present in the porewater. The general interactions and processes contributing to the fate of contaminants and pollutants is shown in Figure 4.1. We will consider these in greater detail in the next few chapters. At this stage we can consider the four main groups of events that fall under a general characterization described in overall terms as *fate description*:

1. **Persistence** — this includes pollutant recalcitrance, degradative and/or intermediate products, and partitioning;
2. **Accumulation** — describes the processes involved in the removal of the contaminant solutes from solution, e.g., adsorption, retention, precipitation, and complexation;
3. **Transport** — accounts for the environmental mobility of the contaminants and includes partitioning, distribution, and speciation;
4. **Disappearance** — this grouping is meant to include the final disappearance of the contaminants. In some instances, the elimination of pollutant toxicity or threat to

human health and the environment of the contaminant (even though it may still be present in the substrate) has been classified under this grouping, i.e., disappearance of the threat posed by the pollutant.

The question frequently asked here is: "Why do we want (need) to know the fate of pollutants?" Of the many answers that come to mind, two very quick ones can be cited:

- For prediction of transport and status of the pollutants resident in the ground over long periods of time — e.g., 25 to 250 years — it is important to be able to say that the contaminants of interest (i.e., pollutants) are properly managed, or will continue to pose a threat because of their continued presence in concentrations or forms deemed to be unacceptable. The question of *risks and risk management* comes immediately to mind.
- Performance and/or acceptance criteria established by many regulatory agencies using the *natural attenuation capability* (also known as *managed natural attenuation*) of soil-engineered and natural soil substrate barriers rely on *pollutant retention* as the operative mechanism for attenuation of pollutants.

The many mechanisms of interaction between contaminants (i.e., non-pollutants and pollutants) and soil fractions do not necessarily assure permanent removal of the contaminant solutes from the transporting fluid phase (leachates). We have seen from Section 2.1.1 and Figure 2.4 that we need to be careful in distinguishing between the many mechanisms or processes contributing to pollution attenuation by the soil-water system. The processes contributing to pollutant attenuation in the soil substrate by *retardation*, *retention*, and *dilution* are not similar, and the end results will also be distinctly different.

The term *attenuation* is most often used in relation to the transport of pollutants in the soil substrate, and generally refers to the reduction in concentration of the pollutant load in the transport process. It does not describe the processes involved. A distinction between processes that result in temporary and permanent sorption of the sorbate (solutes in the porewater) by the soil fractions should be made. The nature and extent of the interactions and reactions established between pollutants and soil fractions (Figure 4.1) will determine whether irreversible or reversible (temporary) sorption of the sorbate occurs, resulting in the pollutant transport profiles shown in the schematic diagram given as Figure 2.5.

Partitioning of pollutants by *retention* mechanisms will result in irreversible sorption of the pollutants by the soil fractions. Desorption or release of the sorbate is not expected to occur. The term *attenuation* has been used by soil scientists to indicate reduction of contaminant concentration resulting from retention of contaminants during contaminant transport in the soil, i.e., chemical mass transfer of contaminants from the porewater to the soil solids. On the assumption that the contaminants held by exchange mechanisms or reactions are the easiest to remove, we can stipulate a threshold which might say, for example, that attenuation occurs when the sorbate (contaminants) will not be extractable when exposed to neutral salts or mild acid solutions.

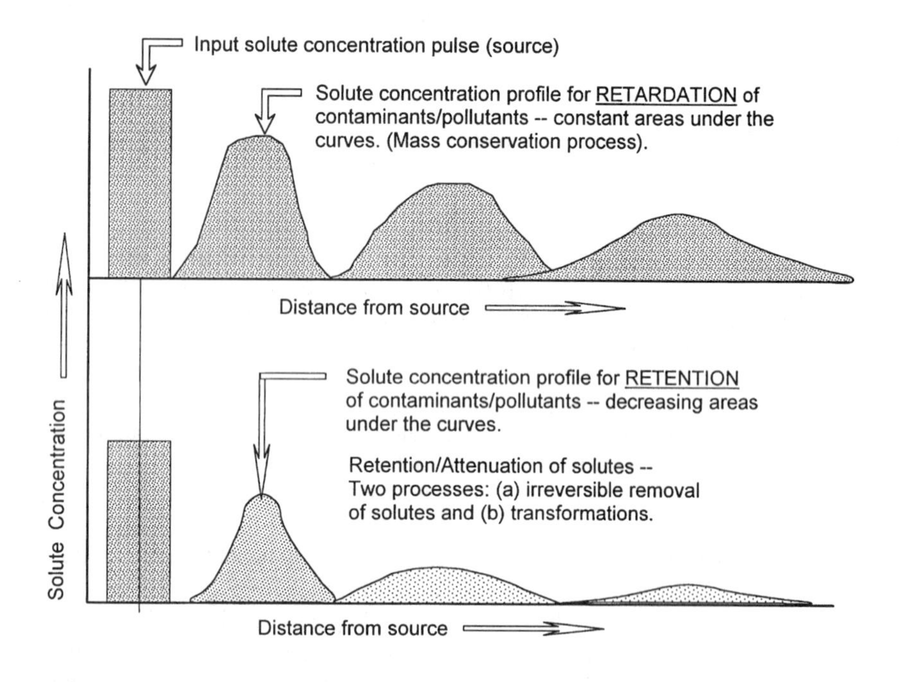

Figure 4.2 Retardation and retention processes. Note that the solute pulse shapes in the top show solute mass conservation, i.e., areas under the pulse curves are all equal to each other.

The term *retardation*, which has been used in literature in the context of contaminant transport in the substrate, refers to a diminished concentration of pollutants in the contaminant load undergoing transport. Attenuation of contaminants by retardation processes or mechanisms differ considerably from attenuation by retention mechanisms. Because retardation mechanisms involve sorption processes that are reversible, release of the sorbate will eventually occur. This will result in delivery of all the pollutants to the final destination. The schematic illustration given in Figure 2.5 portrays the resultant effects between the two kinds of processes. If the pollutant solute pulse (i.e., total pollutant load represented by the rectangular area at the top) is retarded, the area under each of the retardation pulse curves remains constant as the pulse travels downward toward the aquifer. The height of the bell-shaped curves will be reduced, but the base of the bell-shaped curves will be increased, as seen in Figure 4.2. The areas of the curves are similar since the total pollutant load is constant. Eventually, all of the pollutants will be transported to the aquifer. In contrast, the retention pulse shows decreasing areas under the pulse-curves. Partitioning by chemical mass transfer and irreversible sorption decreases the total pollutant load. The pollutant concentration is similarly decreased, and a much lesser amount of pollutants is transported to the aquifer. If proper landfill barrier design is implemented, the pollutant load reaching the aquifer will be negligible.

Failure to properly distinguish between attenuation by retention and retardation mechanisms, especially in respect to pollution of the ground and groundwater and transport modelling for prediction of pollutant plume migration, can lead to severe consequences. Differences in the predicted rate and penetration of a pollutant plume depend not only on the choice of transport coefficients, but also on whether the pollutants are *retained* in the soil through retention mechanisms or *retarded* because of physical interferences and/or sorption processes that are reversible. That being said, it is often not easy to distinguish between these two processes inasmuch as direct mechanistic observations in the field are not always possible. This will be explored in greater detail in the next two chapters.

A proper knowledge of the fate of contaminants is important and necessary for:

- Accurate prediction of the status (nature, concentration, and distribution) of the pollutants in the leachate plume during transport in the substrate — with passage of time;
- Design, specification, construction, and management of proper containment systems;
- Monitoring requirements and processes associated with management of the contaminant plume;
- Structuring of the mitigation and/or remediation technology that would be effective in reducing pollutant concentrations or removal of the pollutants;
- Risk documentation, analyses, and predictions; and
- Regulatory processes associated with the development of documentation regarding mitigation and remediation effectiveness, and safe disposal/containment of waste products on land.

To ensure that the environment and public health are protected, it is necessary to recognize where the various pollutants will be transported within the substrate, and whether the pollutants will be retained within the domain of interest. In addition, it is important to be able to account for the nature, concentration, and distribution of the pollutants within the domain of interest, if we are to implement proper risk management. Accordingly, it is necessary to have knowledge of the various interactions established between pollutants and soil fractions. The outcome of these interactions will determine the fate of the pollutants. The pH and *pE* regimes are known to be influential in the control of the status of a pollutant. Reactions involving electron transfer from one reactant to another will result in the transformation of both the pollutants and soil fractions. Changes in the oxidation states will produce transformed pollutants that can differ significantly in solubilities, toxicities, and reactivities from the original form of the pollutants. Dissolution of the solid soil minerals and/or precipitation of new mineral phases can occur with changes in the oxidation states.

4.1.1 Persistence and Fate

The terms *persistence* and *fate* are often used in conjunction with pollutants and contaminants detected in the substrate. Whereas concern is expressed for where the contaminants from waste materials and waste discharges end up, and whereas it is important to establish that these contaminants do not pose immediate or potential threats to the environment and human health, it is the *pollutant* aspect of the contamination problem that is frequently used in reference to such concerns (see previous chapters). The *fate* of a pollutant is generally taken to mean the destiny of a pollutant, i.e., the final outcome or *state* of a pollutant found in the substrate. The term *fate* is most often used in studies on contaminant transport where concern is directed toward whether a contaminant will be retained (accumulated), attenuated within the domain of interest, or transported (mobile) within the substrate domain of interest.

A pollutant or contaminant in the substrate is said to be *persistent* if it remains in the substrate environment in its original form or in a transformed state that poses an immediate or potential threat to human health and the environment. Strictly speaking, *persistence* is part of *fate*. An organic chemical is said to be a *recalcitrant chemical* or *compound* or labelled as a *persistent organic chemical* or *compound* when the original chemical which has been transformed in the substrate persists as a threat to the environment and human health. A major concern in the use of pesticides, for example, is the persistence of certain pesticides. It is most desirable for the pesticide to be completely degraded and/or rendered harmless over a short space of time.

Persistence is most often used in conjunction with organic chemicals where one is concerned not only with the presence of such chemicals, but also the state of the organic chemicals found in the substrate. This refers to the fact that the chemical may or may not retain its original chemical composition because of transformation reactions, e.g., redox reactions. However, most organic chemicals do not retain their original composition over time in the substrate because of the aggressive chemical and biological environment in the immediate surroundings (microenvironment). Some alteration generally occurs, resulting in what is sometimes known as *intermediate products*. This refers to the production of new chemicals from the original chemical pollutant. It is not uncommon to find several intermediate products along the transformation path of an organic chemical. The reductive dehalogenation of tetrachloroethylene or perchloroethylene (PCE) is a very good example. Tetrachloroethylene CCl_2CCl_2 (perchloroethylene) is an organic chemical used in dry cleaning operations, metal degreasing, and as a solvent for fats, greases, etc. Progressive degradation of the compound through removal and substitution of the associated chlorines with hydrogen will form intermediate products. However, because of the associated changes in the water solubility and partitioning of the intermediate and final products, these products can be more toxic than the original pollutant (tetrachloroethylene, PCE).

4.2 POLLUTANTS OF MAJOR CONCERN

The most common types of pollutants found in contaminated sites fall into two categories: (a) inorganic substances, e.g., heavy metals such as lead (Pb), copper (Cu), cadmium (Cd), etc.; and (b) organic chemicals such as polycyclic aromatic hydrocarbons (PAHs), petroleum hydrocarbons (PHCs), benzene, toluene, ethylene, and xylene (BTEX), etc. Since interactions between the pollutants (and contaminants) will be between the surface reactive groups that characterize the surfaces of both the soil fractions and the pollutants, it is useful to obtain an appreciation of the nature of the broad groups of pollutants, and the various factors that control their interactions in the soil-water system.

4.2.1 Metals

The alkali and alkaline-earth metals are elements of Groups I and II (periodic table). The common alkali metals are Li, Na, and K, with Na and K being very abundant in nature. The other alkali metals in Group IA Rb, Cs, and Fr are less commonly found in nature. The alkali metals are strong reducing agents, and are never found in the elemental state since they will react well with all nonmetals.

Of the metals in Group II (Be, Mg, Ca, Sr, Ba, and Ra), Mg and Ca are the more common ones, and similar to the Group IA metals, these are strong reducing agents. They react well with many nonmetals. While Be, Ba, and Sr are less common, they can be found from various sources, e.g., Be from the mineral beryl, and Ba and Sr generally from their respective sulphates.

Strictly speaking, *heavy metals* (HMs) are those elements with atomic numbers higher than Sr — whose atomic number is 38. However, it is not uncommon to find usage of the term heavy metals to cover those elements with atomic numbers greater than 20 (i.e., greater than Ca). We will use the commonly accepted grouping of HM pollutants, i.e., those having atomic numbers greater than 20. These can be found in the lower right-hand portion of the periodic table, i.e., the *d*-block of the periodic table, and include 38 elements that can be classified into three convenient groups of atomic numbers as follows:

- **From atomic number 22 to 34** — Ti, V, Cr, Mn, Fe, Co, Ni, Cu, Zn, Ga, Ge, As, and Se;
- **From 40 to 52** — Zr, Nb, Mo, Te, Ru, Rh, Pd, Ag, Cd, In, Sn, Sb, and Te; and
- **From 72 to 83** — Hf, Ta, W, Re, Os, Ir, Pt, Au, Hg, Tl, Pb, and Bi.

Most of the metals in this group, which excludes Zn and those metals in Group III to Group V, are *transition metals,* because these are elements with at least one ion with a partially filled *d* sub-shell. It can be said that almost all the properties of these transition elements are related to their electronic structures and the relative energy levels of the orbitals available for their electrons. This is particularly significant in metal classification schemes such as the one proposed by Pearson (1963) (Section 4.3.1).

The more common toxic HMs associated with anthropogenic inputs, landfill and chemical waste leachates and sludges, include lead (Pb), cadmium (Cd), copper (Cu), chromium (Cr), nickel (Ni), iron (Fe), mercury (Hg), and zinc (Zn). Metallic ions such as Cu^{2+}, Cr^{2+}, etc. (M^{n+} ions) cannot exist in aqueous solutions (porewater) as individual metal ions. They are generally coordinated (chemically bound) to six water molecules, and in their hydrated form they exist as $M(H_2O)_x^{n+}$. By and large, M^{n+} is used as a simplified notational scheme. Since M^{n+} coordination with water is in the form of bonding with inorganic anions, replacement of water as the ligand for M^{n+} can occur if the candidate ligand, generally an electron donor, can replace the water molecules bonded to the M^{n+}.

We define *ligands* as those anions that can form coordinating compounds with metal ions. The characteristic feature of these anions is their free pairs of electrons. In this instance, the water molecules that form the coordinating complex are the ligands, and the metal ion M^{n+} would be identified as the central atom. The number of ligands attached to a central metal ion is called the *coordination number.* In general, the coordination number of a metal ion is the same regardless of the type or nature of ligand. The coordination number for Cu^{2+}, for example, is 4 — as found in $Cu(H_2O)_4^{2+}$ and $CuCl^{2-}$. In the case of Fe^{3+}, whose coordination number is 6, we have $Fe(CN)_6^{3-}$ and $Fe(H_2O)_6^{3+}$ as examples. By and large, the common coordination numbers for heavy metals is 2, 4, and 6, with 6 being the most common. Complexes with a coordination number of 2 will obviously have a linear arrangement of ligands, whereas complexes with a coordination number of 4 will generally have tetrahedral arrangement of ligands. In some cases, a square-planar arrangement of ligands is also obtained. In the case of complexes of coordination number 6, the ligands are arranged in an octahedral fashion.

If a ligand only possesses one bonding site, i.e., a ligand atom, the ligand is called an *unidentate ligand.* Ligands that have more than one ligand atom are *multidentate ligands*, although the prefixes bi- and tri- are sometimes used for ligands with two and three ligand atoms, respectively. The complexes formed by metal ions M^{n+} and multidentate ligands are called *chelated complexes*, and the multidentate ligands themselves are most often referred to as chelating agents. Three of the more common chelating agents are EDTA (ethylene-diamine tetraacetate), sodium nitrilo-triacetate (NTA), and sodium tripolyphosphate (TPP).

Some of the more common inorganic ligands that will form complexes with metals include: CO_3^{2-}, SO_4^{2-}, Cl^-, NO_3^-, OH^-, SiO_3^-, CN^-, F^-, and PO_4^{3-}. In addition to anionic-type ligands, metal complexes can be formed with molecules with lone pairs of electrons, e.g., NH_3 and PH_3. Examples of these kinds of complexes are: $Co(NH_3)_6^{3+}$ where the NH_3 is a Lewis base and a neutral ligand, and $Fe(CN)_6^{4-}$ where CN^- is also a Lewis base and an anionic ligand. Complexes formed between soil-organic compounds and metal ions are generally chelated complexes. These naturally occurring organic compounds are humic and fulvic acids, and amino acids.

Some of the HMs can exist in the porewater in more than one oxidation state, depending on the pH and redox potential of the porewater in the microenvironment. For example, selenium (Se) can occur as SeO_3^{2-} with a valence of +4, and as SeO_4^{2-} with a valence of +6. Similarly, we have two possible valence states for the existence of copper (Cu) in the porewater. These are valencies of +1 and +2 for CuCl and CuS, respectively. Chromium (Cr) and iron (Fe) present more than one ionic form for each of their two valence states. For Cr, we have CrO_4^{2-} and $Cr_2O_7^{2-}$ for the valence state of +6, and Cr^{3+} and $Cr(OH)_3$ for the +3 valence state. In the case of Fe we have Fe^{2+} and FeS for the +2 valence state and Fe^{3+} and $Fe(OH)_3$ for the +3 valence state.

Variability in oxidation states is a characteristic of transition elements (i.e., transition metals). Many of these elements have one oxidation state that is most stable, e.g., the most stable state for Fe is Fe(III) and Co(II) and Ni(II) for cobalt and nickel, respectively. Much of this is a function of the electronic configuration in the *d* orbitals. Unpaired electrons which compose one half of the sets in *d* orbitals are very stable. This explains why Fe(II) can be easily oxidized to Fe(III) and why the oxidation of Co(II) to Co(III) and Ni(II) to Ni(III) cannot be as easily accomplished. The loss of an additional electron to either Co(II) and Ni(II) still does not provide for one half unpaired electron sets in the *d* orbitals. This does not mean to say that Co(III) does not readily exist. The complex ion $[Co(NH_3)_6]^{3+}$ has Co at an oxidation state of +3.

4.2.2 Organic Chemical Pollutants

There is a whole host of organic chemicals that find their way into the land environment. These have origins in various chemical industrial processes and as commercial substances for use in various forms. Products for commercial use include organic solvents, paints, pesticides, oils, gasoline, creasotes, greases, etc. are some of the many sources for the chemicals found in contaminated sites. One can find at least a million organic chemical compounds registered in the various chemical abstracts services available, and many thousands of these are in commercial use. It is not possible to categorize them all in respect to how they would interact in a soil-water system. The more common organic chemicals found in contaminated sites fall into convenient groupings which include:

- **Hydrocarbons** — including the PHCs (petroleum hydrocarbons), the various alkanes and alkenes, and aromatic hydrocarbons such as benzene, MAHs (multi-cyclic aromatic hydrocarbons), e.g., naphthalene, and PAHs (polycyclic aromatic hydrocarbons), e.g., benzo-pyrene; and
- **Organohalide compounds** — of which the chlorinated hydrocarbons are perhaps the best known. These include: TCE (trichloroethylene), carbon tetrachloride, vinyl chloride, hexachlorobutadiene, PCBs (polychlorinated biphenyls), and PBBs (poly-brominated biphenyls).
- The other groupings could include oxygen-containing organic compounds such as phenol and methanol, and nitrogen-containing organic compounds such as TNT (trinitrotoluene).

In respect to the presence of these chemicals in the ground, the characteristic of particular interest is whether they are lighter or denser than water, since this influences the transport characteristics of the organic chemical. The properties and characteristics of these pollutants are discussed in detail in considerations of persistence and fate of organic pollutants in Chapter 6.

A well-accepted classification is the NAPL (non-aqueous phase liquids) scheme which breaks the NAPLs down into the light NAPLs identified as LNAPLs, and the dense ones called the DNAPLS. The LNAPLs are considered to be lighter than water and the DNAPLs are heavier than water. The consequence of these characteristics is shown in the schematic in Figure 4.3. Because the LNAPL is lighter than water, the schematic shows that it stays above the water table. On the other hand, since the DNAPL is denser than water, it will sink through the water table and will come to rest at the impermeable bottom (bedrock). Some typical LNAPLs include gasoline,

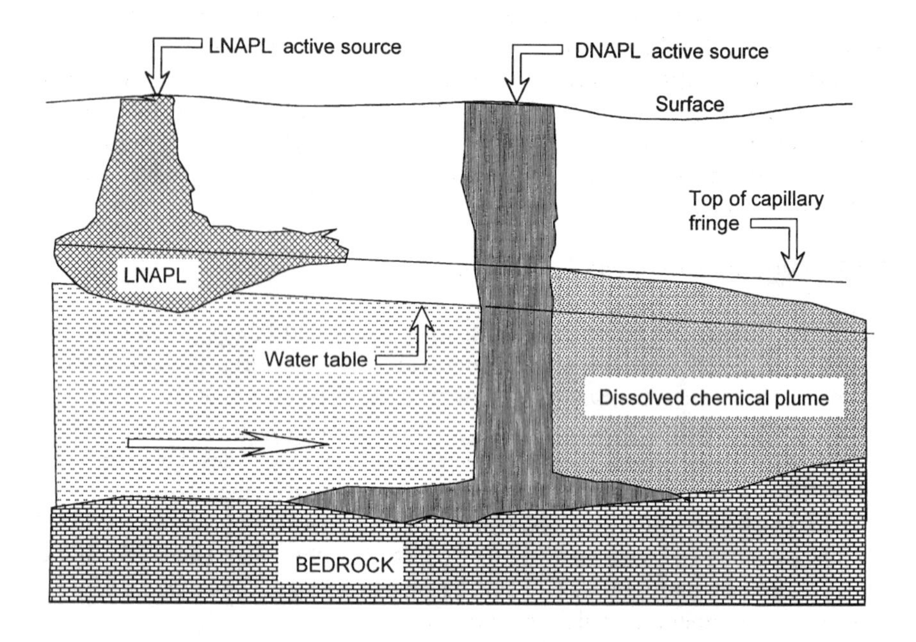

Figure 4.3 Schematic diagram showing LNAPL and DNAPL penetration in substrate. Note influence of water table on extent of LNAPL penetration.

heating oil, kerosene, and aviation gas. DNAPLs include the organohalide and oxygen-containing organic compounds such as 1,1,1-trichloroethane, creasote, carbon tetrachloride, pentachlorophenols, dichlorobenzenes, and tetrachloroethylene.

4.3 CONTROLS AND REACTIONS IN POREWATER

The presence of naturally occurring salts in the porewater (Groups I and II in the periodic table) together with the inorganic and organic pollutants result in a complex aqueous chemical regime. The transport and fate of pollutants are as much affected by the surface reactive groups of the soil fractions as by the chemistry of the porewater. At equilibrium, the chemistry of the porewater is intimately connected to the chemistry of the pollutants and the surfaces of the soil fractions. Evaluation of the interactions among contaminants, pollutants, and soil fractions cannot be fully realized without knowledge of the many different sets of chemical reactions occurring in the porewater. Included in these sets of reactions are the biologically mediated chemical processes and reactions that occur because of the presence of microorganisms and their response to the microenvironment.

Figure 2.2 showed a highly simplistic picture of the interaction between a soil fraction and a contaminant. As stated previously, the nature of these interactions is determined by the characteristics of the interacting surfaces, and can be physical chemical in nature. Chemical interactions between the pollutants and soil fractions are by far the most significant. We would thus expect that the chemistry of the surfaces of these interacting elements, and the environment within which they reside, would be important factors that will control the fate of the pollutants. The pH of the soil-water system and the various other dissolved solutes in the porewater influence the various interaction mechanisms. Bonding between pollutants and soil fractions, acid-base reactions, speciation, complexation, precipitation, and fixation are some of the many manifestations of the interactions.

4.3.1 Acid-Base Reactions — Hydrolysis

Hydrolysis falls under the category of acid-base reactions, and in its broadest sense refers to the reaction of H^+ and OH^- ions of water with the solutes and elements present in the water. In general, hydrolysis is a neutralization process. In the context of a soil-water system, it is useful to bear in mind that many soil minerals, for example, are composed of ionized cations and anions. These may be strongly or weakly ionized, the result of which will produce resultant pH levels in the soil-water system that can vary from below neutral to above neutral pH values. Abrasion pH values from neutral to pH 11 have been reported for some silicate rock-forming minerals such as feldspars, amphiboles, and pyroxenes which consist of strongly ionized cations and weakly ionized anions (Keller, 1968). For hydrolysis reactions to continue, the reaction products need to be removed if the system is to continue the reactions. In terms of pollutants and soil-water systems, this means processes associated with precipitation, complexation, and sorption will remove the reaction products. Fresh input (from transport) of pollutants will serve to continue the hydrolysis reactions.

Water is both a *protophillic* and a *protogenic* solvent, i.e., it is *amphiprotic* in nature. It can act either as an acid or as a base. It can undergo self-ionization, resulting in the production of the conjugate base OH^- and conjugate acid H_3O^+. For strictly aqueous solutions, the concept of acids and bases proposed by Arrhenius has been shown to be useful, i.e., we define an *acid* as a substance which dissociates to produce H^+ ions. If dissociation in an aqueous solution produces OH^- ions, the substance is identified as a *base*. Since soil solids and water form the soil-water system, and since pollutants consist of both inorganic and organic substances, it is necessary to use the broader concepts of acids and bases in describing the various reactions and interactions occurring in a soil-water-pollutant system.

The Brønsted-Lowry concept considers an *acid* as a substance that has a tendency to lose a proton (H^+), and, conversely, a *base* is considered as a substance that has a tendency to accept a proton. In the Brønsted-Lowry acid-base scheme, an *acid* is a *proton donor* (*protogenic* substance) and a *base* is a *proton acceptor* (*protophillic* substance). Substances that have the capability to both donate and accept protons (i.e., both protogenic and protophillic), such as water and alcohols, are called *amphiprotic* substances.

Acid-base reactions involve proton transfer between a proton donor (acid) and a proton acceptor (base). The transfer is called a *protolytic* reaction and the process is called *protolysis*. The self-ionization of water, for example, is called *autoprotolysis*, and *neutralization* is the reverse of autoprotolysis. All bases have a lone pair of electrons to share with a proton. The donation of the electron pair in covalent bonding to an acid that accepts the electron pair will leave the electron donor (base) electron-deficient. This brings us to the broader concept of acids and bases used by Lewis (1923). He defined an acid as a substance that is capable of accepting a pair of electrons for bonding, and a base as a substance that is capable of donating a pair of electrons. As with the donor-acceptor terminology, *Lewis acids* are electron acceptors, and *Lewis bases* are electron donors. As an example, all metal ions M^{nx} are Lewis acids, and in the previous discussion on heavy metals and complexes formed with ligands, we see that the HMs are bonded with Lewis bases. This is explained by the fact that Lewis acids can accept and share electron pairs donated by Lewis bases. Whilst Lewis bases are also Brønsted bases, Lewis acids are not necessarily Brønsted acids since Lewis acids include substances that are not proton donors. However, the use of the Lewis acid-base concept permits us to treat metal-ligand bonding as acid-base reactions.

Pearson (1963) has classified Lewis acids and bases according to their mutual behaviour into categories of *hard* and *soft* acids and bases, based on demonstrated properties:

- **Hard acids** — generally small in size with high positive charge; high electronegativity; low polarizability; and no unshared pairs of electrons in their valence shells.
- **Soft acids** — generally large in size with a low positive charge; low electronegativity; high polarizability; and with unshared pairs of electrons in their valence shells.
- **Hard bases** — usually have high electronegativity; low polarizability; and difficult to oxidize.
- **Soft bases** — usually have low electronegativity; high polarizability; and easy to oxidize.

Hydrolysis reactions of metal ions can be expressed as:

$$MX + H_2O \rightarrow MOH + H^+ + X^- \tag{4.1}$$

and are influenced by: (a) pH of the active system; (b) type, concentration, and oxidation state of the metal cations; (c) redox environment; and (d) temperature. High temperatures favour hydrolysis reactions, as do low organic contents, low pH values, and low redox potentials.

A sense of the degree of dissociation of a compound is obtained by a knowledge of the dissociation constant *k*. The *pk value* is commonly used to express this dissociation in terms of the negative logarithm (to base 10) of the dissociation constant, i.e., $pk = -log(k)$. The smaller the *pk* value, the higher the degree of ionic dissociation and hence the more soluble the substance. A knowledge of relative values *pk* between compounds will tell us much about the transport and adsorption of chemical species in the ground. The *pk* value can also be used to indicate the strength of acids and bases. Strong acids are strong proton donors. Weak acids do not provide much proton donor capability, i.e., they do not favor the formation of H^+ ions, and will consequently show higher pH values than strong acids. In respect to heavy metals, for example, most highly charged cationic metals have low *pk* values and are strongly hydrolyzed in aqueous solution. *pk* values can be determined using the Henderson-Hasselbalck relationship:

$$pk = pH - \log_{10}\frac{unprotonated\,form\,(base)}{protonated\,form\,(acid)} \tag{4.2}$$

Hydrated metal cations can act as acids or proton donors, with separate *pk* values for each. In the context of interaction with clay particles in a soil-water system, these *pk* values decrease with dehydration of the soil. Water molecules are strongly polarized by the exchangeable metal cations on the surfaces of clay particles. These strongly polarized water molecules contribute considerably to the proton donating process of clay particles, as witness the observations that the acidity of this water is greater than what might be expected from considerations of the *pk* values of the hydrated metal cations in water (Mortland and Raman, 1968). The hydrolysis properties of the cations appear to be influenced by the effect of exchangeable cation on the protonation process.

4.3.2 Oxidation-Reduction (Redox) Reactions

In addition to the considerations of acid-base reactions given in the previous section, we need to note that the porewater in soils also provides the medium for oxidation-reduction reactions which can be abiotic and/or biotic. Microorganisms play a significant role in catalyzing redox reactions. The bacteria in the soil utilize oxidation-reduction reactions as a means to extract the energy required for growth, and as such are the catalysts for reactions involving molecular oxygen and soil organic matter and organic chemicals. Since oxidation-reduction reactions involve the transfer of electrons between the reactants, the activity of the electron e^- in the chemical system plays a significant role. A fundamental premise in respect to chemical reactions is that these reactions are directed toward establishing a greater stability of the outermost electrons of the reactants, i.e., electrons in the outermost shell of the substances involved. There is a link between redox reactions and acid-base reactions. Generally speaking, the transfer of electrons in a redox reaction is accompanied by proton transfer. The loss of an electron by iron(II) at pH 7 is accompanied by the loss of three hydrogen ions to form highly insoluble ferric hydroxide (Manahan, 1990), according to the following:

$$Fe(H_2O)_6^{2+} \rightarrow Fe(OH)_3(s) + 3H^+ + e^- \tag{4.3}$$

For inorganic solutes, redox reactions result in the decrease or increase in the oxidation state of an atom. This is significant in that some ions have multiple oxidation states, and thus impact directly on the fate of the inorganic pollutant with such a characteristic. Organic chemical pollutants, on the other hand, show the effects of redox reactions through the gain or loss of electrons in the chemical. In terms of relative importance, it is generally assumed that biotic redox reactions are of greater significance than abiotic redox.

There are two classes (each) for electron donors and electron acceptors of organic chemical pollutants. In the case of electron donors, we have (a) electron-rich π-cloud donors which include alkenes, alkynes, and the aromatics, and (b) lone-pair electron donors which include the alcohols, ethers, amines, and alkyl iodides. For the electron acceptors, we have (a) electron-deficient π-electron cloud acceptors which include the π-acids, and (b) weakly acidic hydrogens such as *s*-triazine herbicides and some pesticides.

The *redox potential Eh* is considered to be a measure of electron activity in the porewater. It is a means for determining the potential for oxidation-reduction reactions in the pollutant-soil system under consideration, and is given as:

$$Eh = pE\left(\frac{2.3RT}{F}\right) \tag{4.4}$$

where E = electrode potential, R = gas constant, T = absolute temperature, and F = Faraday constant.

The electrode potential E is defined in Equation 4.4 in terms of the half reaction:

$$2H^+ + 2e^- \rightleftharpoons H_2(g) \tag{4.5}$$

When activity of H^+ = 1, and pressure H_2 *(gas)* = 1 atmosphere, then E = 0.

The expression pE is a mathematical term that represents the negative logarithm of the electron activity e^-. At a temperature of 25°C, the relationship between Eh and pE is:

$$\begin{aligned} Eh &= 0.0591\,pE \qquad (4.6) \\ &= E^o + \left(\frac{RT}{nF}\right)\ln\frac{a_{i,\,ox}}{a_{i,\,red}} \end{aligned}$$

where E^o = standard reference potential, n = number of electrons, and the subscripts for a refer to the activity of the ith species in the oxidized (ox) or reduced (red) states. The redox capacity measures the maximum amount of electrons that can be added or removed from the soil-water system without a measurable change in the Eh or pE. This concept corresponds exactly to the *buffering capacity* of soils which refers to a measure of the amount of acid or base that can be added to a soil-water system without any measurable change in the system pH. The factors that affect the redox potential Eh include pH, oxygen content or activity, and water content of the soil.

4.3.3 *Eh*-pH relationship

Without taking into account the presence of soil fractions, and considering only the porewater as a fluid medium, the stability of inorganic solutes in the porewater is a function of several factors. Amongst these, the pH, *Eh*, or *pE* of the porewater, the presence of ligands, temperature, and concentration of the inorganic solutes are perhaps the most significant. The influence of all of these can be calculated using the Nernst equation, similar in form to Equation 4.6. Thus, if *A* and *B* represent the reactant and product, respectively, we will have:

$$pE = 16.92Eh \qquad (4.7)$$

$$= E^o + \left(\frac{RT}{nF}\right)\ln\frac{[A]^a[H_2O]^w}{[B]^b[H^+]^h}$$

where the superscripts a, b, w, and h in the equation refer to number of moles of reactant, product, water, and hydrogen ions, respectively. The stable product for a given set of reactants or the valence state of the reactants will be seen to be a function of the pH-*pE* status. Using information from Manahan (1990), Sawyer et al. (1994), and Fetter (1993), Figure 4.4 shows a simplified *pE*-pH diagram for an iron (Fe)-water system for a maximum soluble iron concentration of 10^{-5} *M*.

The uppermost sloping boundary defines the limit of water stability, above which the water is oxidized. Likewise, the lowest sloping boundary marks the limit of water stability below which the water is reduced. The redox reactions are given as follows:

$$2H_2O \rightleftharpoons O_2(g) + 4H^+ + 4e^- \quad (oxidation)$$

$$2H_2O + 2e^- \rightleftharpoons H_2(g) + 2OH^- \quad (reduction) \qquad (4.8)$$

The *pE*-pH diagram provides a quick view of the various phases of Fe. For example, we see that at a *pE* value of 4, Fe exists as Fe^{3+} at the lower pH values. Staying with a *pE* value of 4 and continuing with increases in pH, we note that as we approach a pH of about 6.4 and beyond, precipitation occurs, resulting in the formation of Fe(III) hydroxides ($Fe(OH)_3$). A decrease in *pE* at the higher pH values will result in precipitates of Fe(II), as seen in the diagram. Similar diagrams can be constructed for other inorganic pollutants. The interested reader should consult textbooks on aquatic chemistry, geochemistry, and soil chemistry for more details.

4.4 PARTITIONING AND SORPTION MECHANISMS

The *partitioning* of contaminants (pollutants and non-pollutants) refers to processes of chemical and physical mass transfer (or removal) of the contaminants from

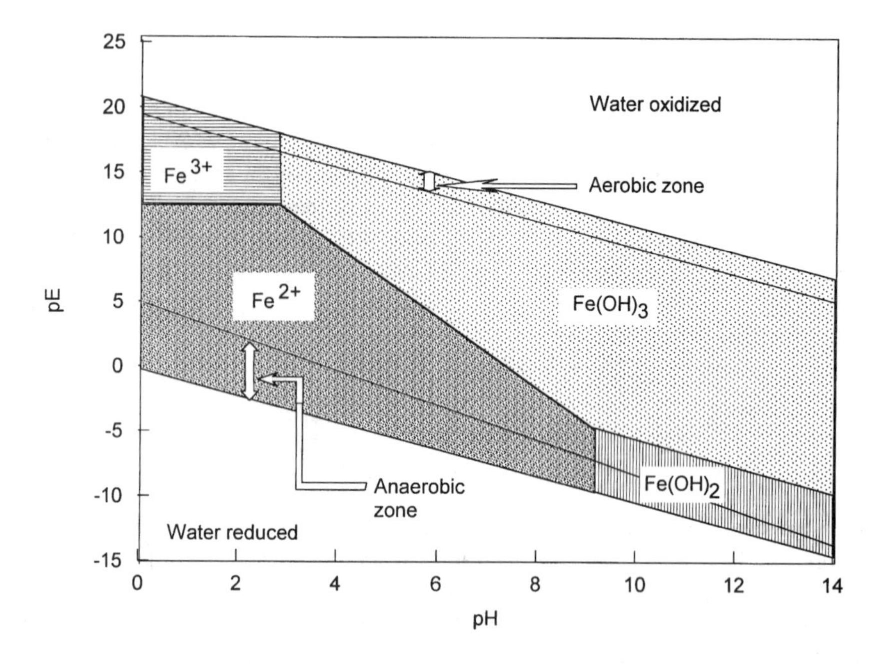

Figure 4.4 *p*E-pH diagram for Fe and water with maximum soluble Fe concentration of 10^{-5} *M*. Note that the zone between the *aerobic* and *anaerobic* zones is the *transition* zone.

the porewater to the surfaces of the soil fractions. We refer to contaminants (pollutants and non-pollutants) partitioned onto soil fractions' surfaces as *sorbate*, and to the soil fractions responsible for this partitioning as the *sorbent*. Partitioning, as a process or phenomenon, is most generally associated with considerations of transport of pollutants in soils.

We use the term *sorption* to refer to the adsorption processes responsible for the partitioning of the dissolved solutes in the porewater to the surfaces of the soil fractions. The dissolved solutes include ions, molecules, and compounds. It is often not easy to fully distinguish amongst all the processes that contribute to the overall adsorption phenomenon. Hence the term *sorption* is used to indicate the general transfer of dissolved solutes from the aqueous phase to the interfaces of the various soil fractions via mechanisms of physical adsorption, chemical adsorption, and precipitation. Adsorption reactions are processes by which contaminant solutes in solution become attached to the surfaces of the various soil fractions. These reactions are basically governed by the surface properties of the soil fractions, the chemistry of the pollutants and the porewater, and the *pE*-pH of the environment of interaction. The various sorption mechanisms can include both short-range chemical forces such as covalent bonding, and long-range forces such as electrostatic forces.

4.4.1 Molecular Interactions and Bondings

Sorption processes involving molecular interactions are Coulombic, and are interactions between nuclei and electrons. These are essentially electrostatic in nature. The major types of interatomic bonds are ionic, covalent, hydrogen, and van der Waals. Ionic forces hold together the atoms in a crystal. The bonds formed from various forces of attraction include:

- **Ionic** — Electron transfer occurs between the atoms, which are subsequently held together by the opposite charge attraction of the ions formed.
- **Covalent** — Electrons are shared between two or more atomic nuclei.
- **Coulombic** — This involves ion-ion interaction.
- **van der Waals** — This involves dipole-dipole (Keesom); dipole-induced dipole (Debye); instantaneous dipole-dipole (London dispersion).
- **Steric** — This involves ion hydration surface adsorption.

Forces of attraction between atoms and/or molecules originate from several sources, the strongest of which is the Coulombic or ionic force between a positively charged and a negatively charged atom. This force decreases as the square of the distance separating the atoms, and is an important force in developing sorption between charged contaminants and charged (reactive) surfaces of the soil fractions. Interactions between instantaneous dipoles, and dipole-dipole interactions produce forces of attractions categorized as *van der Waals* forces. The three dominant types, as listed above are: (a) Keesom — forces developed as a result of dipole orientation; (b) Debye — forces developed due to induction; and (c) London dispersion forces. For non-polar molecules (e.g., organic chemicals) this is frequently the most common type of bonding mechanism established with the mineral soil fractions.

Soil-organic matter in soils can form hydrogen bonds with clay particles. These are electrostatic or ionic bonds. The bonding between the oxygen from a water molecule to the oxygen on the clay particle surface is a strong bond in comparison with other bonds between neutral molecules. This mechanism of bonding is important in (a) bonding layers of clay minerals together; (b) holding water at the clay surface; and (c) bonding organic molecules to clay surfaces. Electrical bonds are formed between the negative charges on clay mineral surfaces and positive charges on the organic matter. They can also be formed between negatively charged organic acids and positively charged clay mineral edges.

Whilst organic anions such as those in organic chemicals are normally repelled from the surfaces of negatively charged particles, some adsorption can occur if polyvalent exchangeable cations are present. Bonding with clay mineral particle surfaces will be via polyvalent bridges. The sorption mechanism can be in the form of (a) anion associated directly with cation, or (b) anion associated with cation in the form of a water bridge, referred to as a *cation bridge*. The process essentially consists of replacement of a water molecule from the hydration shell of the exchangeable cation by an oxygen or an anionic group, e.g., carboxylate or phenate of the organic polymer. Charge neutrality at the surface is established by the ion formerly satisfying the charge of the organic group entering the exchange complex of the clay. Because positive sites normally exist in aluminum and iron hydroxides, at least below pH 8 (Parks, 1965), organic anions can be associated with the oxides by simple Coulombic attraction. The adsorption of the organic anion is readily reversible by exchange with chloride or nitrate ions. In addition to anion exchange reactions, specific adsorption of anions by these (humic) materials normally occurs, i.e., the anions penetrate into the coordination shells of iron or aluminum atoms in the surface of the hydroxide. This type of specific adsorption is generally called *ligand exchange*. Unlike anion exchange reactions, the specifically adsorbed anions cannot be displaced from the complex.

4.4.2 Cation Exchange

Cation exchange in soils occurs when positively charged ions (contaminant ions and salts) in the porewater are attracted to the surfaces of the clay fractions. The process is set in motion because of the need to satisfy electroneutrality and is stoichiometric. Electroneutrality requirements necessitate that replacing cations must satisfy the net negative charge imbalance shown by the charged clay surfaces. In terms of the DDL model, this means that the cations leaving the diffuse ion-layer must be replaced by an equivalent amount of cations if the negative charges from the clay particle surfaces are to be balanced. The replaced cations are identified as *exchangeable cations*, and when they possess the same positive charge and similar geometries as the replacing cations, the following relationship applies: $M_s/N_s = M_o/N_o = 1$, where M and N represent the cation species and the subscripts s and o represent the surface and the bulk solution. Exchangeable cations are identified as such because one cation can be readily replaced by another of equal valence, or by two of one half the valence of the original one. This is highly significant when it comes to prediction of partitioning of pollutants. Thus, for example, if the substrate soil material contains sodium as an exchangeable cation, cation exchange with an incoming lead chloride ($PbCl_2$) leachate would occur according to the following:

$$Na_2\ clay + PbCl_2 \rightleftharpoons Pb\ clay + 2\ NaCl$$

The quantity of exchangeable cations held by the soil is called the *cation-exchange capacity* (CEC) of the soil, and is generally equal to the amount of negative charge. It is expressed as milliequivalents per 100 g of soil (meq/100 g soil). The predominant exchangeable cations in soils are calcium and magnesium, with potassium and sodium being found in smaller amounts. In acid soils, aluminium and hydrogen are the predominant exchangeable ions. Extensive leaching of the soil will remove the cations that form bases (calcium, sodium, etc.), leaving a clay with acidic cations, aluminium, and hydrogen.

We can determine the relative energy with which different cations are held at the clay surface by assessing the relative ease of replacement or exchange by a chosen cation at a chosen concentration. Because the valency of the cation has a dominant influence on its ease of replacement, the higher the valency of the cation, the greater is the replacing power of the ion. Conversely, the higher the valency of the cation at the surface of the clay particles, the harder it is to replace. For ions of the same valence, increasing ion size endows it with greater replacing power. There are some minor exceptions to this simple rule. The best example of this exception is potassium, which is a monovalent cation. It has a high replacing power, and is strongly held because it fits nicely into the hexagonal holes of the silica sheet of the layer lattice structure of clay minerals. The result is that potassium will replace a divalent ion much more easily than will monovalent sodium.

Some representative cations arranged in a series that portrays their relative replacing power can be shown as:

$$Na^+ < Li^+ < K^+ < Rb^+ < Cs^+ < Mg^{2+} < Ca^{2+} < Ba^{2+} < Cu^{2+} < Al^{3+} < Fe^{3+} < Th^{4+}$$

The positions shown above are generally the more likely replacement positions, and are to a very large extent dependent on the size of the hydrated cation. In heterovalent exchange, the selective preference for monovalent and divalent cations is dependent on the magnitude of the electric potential in the region where the greatest amount of cations are located. Changes in the relative positions can occur in the above (lyotropic) series depending on the kind of clay and ion which is being replaced. The number of exchangeable cations replaced obviously depends upon the concentration of ions in the replacing solution (contaminant leachate). If a clay containing sodium cations is contacted by a contaminant leachate containing divalent ions, exchange will take place until, at equilibrium, a certain percentage of the exchangeable ions will still be sodium and the remainder will be the divalent contaminant ion (e.g., Pb^{2+}, Cd^{2+}, etc.). The proportion of each exchangeable cation to the total CEC, as the outside ion concentration varies, is given by the exchange-equilibrium equations. Of the several equations that have been derived with different assumptions about the nature of the exchange process, perhaps the simplest useful equation is that used first by Gapon:

$$\frac{M_e^{+m}}{N_e^{+n}} = K\frac{\left[M_o^{+m}\right]^{\frac{1}{m}}}{\left[N_o^{+n}\right]^{\frac{1}{n}}} \tag{4.9}$$

where:

- superscripts m and n refer to the valence of the cations;
- subscripts e and o refer to the exchangeable and bulk solution ions;
- constant K is a function of specific cation adsorption and nature of the clay surface. K decreases in value as the surface density of charges increases.

4.4.3 Physical Adsorption

Physical adsorption of pollutants in the porewater (or from incoming leachate) by the soil fractions occurs as a result of the attraction of the pollutants to the surfaces

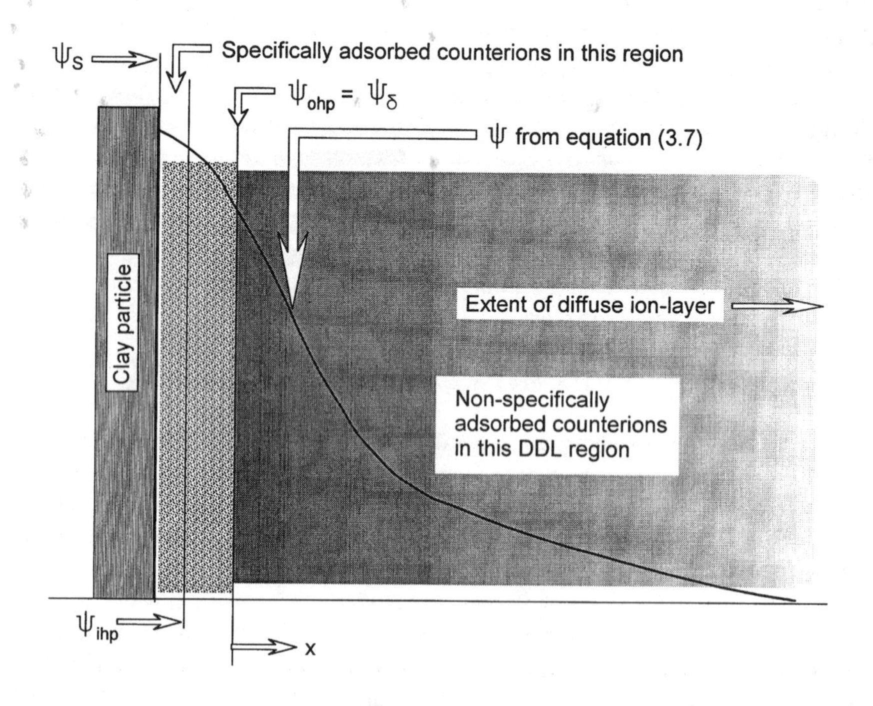

Figure 4.5 Specifically and non-specifically adsorbed counterions in DDL model.

of the soil fractions. This is in response to the charge deficiencies of the soil fractions (i.e., clay minerals). As mentioned previously, the counterions are drawn to the soil fractions (primarily clay minerals) because of the need to establish electroneutrality. Cations and anions are specifically or non-specifically adsorbed by the soil solids, as shown in Figure 4.5, depending on whether they interact in diffuse ion-layer or in the Stern layer. The counterions in the diffuse ion-layer will reduce the potential ψ, and are generally referred to as *indifferent ions*. They are *non-specific,* and they do not reverse the sign of ψ.

Non-specific adsorption refers to ions that are held primarily by electrostatic forces. Sposito (1984) uses this term to refer to outer-sphere surface complexation of ions by the functional groups exposed on soil particles. Calculations for the concentration of ions held as non-specific ions at distances of x away from the particle surface, in the diffuse ion-layer, can be made using the relationship shown as Equation 3.7, on the assumption that the soil solids can be approximated by the parallel-plate model. Examples of non-specific adsorption are the adsorption of alkali and alkaline earth cations by the clay minerals. If we consider cations as point charges, as assumed in the DDL model discussed in the previous chapter, the adsorption of cations would be related to their valence, crystalline unhydrated and hydrated radii. Cations with smaller hydrated size or large crystalline size would be preferentially adsorbed, everything else being equal. Cation exchange involves those cations associated with the negative charge sites on the soil solids, largely through electrostatic forces. It is important to note that ion exchange reactions occur with the various soil fractions, i.e., clay minerals and non-clay minerals.

4.4.4 Specific Adsorption

Specific adsorption of contaminants and pollutants occurs when their respective ions are adsorbed by forces other than those associated with the electric potential within the Stern layer, as shown in Figure 4.5. Sposito (1984) refers to specific adsorption as the effects of inner-sphere surface complexation of the ions in solution by the surface functional groups associated with the soil fractions. The specifically adsorbed ions can influence the sign of ψ, and are referred to as *specific ions*. Cations specifically adsorbed in the inner part of the Stern layer will lower the point of zero charge (Arnold, 1978). Specific adsorption of anions on the other hand will tend to shift the point of zero charge (zpc) to a higher value.

4.4.5 Chemical Adsorption

Chemical adsorption or *chemisorption* refers to high affinity, specific adsorption which occurs in the inner Helmholtz layer (see Figures 4.5 and 3.12) through covalent bonding. In specific cation adsorption, the ions penetrate the coordination shell of the structural atom and are bonded by covalent bonds via *O* and *OH* groups to the structural cations. The valence forces bind atoms to form chemical compounds of definite shapes and energies. The chemisorbed ions can influence the sign of ψ, and are called *potential determining ions* (pdis). To that extent, chemisorbed ions are also referred to as *high affinity specifically sorbed ions*. It is not always easy to distinguish the interaction mechanisms associated with chemical adsorption from electrostatic positive adsorption. Due to the nature of the adsorption phenomenon, we would expect that higher adsorption energies would be obtained for reactions resulting in chemical adsorption. These reactions can be either endothermic or exothermic, and usually involve activation energies in the process of adsorption, i.e., the energy barrier between the molecule/ion being adsorbed and the soil solid surface must be surmounted if a reaction is to occur. Strong chemical bond formation is often associated with high exothermic heat of reaction, and the first layer is chemically bonded to the surface with additional layers being held by van der Waals forces.

The three principal types of chemical bonds between atoms are:

- **Ionic** — Where electron transfer between atoms results in an electrostatic attraction between the resulting oppositely charged ions;
- **Covalent** — More or less equal sharing of electrons exists between the partners;
- **Coordinate-covalent** — The shared electrons originate only from one partner.

4.4.6 Physical Adsorption of Anions

The soil fractions that have positive charge sites are primarily the oxides and edges of some clay minerals. Physical adsorption of anions is thus considerably less than the adsorption capacity for cations. The capacity for adsorption of anions is influenced by the pH of the soil-water system and the electrolyte level, and selectivity for anion sorption is greater in comparison to cation sorption as previously described. Experimental evidence shows the following preference:

$$Cl \simeq NO_3 < SO_4 \ll PO_4 < SiO_4$$

4.5 pH ENVIRONMENT, SOLUBILITY, AND PRECIPITATION

We have seen in the example given in Figure 4.4 that the various changes in both pH and *pE* affect the speciation of Fe. In general, the pH of the microenvironment in a representative elementary volume which encompasses soil solids and porewater is a significant factor in the environmental mobility of heavy metal pollutants. To a very large extent, this is because of the influence of pH on the solubility of the heavy metal complexes. Nyffeler et al. (1984) show that the pH at which maximum adsorption of metals occurs can be expected to vary according to the first hydrolysis constant of the metal (cationic) ions.

Under slightly alkaline conditions, precipitation of heavy metals as hydroxides and carbonates can occur. The process requires the ionic activity of the heavy metal solutes to exceed their respective solubility products. The precipitation process, which is mostly associated with the heavy metal pollutants, results in the formation of a new substance in the porewater by itself or as a precipitate attached to the soil solids. The process itself is the converse of dissolution. This occurs when the transfer of solutes from the porewater to the interface results in accumulation of a new substance in the form of a new soluble solid phase. Generally speaking, there are two stages in precipitation: nucleation and particle growth. Gibbs' phase rule restricts the number of solid phases that can be formed.

Since the various sorption mechanisms and precipitation all result in the removal of pollutants (heavy metals in this case) from the porewater, it is not easy to distinguish the various processes responsible for the removal, e.g., (a) net accumulation of contaminants by the soil fractions, and (b) formation of new precipitated solid phases. One of the reasons why a distinction between these two processes is not always easy to obtain is because the chemical bonds formed in both processes are nearly similar (Sposito, 1984). The primary factors that influence formation of precipitates include the pH of the soil and porewater, type and concentration of heavy metals, availability of inorganic and organic ligands, and precipitation pH of the heavy metal pollutants. Figure 4.6 shows the solubility-precipitation diagram for a metal hydroxide complex. The left-shaded area marked as *soluble* identifies the zone where the metals are in soluble form with positively charged complexes formed with inorganic ligands. The right-shaded *soluble* area contains the metals in soluble form with negatively charged compounds. The *precipitation region* shown between the two shaded areas denotes the region where the various metal hydroxide species exist. The boundaries are not distinct separation lines. Transition between the two regions or zones occurs in the vicinity of the boundaries, and will overlap the boundaries throughout the entire pH range.

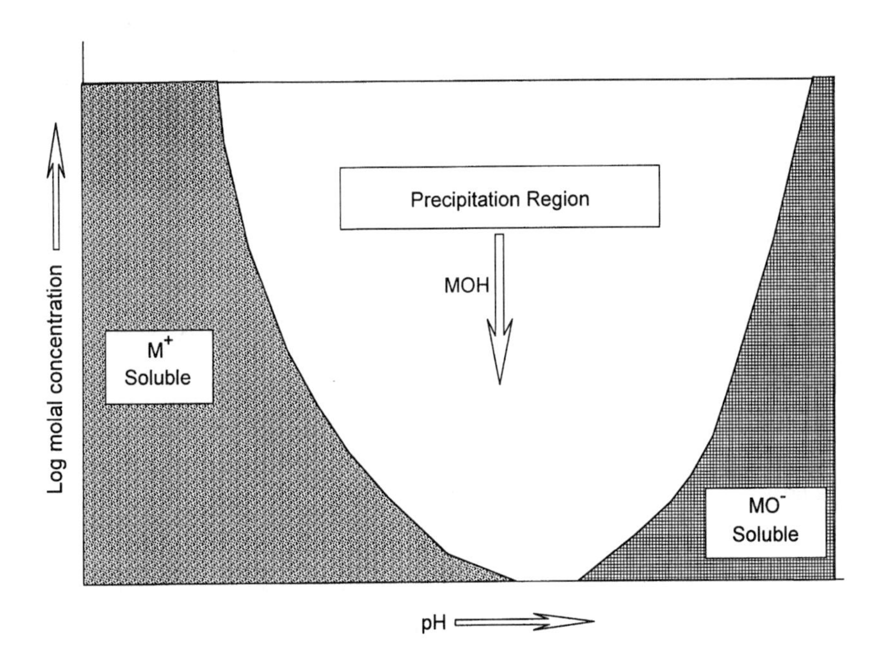

Figure 4.6 Solubility-precipitation diagram for a metal hydroxide complex.

The solubility-precipitation diagram (Figure 4.6) gives us the opportunity to better appreciate the state or fate of metal pollutants in soils, in relation to both the varying nature of the pH environment and the sorption characteristics resulting therefrom. If, for example, a heavy metal contaminant (Pb) was introduced into a soil solution as a $PbCl_2$ salt, the left-shaded area containing soluble metal ions will show that a significant portion of the metal ions would be sorbed by the soil particles, and that the ions remaining in solution would either be hydrated or would form complexes, giving one Pb^{2+}, $PbOH^+$, and $PbCl^+$. In the right-shaded area, one would obtain PbO_2H^-, and PbO_2^{2-}. The total amount of Pb sorbed by the soil particles (in the left-shaded area) would vary with the level of pH, and with the maximum amount sorbed as the pH comes close to the precipitation pH of the metal.

Precipitation of heavy metals in the porewater can be examined by studying the precipitation behaviour of these metals in aqueous solutions. The heavy metal precipitation information presented in Figure 4.7 using data obtained from MacDonald (1994) shows that the transition from soluble forms to precipitate forms occurs over a range of pH values for three heavy metals. The results show that onset of precipitation can be as early as pH of about 3.2 in the case of the single heavy metal species (Pb). Precipitation occurs as a continuous process from an onset at some early pH to about a pH of 7 for most of the metals. The presence of other heavy metals is seen in the results of the mixtures. A good example of this is shown by the onset of precipitation of Zn which appears to be at about pH 6.4 for the single

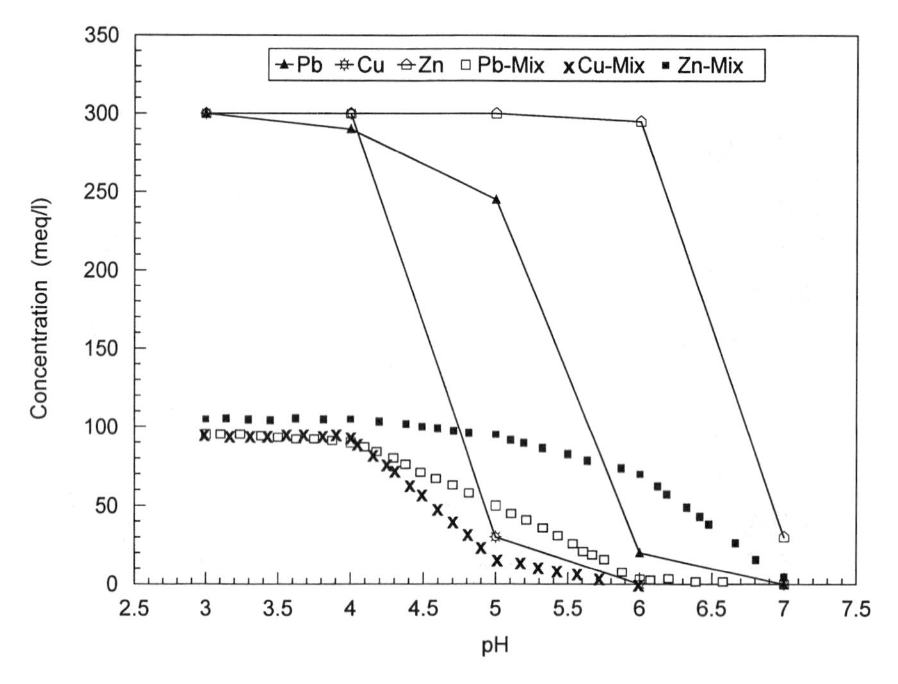

Figure 4.7 pH effect on precipitation of three heavy metals, Pb, Cu, and Zn. Bottom points show precipitation of the individual metals from metal nitrate solution with equal proportions of each metal (100 meq/each). Top curves are for single solutions of each heavy metal with 300 meq/l concentration each.

component species. When other heavy metals are present, as identified by the Zn-mixture curve, the onset of precipitation for Zn is reduced from pH 6.4 to about pH 4.4, a significant drop in the precipitation pH value. The precipitation characteristics for Pb and Cu do not appear to be significantly affected by the presence of other heavy metal pollutants.

4.6 NATURAL SOIL ORGANICS AND ORGANIC CHEMICALS

The close similarity in the chemical structures between natural soil organics (NOCs) and synthetic organic compounds provides opportunities for soil microorganisms, of which should be present in the soil substrate, in a contaminated site, to metabolize (biodegrade) the synthetic organic compounds (SOCs). Table 4.1 shows some of the NOCs and SOCs reported by Hopper (1989).

4.7 SOIL SURFACE SORPTION PROPERTIES — CEC, SSA

We have discussed the surface properties of soils in Chapter 3. For this section, we wish to examine two particular soil surface features that are important in the

Table 4.1 Closely Similar Types of Natural and Synthetic Organic Materials

Natural Soil Organics	Synthetic Organic Compounds
Aromatic NOCs	***Aromatic SOCs***
Phenylalanine	Benzenes, toluenes
Vanillin	Xylenes
Lignin	Chlorophenols
Tannins	PAHs, phenols, napthalenes, phthalates
NOCs (Sugar)	***SOCs (Sugar)***
Glucose	Cyclohexane
Cellulose	Cyclohexanol
Sucrose	Chlorocyclohexanes
Pectin	Heptachlor
Starch	Toxaphene
Aliphatic NOCs	***Aliphatic SOCs***
Fatty acids	Alkanes
Ethanol	Alkenes
Acetate	Chloroalkenes
Glycine	Chloroalkanes
Cyanides	Cyanides, nitriles, paraffins

characterization of the sorption of pollutants by soil fractions (see Figure 4.8) These are (a) cation exchange capacity (CEC) of the soil, and (b) the specific surface area (SSA) of the soil. These are surface characteristics associated with the types of soil fractions as seen, for example, in Table 2.1 for the different types of clay minerals.

Characterization of the soil surface sorption properties is useful because it provides some insight into the sorption capability of the soil. Except for non-reactive soil surfaces, we can conclude that the measured values for the specific surface area of soils are operationally defined. This means that the measured values (data) obtained are dependent on the method used to measure the surface area of the soil sample. It is important to realize that we do not directly measure the surface area of any sample of soil. Instead, we deduce or calculate the surface area of the sample being tested from indirect laboratory measurements. Along the same lines, we can say that to a lesser extent, the values obtained from laboratory "measurements" for CEC are also operationally defined.

4.7.1 Soil Surface Area Measurements

Laboratory techniques for determination of the surface area of a soil can be made directly either by visual measurements using electron microscopy, or by indirect procedures that have one common feature, i.e., measurement of the amount of material adsorbed onto the surfaces of the soil fractions in the soil. The direct visual technique requires observations on samples to obtain an appreciation of the nature

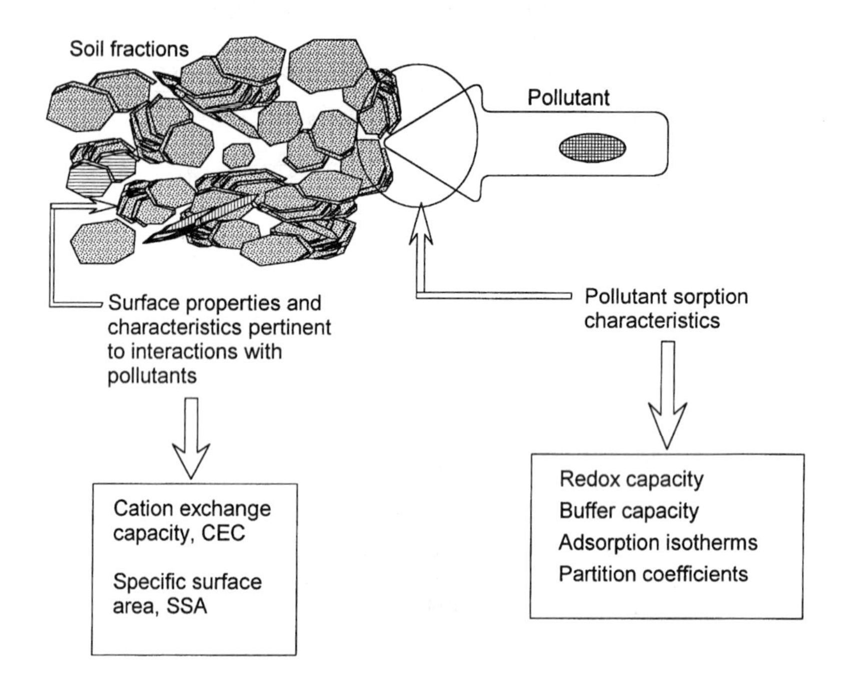

Figure 4.8 Some simple properties and characteristics of soils pertinent to sorption and partitioning of pollutants.

of the soil particles, and implementation of calculation procedures such as those given in Sposito (1984) to determine the surface areas.

What distinguishes one indirect technique from another is the adsorbate (material adsorbed onto the surfaces) used. The types of materials (adsorbates) fall into two groups: (a) gaseous or vapour phase, and (b) liquid phase. In addition to the preceding, another distinguishing feature is the method of preparation of the soil sample for surface area measurement. In particular, the use of a gaseous phase as an adsorbate requires that all the surfaces of all the particles must be totally dry.

Regardless of whether one uses a gaseous or liquid phase, the adsorptive techniques require that:

1. All the surfaces of all of the soil particles are available for adsorption of the adsorbate. This means that all the soil particles in any soil sample must be made available. Techniques for dispersion of soil particles include the ultrasonic dispersion of particles and dispersing agents in soil solutions. Coating of soil particles by amorphous materials and cementation bonds between particles are some of the problems that need particular attention.
2. Physico-chemical reactions between the adsorbate and soil particles occur so that a uniform adsorption coating of all the external surfaces of all the particles will result. The penetration of the adsorbate into interlayers of layer-lattice clay minerals is a decided problem, particularly in respect to the smectite-type minerals. Sposito (1984) compares both positive and negative adsorption methods for determination

of SSA and has shown that considerable differences in determined SSA are due not only to the two types of adsorption techniques, but also to the nature of the adsorbate used in both cases. Much of the problem lies in the reactions established at the surfaces and interlayer penetrations.

3. There exists a means to determine and quantify the monolayer adsorption coating. The problems encountered here are similar to the ones in respect to the previous point #2.

The term *operationally defined* SSA is used in recognition of the fact that the measured (or calculated) values for SSA are dependent on test technique. Citations of methods for determination of SSA should always be made in conjunction with the reporting of the SSA values. The ethylene glycol mono-ethyl ether (EGME) method described by Eltantawy and Arnold (1973) and Carter et al. (1986) is commonly used at present for routine determinations. The use of other tested and proven adsorbates, however, are equally useful and valid, provided that reporting of the technique and adsorbate are provided with the test results.

4.7.2 Cation Exchange Capacity, CEC

The CEC is related to the specific surface area (SSA) of a soil and the surface charge density σ_s by the following relationship:

$$CEC = 10^5 \, \sigma_s \, (SSA)$$

If σ_s and SSA are expressed in units of m²/kg and keq/m², CEC will be obtained as meq/100 g soil.

Similar sets of concern attend the measurement and reporting of the cation exchange capacities of soils. Since the intent of the cation exchange capacity measurement is the quantitative determination of the easily exchangeable cations in a soil, the same three main points of concern expressed for SSA measurements apply — with application to adsorption phenomena. These are:

1. All the surfaces of all of the soil particles and all the sorption sites are available for sorption of cations. This assumes that a strong cation will be used to saturate the soil. Techniques used for dispersion of soil particles should not interfere with effective cation saturation.
2. Chemical reactions — Cation sorption should occur on all available sites. The problem of saturation (and subsequent exchange) in the interlayers of layer-lattice clay minerals is particularly difficult in the case of smectites. In addition, reactions between the saturating cation solution and soil fractions will lead to serious errors in measurements. The use of ammonium acetate (NH_4OAC) as a saturation fluid, for example, for soils with significant amounts of carbonates, can cause dissolution of $CaCO_3$ and gypsum, resulting in an excess extraction of Ca^{2+} by NH^{4+}. Since this technique is designed to provide the opportunity for measurement of the Na^+, K^+, Ca^{2+}, and Mg^{2+} in the supernatant, this excess Ca^{2+} will provide for a higher measured CEC value.
3. There needs to be a means to determine that the saturation and replacement of the cations have been effectively accomplished.

In addition to the above, the composition or the nature and distribution of the soil fractions can be very significant factors. Amorphous materials and natural soil organics, because of their compositional features, present difficulties in obtaining CEC values. In the case of the oxides and allophanes, for example, composition and surface features are highly variable. Measured values for CEC can range from 15 to 24 meq/100 g soil for Fe oxides, from 10 to 18 meq/100 g soil for Al oxides, and from 20 to 30 meq/100 g for allophanes. The corresponding measured SSAs for these materials range from 300 to 380 m^2/g soil, 200 to 300 m^2/g soil and 450 to 550 m^2/g soil, respectively.

The surface functional groups of natural soil organics (SOM — soil organic material), in addition to composition and surface features, exert controlling influence on the CEC measured. The nature and distribution of the oxygen-containing functional groups of SOM are dependent on the composition of the material. Since there is a very high degree of variability in nature and composition of the material, it is clear that reported values for CEC and SSA must be considered to be estimates or representative values. The carboxyl and phenolic functional groups appear to be the major contributors to the CEC of these soils. As with the SSA reporting requirement, it is necessary to report the technique used for determination of CEC values.

Soil fractions that have charge characterization dependent on the pH of the system require attention to the pH when tests for determination of CEC are conducted. Because of the pH dependency of the net surface charges developed (see Chapter 3, Section 3.3), values of CEC will vary depending on the pH of the system. The soils that are most likely candidates for such attention include kaolinites amongst the clay minerals, natural soil organics, and the various oxides or amorphous materials. In kaolinites, for example, the values of CEC can vary by a factor of 3 between the CEC at a pH of 4 (CEC = 2) to a pH of 9 (CEC = 6). The variations would be expected to be higher for the oxides. This is because the proportion of pH-dependent charges are much higher for the oxides, in comparison to the proportion of pH-dependent edge surface charges to planar surface charges in the kaolinites. It is useful to conduct tests for CEC measurements over a range of pH values for the system. In addition, it would also be useful to conduct corresponding tests for AEC (anion exchange capacity) measurement.

4.8 POLLUTANT SORPTION CAPACITY CHARACTERIZATION

The chemical buffering capacity of soils was briefly mentioned in Chapter 1 as a consideration under the sets of concerns dealing with chemical stresses to the land environment. The soil buffer capacity can be determined experimentally by titration of the soil with a strong acid or base (see Figure 4.9). The buffer capacity of soil is defined as the number of moles of H^+ or OH^- that must be added to raise or lower the pH of 1 kg of soil by 1 pH unit. The buffer capacity of soil is the reciprocal of the slope of the titration curve of soil. The titration results of Phadungchewit (1990) are shown in Figure 4.9 to provide the basis for creating the buffer capacity curves, which would characterize the buffering capacity of soils. The results indicate that the illite soil has the highest capability to accept increasing acid inputs with the least

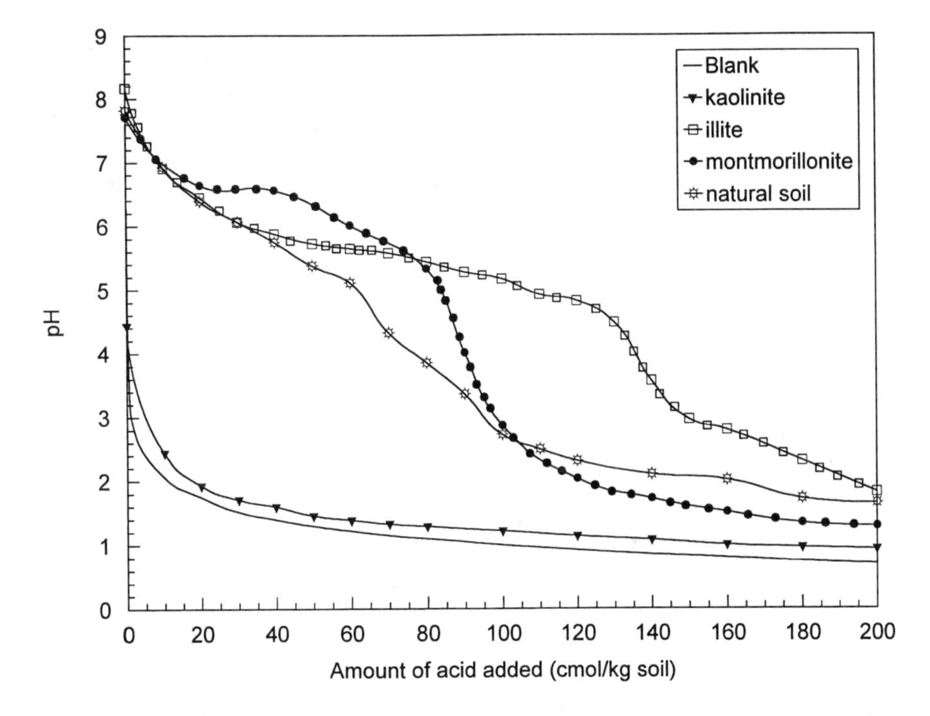

Figure 4.9 pH titration curves for four soils and a blank. (Data from Phadungchewit, 1990.)

change in pH, i.e., it shows the least change in pH when more acid is introduced into the soil solution. In contrast, we can see that the kaolinite soil is not capable of accommodating much acid input. Its titration performance mimics the blank control test results.

From the titration curves of Figure 4.9, the soil buffer capacity can be determined from the negative inverse slope of the curves, and plotted in relation to pH. If we express the buffering capability β of a soil in terms of changes in the amount of hydroxyl ions (OH^-) or hydrogen ions (H^+) added to the system, and in respect to the pH changes resulting therefrom, we can obtain some measure of quantification of the ability of a soil to perform as a chemical buffer. The relationship is expressed as follows:

$$\beta = \frac{dOH^-}{dpH} = \frac{dH^+}{dpH} \tag{4.10}$$

The β (buffering capacity) curves shown in Figure 4.10 provide a better picture of the capability of the soils to perform as chemical buffers, a fact that can be deduced from the titration results shown previously in Figure 4.9. By expressing the results in the form of β in relation to pH, a clearer picture is obtained in respect to the buffering potential differences between the various soils tested. The pH range where chemical buffering works well for each soil is very evident. When the pH of the soil-water system is greater than 4, the buffer capacity of illite is higher than

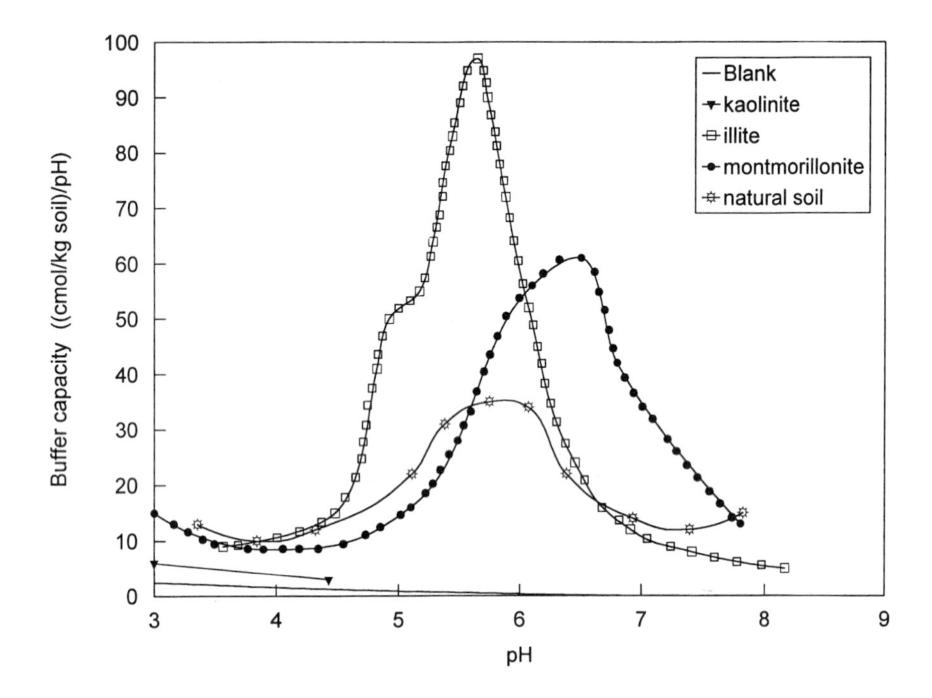

Figure 4.10 Buffering curves for the soils shown in Figure 4.9.

montmorillonite and the natural clay soil, indicating that it has a higher resistance to pH changes than the other soils. Although the illite soil has a smaller CEC than the montmorillonite soil, the high resistance in pH change in illite is not only due to adsorption of H^+ onto the exchange sites, but includes the neutralization of H^+ by the carbonates in the soils.

4.8.1 Adsorption Isotherms

Characterization of the adsorption capacity of soils (in respect to candidate pollutants) is generally performed using batch equilibrium testing procedures. These tests on replicate soil solutions of constant soil-aqueous proportions are conducted with progressively higher concentrations of the candidate pollutant in the aqueous phase, as shown in Figure 4.11. The procedure characterizes adsorption of the candidate pollutant by a completely dispersed soil particle system, i.e., all the soil particle surfaces are exposed and available for sorption of the pollutant. Since most natural soils contain various kinds of soil fractions, and these have different individual specific surface areas, it is clear that the test results obtained are the result of the average effect of all the sorption surfaces. The results obtained can be expressed in a graphical form as shown in Figure 4.12. Changes in the character of the adsorption isotherm (for a particular pollutant or sorbate) can be obtained by changing the proportions of various soil fractions — as might be desired for "designer soils" for engineered barrier systems.

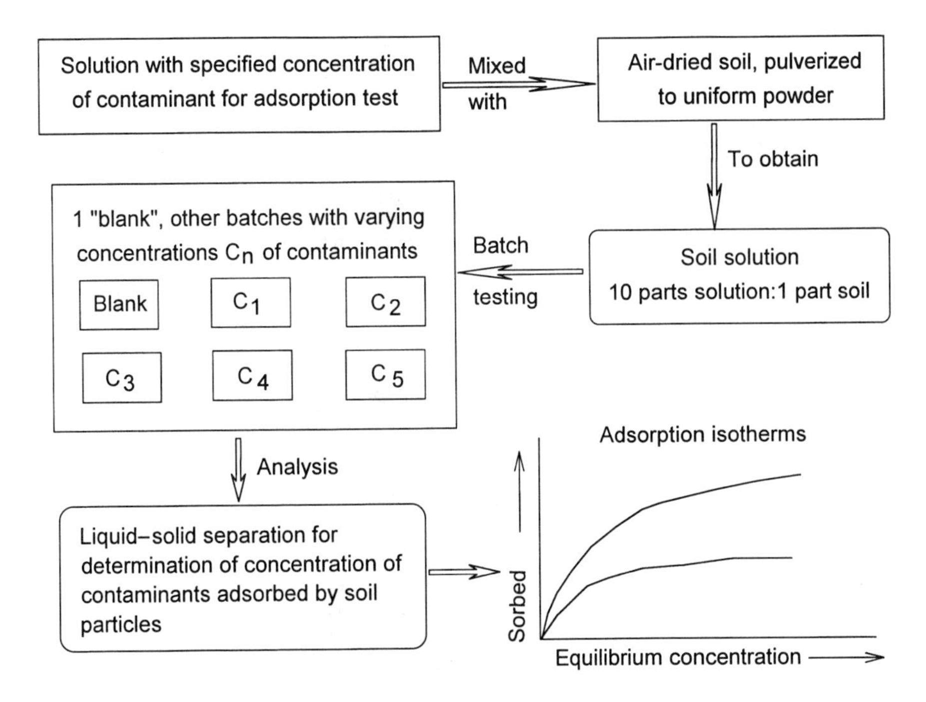

Figure 4.11 Batch equilibrium test procedure for determination of adsorption isotherms.

The ordinate (Figure 4.12) describes the concentration of the heavy metal pollutant (Pb) removed from the aqueous phase of the soil solution (identified as Pb retained by the soil solids), whilst the abscissa expresses the concentration of the pollutant in the aqueous phase of the soil solution (called the solution). We can produce two separate relationships: (a) a direct relationship between the Pb retained and the initial concentration of Pb in the soil solution, and (b) a relationship between the Pb retained and the equilibrium concentration of Pb in the soil solution. This *equilibrium concentration* is what is left in equilibrium in the aqueous phase of the soil solution. This equilibrium concentration expression (i.e., method b) is the preferred method of expression in the derivation of adsorption isotherms. Using this preferred method, the curve drawn through the filled circles in Figure 4.12 represents the adsorption isotherm for the illite soil in respect to Pb. The adsorption isotherms (in respect to Pb) for two other soils are shown in Figure 4.13.

Several pertinent points need to be noted and emphasized if adsorption isotherms are to be used for assessment of soil sorption capability.

1. The units used for adsorbed concentration (ordinate) and the equilibrium concentration (abscissa) must be consistent with each other. This is important in the subsequent use of the isotherm for determination of the partition coefficient k_d, as will be seen later.
2. We can control the pH of each batch of soil solution by adding buffering agents. However, when we choose to do so, we must be aware of the fact that the adsorption

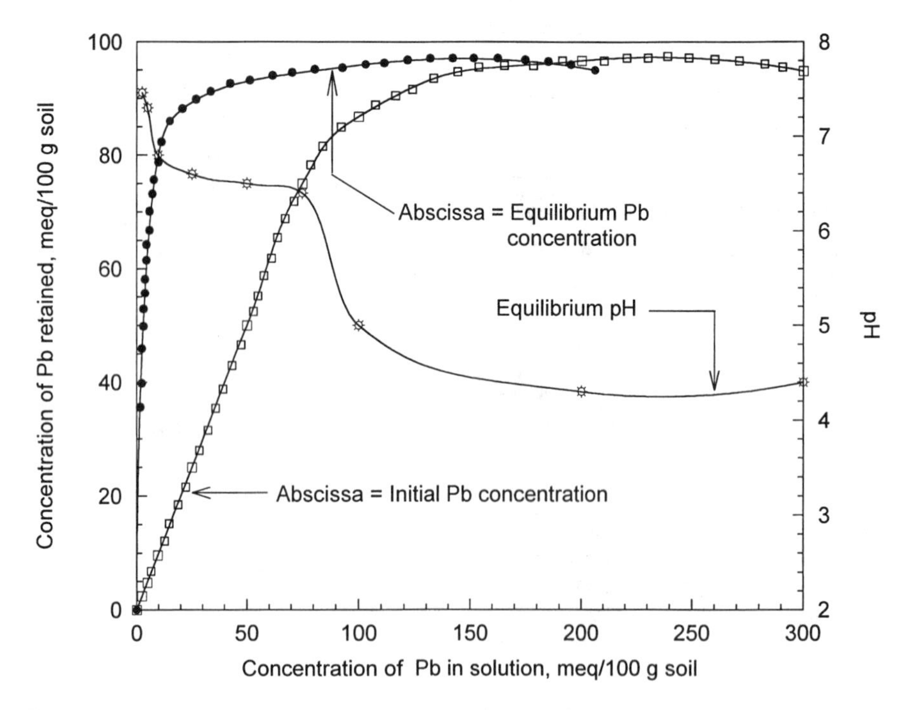

Figure 4.12 Adsorption isotherms for Pb by illite soil — in relation to Pb concentration in the aqueous phase. Note that differences in choice of expression of Pb concentration in the aqueous phase will yield different "slopes" of the adsorption isotherm. (Data from MacDonald, 1994.)

characteristics can also change. The amount and the nature of change will depend on both the type of soil and the pollutant.

3. The adsorption isotherms will likely be different when multiple pollutants are used. In addition, the types of conjugate ions used for the pollutant(s) will also influence the character of the adsorption isotherm.

Figure 4.14 shows some of the typical shapes of adsorption isotherms. These have been classified according to their shapes and have been characterized as, for example, *constant-type* (linear adsorption curve), and *Freundlich-type* isotherms, as seen in the figure. Both the *linear* and *Freundlich-type* isotherms will predict continuous (no limit) adsorption and should therefore be used with knowledge of the extrapolation limits, usually based on experimental information. Relationships describing these isotherms have been developed as follows:

$$\begin{aligned} &\textit{Linear} && c^* = k_1 c + k_2 \\ &\textit{Freundlich} && c^* = k_1 c^{k_2} \qquad (4.11) \\ &\textit{Langmuir} && c^* = \frac{k_1 c}{1 + k_2 c} \end{aligned}$$

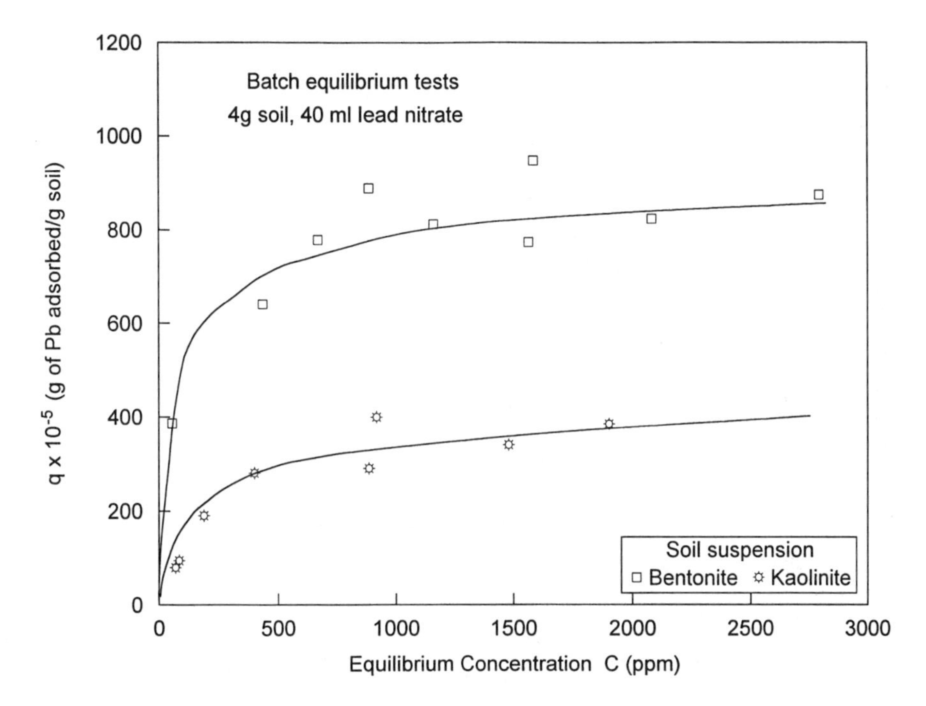

Figure 4.13 Adsorption isotherms for bentonite and kaolinite.

where c^* = concentration of adsorbed contaminants or pollutants, and k_1, k_2 are rate coefficients determined from laboratory tests.

4.8.2 Distribution Coefficient k_d

If we take the *Freundlich* relationship shown in Equation 4.11 and express it in logarithmic format, we obtain:

$$\log c^* = \log k_1 = k_2 \log c \tag{4.12}$$

A plot of the above equation with log c^* as the ordinate and log c as the abscissa will show an intercept on the ordinate of log k_1 and a straight line with a slope of k_2. It is this slope k_2 that has been defined as the *distribution coefficient* k_d (see Figure 4.15). This coefficient finds usefulness in the evaluation and/or prediction of transport of contaminants or pollutants, as will be seen later.

The distribution coefficient k_d is used to describe contaminant or pollutant partitioning between liquid and solids. We can develop various k_d values for different kinds of soils in relation to different pollutants and under various conditions. The primary requirement is the availability of batch equilibrium test data. The debate concerning the use of batch equilibrium testing and leaching column testing to obtain partitioning and distribution coefficients should be noted at this point. The essence

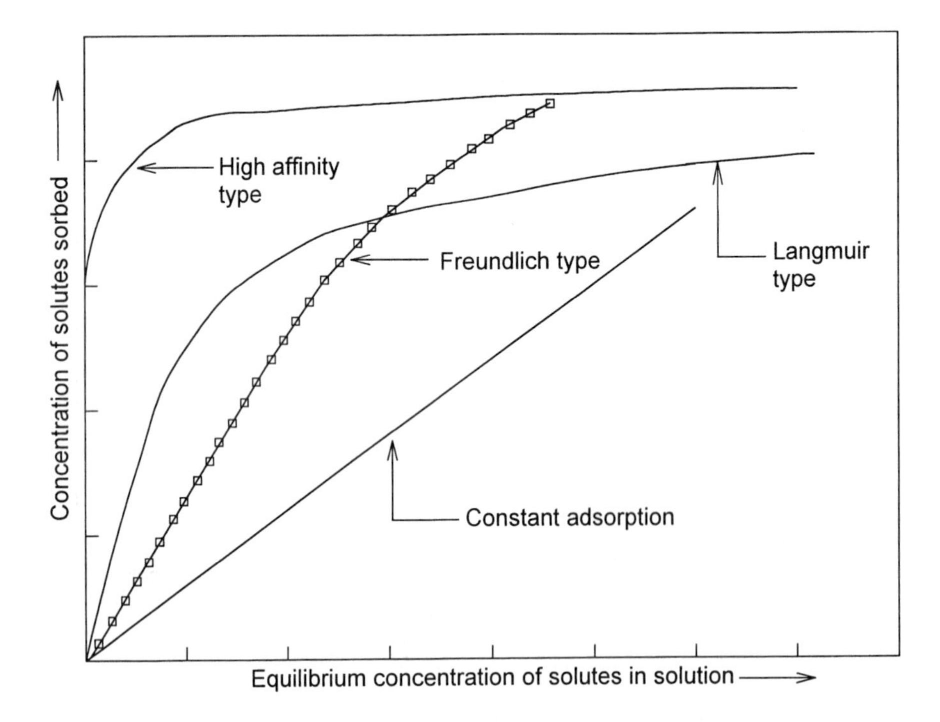

Figure 4.14 Typical adsorption isotherms.

of the debate focuses around the nature of the test sample used to determine partitioning of the pollutant under investigation — i.e., the difference between soil solutions used for batch equilibrium tests and intact soil samples used for leaching column tests. Section 5.3 considers this set of issues in greater detail.

However, regardless of the nature of the soil samples used to determine pollutant partitioning, the partitioning tests assume that chemical equilibrium conditions are attained within a 24-hour period. Thus, the use of k_d should only be applied to those circumstances where reactions that cause the partitioning are fast and reversible. In that sense, this distribution coefficient has been found to be more useful in the evaluation of partitioning of inorganic pollutants. However, certain studies have reported linear adsorption characteristics for low concentrations of hydrophobic organic chemicals such as PAHs and some substituted aromatic compounds.

4.8.3 Partitioning and Organic Carbon Content

We have shown in Section 4.6 that there is considerable similarity between the composition of natural organic matter in soils and synthetic organic chemicals. It is therefore not surprising that the primary mechanism of organic adsorption is the hydrophobic bond established between the synthetic chemical and natural organic matter. The amount of sorbed chemical can be estimated as follows (Karickhoff, 1984):

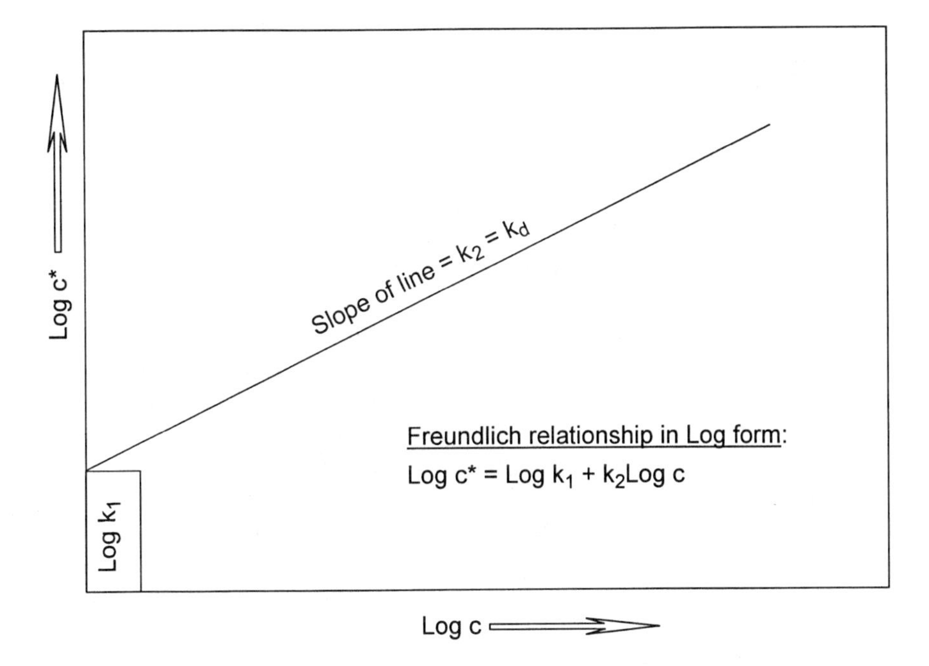

Figure 4.15 Determination of k_d.

$$C_S = k_p C_A = k_{oc} f_{oc} \tag{4.13}$$

where C_S and C_A represent the sorbed concentration of chemical and equilibrium aqueous concentration of the chemical, respectively, $k_p = k_{oc} f_{oc}$ = partition coefficient, f_{oc} is the fractional organic carbon content of the soil and k_{oc} is the proportionality constant of the chemical. When $f_{oc} > 0.001$, good predictions are obtained.

Correlation equations relating k_{oc} to more commonly available chemical properties such as solubility and octanol-water partition coefficient k_{ow} can be obtained (Chiou et al., 1982). The relationships are regression equations obtained from various data and are usually expressed in a log-log form as follows:

$$\log k_{oc} = a + b \log S = m + n \log k_{ow} \tag{4.14}$$

where *a*, *b*, *m*, and *n* are constants. The octanol-water partition coefficient k_{ow} is defined as the ratio of the equilibrium concentration of "C" of the dissolved candidate substance in the two immiscible solvents (*n*-octanol and water), i.e., $k_{ow} = C_{octanol}/C_{water}$. Chemicals with k_{ow} values of about less than 10 are considered to be relatively hydrophilic. They tend to have high water solubilities and low soil adsorption coefficients. At the other end of the scale, chemicals with k_{ow} values in excess of 10^4 are considered to be very hydrophobic.

Studies of the correlation of k_p with soil and pollutant factors have revealed good correlation between sorption and organic carbon content over a wide range of organic carbons content, from ~0.1% to nearly 20% of the soil. Typical values for k_{oc} and k_{ow} can be found for a considerable variety of organic chemicals in the various handbooks on environmental data, e.g., Verschueren (1983), Montgomery and Welkom (1991), and Mackay et al. (1992).

4.9 INTERACTIONS AND POLLUTANT TRANSPORT PREDICTIONS

Transport Prediction Models (TPM) are essential tools in the decision-making process for regulatory agencies and practitioners in: (a) assessment and evaluation of attenuation competence of soil-engineered barriers or soil substrate; (b) prediction of continued progress of pollutant plumes; and (c) risk assessment. These models are designed to provide analyses and predictive performance characteristics of pollutant plumes in transport through the soil medium. The problems addressed include those dealing with evaluation of the capability of clay soils to function as natural attenuation barrier systems over the period of leachate generation and transport in the substrate, generally anywhere from 1 to 50 years and more. An important factor in successful development of TPMs is their ability to represent the many interacting relationships which govern transport of pollutants in the soil-water system. Figure 4.16 illustrates the uses of TPMs and the requirements for proper development and application of the TPMs.

It is not the mandate or purpose of this book to deal with the development of TPMs. The reader should consult the more specialized textbooks on transport modelling for detailed treatment of the analytical and numerical techniques, e.g., Aral (1989), Kinzelbach (1986), and Crank (1975). Instead, the focus on TPMs is in the direction of the capability of the TPMs to fully accommodate or account for the phenomena resulting from the various interactions between pollutants and soil fractions. Since the purpose of analytical/computer models is to represent the physical situation by mathematical relationships, it is not always clear that: (a) the various processes contributing to the physical situation at hand, such as those represented in Figure 4.1 have been properly recognized; and (b) the various driving forces responsible for the resultant fluxes have been well considered.

Prediction of transport of pollutants represented, for example, by pollutant plumes most generally rely on analytical models of saturated transport, i.e., transport in fully saturated soil media. Without considering storage and radioactive decay, the relationships are most often cast in terms of the advection-diffusion relationship as, for example:

$$\frac{\partial c}{\partial t} = D_L \frac{\partial^2 c}{\partial x^2} - v \frac{\partial c}{\partial x} - \frac{\rho}{n \rho_w} \frac{\partial c^*}{\partial t} \tag{4.15}$$

where c = concentration of contaminant of concern, t = time, D_L = diffusion-dispersion coefficient (most often called *diffusion* coefficient), v = advective velocity, x = spatial coordinate, ρ = bulk density of soil media, ρ_w = density of water, n = porosity of soil media, and c^* = concentration of contaminants adsorbed by soil

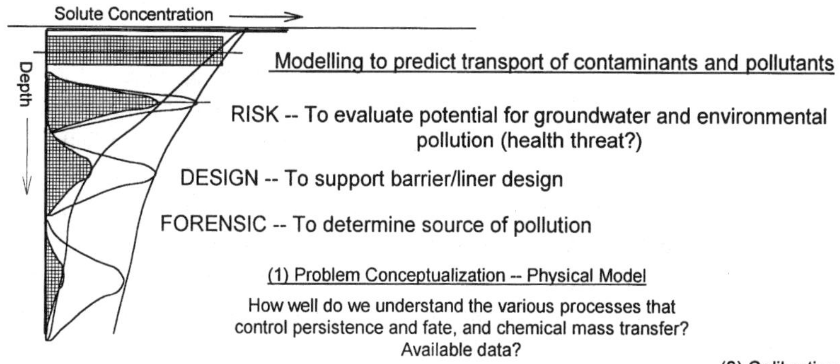

Figure 4.16 Modelling objectives and requirements.

fractions. Note that for the purposes of discussions concerning the TPMs, we will use the more general term *contaminants*, to include both non-pollutants and pollutants. The c^* term used in Equation 4.15 is of particular interest since this is the parameter that reports on the partitioning of contaminants between soil fractions and pore fluid. The examples of some common types of adsorption isotherms shown previously as Figures 4.13 and 4.14 say very little about the process of transfer of contaminants from the fluid phase onto the soil solids. These isotherms do not provide information which allows one to directly distinguish between attenuation and retardation mechanisms responsible for partitioning. Furthermore, we need to be reminded that if batch equilibrium adsorption isotherms are used as the source for determination of k_d, one would be dealing with soil suspensions and that transfer of information to represent comparable behaviour in compact samples may not be a one-to-one relationship.

To determine c^* in Equation 4.15, we can use the partition coefficient k_d defined previously. The common procedure used in most instances is to employ the linear (constant) adsorption isotherm shown in Figure 4.14 and the relationship $c^* = k_d c$. When this is substituted into Equation 4.15, the readily recognized popular relationship shown as Equation 4.16 is obtained as follows:

$$R\frac{\partial c}{\partial t} = D_L\frac{\partial^2 c}{\partial x^2} - v\frac{\partial c}{\partial x} \tag{4.16}$$

where $R = \left[1 + \frac{\rho}{n\rho_w}k_d\right]$.

When a non-linear adsorption isotherm is used in place of the linear isotherm, k_d will not be a constant, and under such circumstances, it is necessary to determine the functional for c^*. The specification of a proper k_d must recognize that this is a function of concentration of the species under consideration, ionic strength, presence of other pollutants, temperature, etc. The predictions made from the relationships will inform one of the concentration of the specific target contaminants used in the model in relation to time and spatial distance. The predicted distribution and attenuation of contaminants can only be considered to be as accurate as the source and quality of the inputs provided for determination of c^*. This will be discussed in the next two chapters when the mechanisms of partitioning and the distribution of partitioned contaminants are examined.

4.9.1 Transport and Partitioning in the Vadose Zone

Transport of pollutants in the vadose zone is not uncommon if contaminated sites are located above the aquifer, and in the case of landfills this is almost inevitable if proper design and placement are to be achieved. Prudent landfill practice is to place landfills in regions where the aquifer is some distance below the bottom of the landfill. The partitioning of contaminants (i.e., the fate of contaminants) during transport of leachate through the unsaturated soil-engineered barrier and particularly through the soil substrate that characterizes the vadose zone is not well understood or studied. The relationships given previously as Equations 4.15 and 4.16 refer to transport of contaminants in the saturated zone, and cannot be readily applied to vadose zone transport. Because moisture transport in the unsaturated soil substrate is most often by diffusive means, and because the moisture acts as the carrying agent for contaminants, vadose zone transport models need to account for water content distribution in the soil.

Moisture movement responds to the soil-water potential ψ (see Section 3.6 in Chapter 3) and in the absence of hydraulic heads, generally moves by diffusive means. Considering gravitational flow to be insignificant, the governing relationship for one-dimensional transport of moisture is given as:

$$\frac{\partial \theta}{\partial t} = \frac{\partial}{\partial x}\left(k(\theta)\frac{\partial \psi_\theta}{\partial x}\right) \tag{4.17}$$

where θ = volumetric water content, ψ_θ = soil-water potential, and k = Darcy permeability coefficient dependent upon the volumetric water content. Since we expect the contaminants to be carried with the water phase, a common procedure in casting the contaminant transport relationship for unsaturated flow is to associate the concentration of contaminant of concern, c, with the volumetric water content θ. Accordingly, one obtains the following:

$$\frac{\partial \theta c}{\partial t} = \frac{\partial}{\partial x}\left(k(\theta, c)\frac{\partial \psi_c}{\partial x}\right) \tag{4.18}$$

where the chemical potential $\psi_c = \psi(\theta, c)$, and is written with specific reference to the target contaminants, and where the Darcy permeability coefficient k is written in respect to dependency on both the volumetric water content θ and the concentration c of the target contaminants. Of the many simplifying assumptions that can be made, the simplest would be: (a) to assume k to be a function only of θ, and (b) to consider $\psi_c = \psi_\theta$ = single-valued function of θ. We obtain thereby

$$\frac{\partial \theta c}{\partial t} = \frac{\partial}{\partial x}\left(D_c(\theta)\frac{\partial c}{\partial x}\right) - \frac{\partial(\rho^* c^*)}{\partial t} \tag{4.19}$$

where D_c = diffusion coefficient of the target contaminants, and ρ^* = bulk density of soil divided by the density of water. To solve the above equation, it is necessary to determine D_c.

4.9.2 Diffusion Coefficient D_c and D_o

The diffusion coefficient of a target contaminant D_c is most often considered as being equivalent or equal to the effective molecular diffusion coefficient. The following treatment of this issue derives from Yong et al. (1992a). In dilute solutions of a single ionic species, the diffusion coefficient of that single species is termed as the infinite solution diffusion coefficient D_o. The studies of molecular diffusion given by both Nernst (1888) and Einstein (1905) show the complex interdependencies that govern this coefficient. The studies on the subject initially dealt with the movement of suspended particles controlled by the osmotic forces in the solution. The three expressions most often cited are:

Nernst-Einstein
$$D_o = \frac{uRT}{N} = uk'T \tag{4.20}$$

Einstein-Stokes
$$D_o = \frac{RT}{6\pi N\eta r} = 7.166 \times 10^{-21}\frac{T}{\eta r} \tag{4.21}$$

Nernst
$$D_o = \frac{RT\lambda^o}{F^2|z|} = 8.928 \times 10^{-10}\frac{T\lambda^o}{|z|} \tag{4.22}$$

where u = absolute mobility of the solute, R = universal gas constant, T = absolute temperature, N = Avogadro's number, k' = Boltzmann's constant, λ^o = conductivity of the target ion or solute, r = radius of the hydrated ion or solute, η = absolute viscosity of the fluid, z = valence of the ion, and F = Faraday's constant.

Figure 4.17 shows the variation in results obtained using the Ogata and Banks (1961) solution for an initial chloride concentration of 3049 ppm as the input source. All the infinite solution diffusion models (i.e., D_o models) depend on such factors

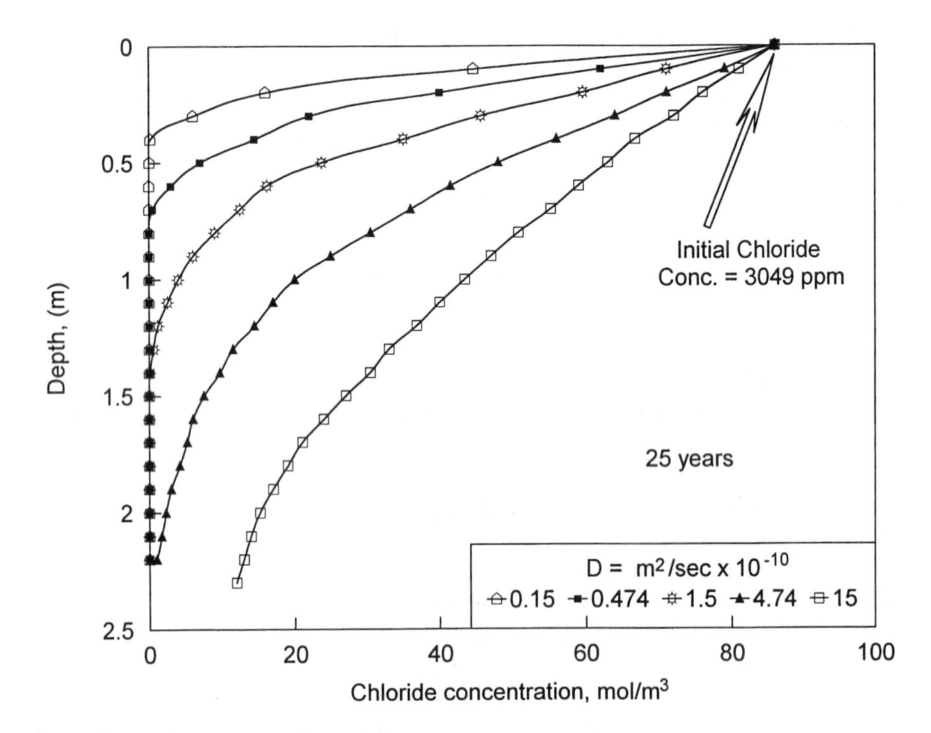

Figure 4.17 Variation of *D* coefficient and its effect on prediction of chloride concentration profiles after 25 years of continuous input of chloride at 3049 ppm.

as ionic radius, absolute mobility of the ion, temperature, viscosity of the fluid medium, valence of the ion, equivalent limiting conductivity of the ion, etc. Li and Gregory (1974) and Lerman (1979), amongst others, have compiled values for D_o for various sets of conditions. Robinson and Stokes (1959) provide an exhaustive summary of experimental values for λ^o for major ions at infinite dilution (water) at various temperatures.

Calculations made using the Nernst-Einstein relationship given as Equation 4.20 show some significant variations in the magnitude of D_o in relation to temperature. A sample of some of these are shown in Figure 4.18, using information from Robinson and Stokes (1959). Considering the large temperature range in field situations, it is important to take into account the temperature factor in determination of the diffusive transport of solutes. A good discussion of the effects of varied contaminant solutes on diffusion coefficients can be found in Robinson and Stokes (1959), Jost (1960), Li and Gregory (1974), and Lerman (1979). Crank (1975) provides an excellent and thorough treatment of the mathematics of the many kinds of diffusion problems, and also the various solutions for cases and conditions required to address variable and transient terms.

The choice of a transport coefficient *D* to be used for Equations 4.16 and 4.19 and other relationships describing the transport of contaminants depends to some extent on whether one concludes that the process of transport is driven by advective

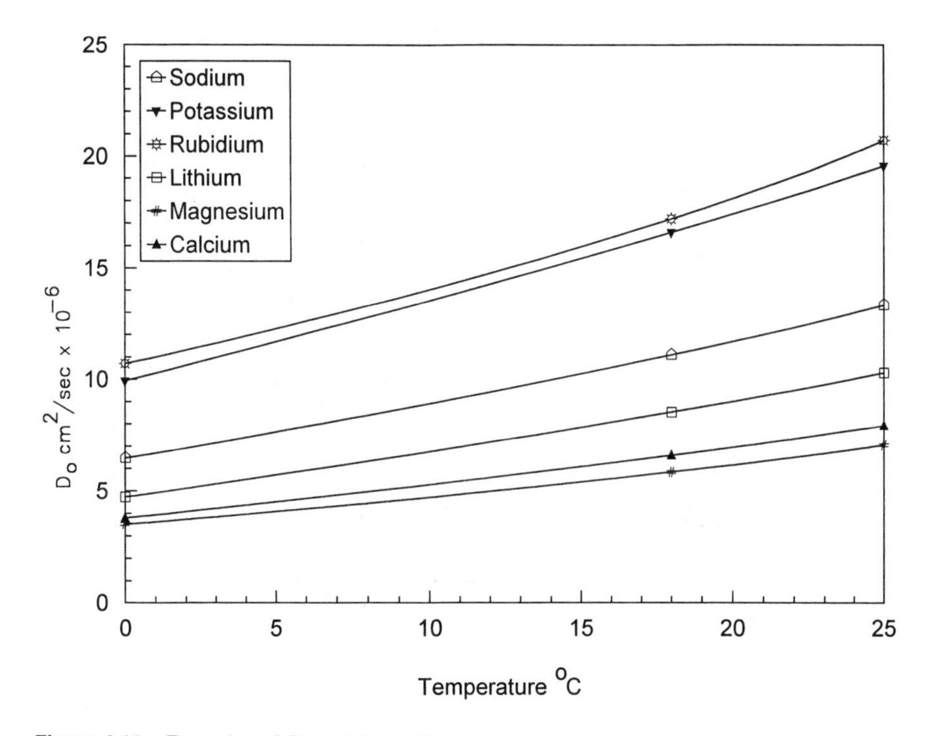

Figure 4.18 Examples of D_o variation with temperature.

forces or more by diffusion-dominated mechanisms. Because the prediction of the manner, rate, amount, and spatial distribution of contaminants in any time interval is conditioned by the choice of transport coefficient D used, the significance of a proper choice of D cannot be overstated. Defining the dimensionless Peclet number as $P_e = v_L \, d / D_o$, where D_o represents the diffusion coefficient in an infinite solution and v_L is the longitudinal flow velocity (advective flow), the information reported by Perkins and Johnston (1963) show that for $P_e < 1$, diffusion (movement) of the contaminant solutes in the contaminant plume travels faster than the advective flow of water. For $P_e > 10$, advective flow constitutes the dominant flow mechanism for the movement of solutes. Between the values of 1 and 10, there is a gradual change from diffusion-dominant to advection-dominant transport as a means of movement of solutes (see Figure 4.19). Since clay-engineered liner/barrier systems are designed to have hydraulic conductivities of 10^{-8} m/s and less, the dominant process for transport of solutes through the barrier system is diffusion driven.

Writing the longitudinal diffusion coefficient D_L as

$$D_L = D_m + \alpha \, v_L$$

where D_m = molecular diffusion = $D_o \, \tau$
and α = dispersivity parameter, and τ = tortuosity factor,

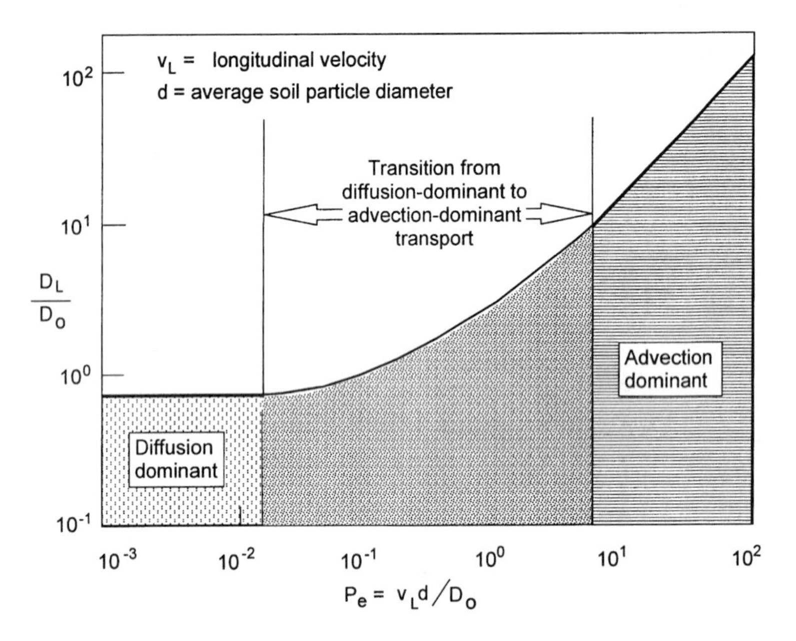

Figure 4.19 Diffusion-dominant and advection-dominant transport ranges in relation to Peclet number. (Adapted from Perkins and Johnston, 1963.)

we will see from Figure 4.19 that in the diffusion-dominant transport region, we can safely neglect the αv_L term since v_L is vanishingly small. Under those circumstances, the diffusion-dominant transport region, we will have $D_L = D_o \tau$. In the advection-dominant transport region, if we consider diffusion transport to be negligible, $D_L = \alpha v_L$. In the transition region, the relationship for D_L will be given as $D_L = D_o \tau + \alpha v_L$.

4.9.3 Soil Structure and Diffusion Coefficients

The preceding section has assumed that the influence of soil structure on the movement of pollutant solutes can be represented by a homogeneous and uniform medium. The fabric and soil structure shown in schematic form in the preceding chapters indicate to us that this is not so. While it may be necessary for purposes of modelling to represent the soil as a homogeneous uniform medium for determination of representative transport properties, such an assumption may not serve the user well in predictions of transport. Pollutant transport through the macropores and micropores will be controlled by mechanisms and processes at the molecular level. As will be discussed in the next two chapters, the interactions between the pollutants and soil fractions will change the properties and characteristics of the soils, particularly if they are pH and/or pE sensitive. The role of soil structure in the transport of pollutants cannot be easily dismissed or ignored.

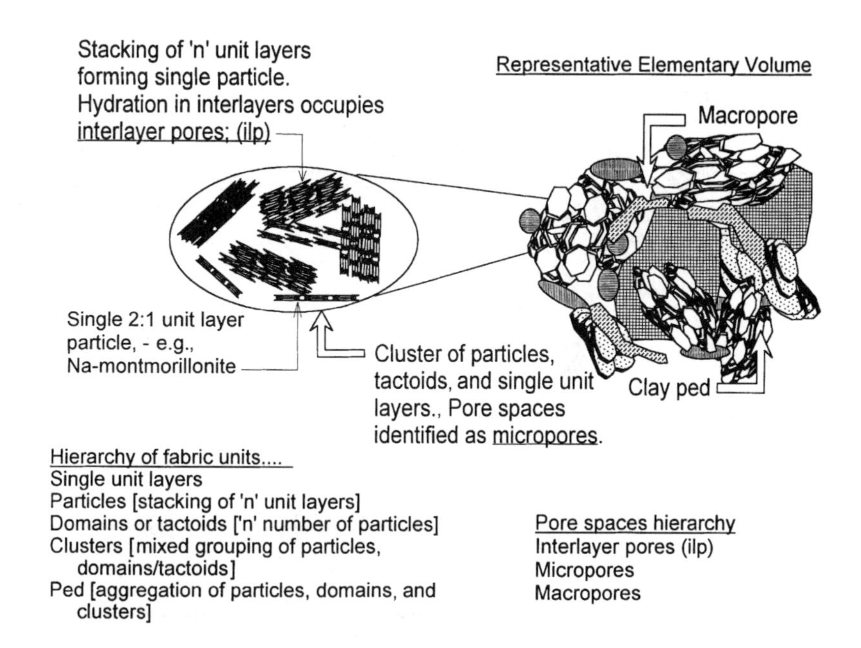

Figure 4.20 Soil structure and hierarchy of fabric units and pore spaces.

In soils with smectites or montmorillonites as a contributing soil fraction, we need to also pay attention to the pore spaces represented as interlayer separations (see Figure 2.10). The hierarchy of pore spaces shown in Figure 4.20 begins with the *ilp* representing the interlayer pores (interlayer separations). This is the lowest order of pore spaces. The next highest level is the micropores, which exist as pore spaces between individual soil particles. These are the pore spaces in the ped fabric units. The pore spaces or void volumes between peds and larger discrete particles are macropores. Not shown in the diagram are pore spaces between larger discrete fine-to-medium silt-sized particles. The combination of all of these contribute to the total void volume in a soil sample. To represent these highly contrasting void volumes as homogeneous and uniform can lead to serious errors in interpretation of the physics of pollutant transport.

Figure 4.21 shows the basic elements of diffusion transport of a pollutant solute in a clay soil. For simplicity in representation, the soil is assumed to be a non-swelling soil, i.e., no smectitic soil materials are in the soil. Interlayer pores are therefore not a consideration in Figure 4.21. The further assumption made in the sketch is that advective flow is not a major consideration, i.e., the Peclet number is less than one ($P_e < 1$). Pollutant transport in the macropores between peds and discrete particles will be by diffusion so long as flow in the macropore obeys $P_e < 1$. Pollutant movement in the micropores takes two forms: (a) movement in the diffuse ion-layer regime, and (b) movement in the Stern layer region. The coefficients of diffusion that best represent the different diffusive rates in these regions are D_1 for diffusion in the macropores, D_2 for diffusion in the diffuse-ion regime, and D_3 for diffusion in the Stern layer region. For the more compact ped units, the thickness

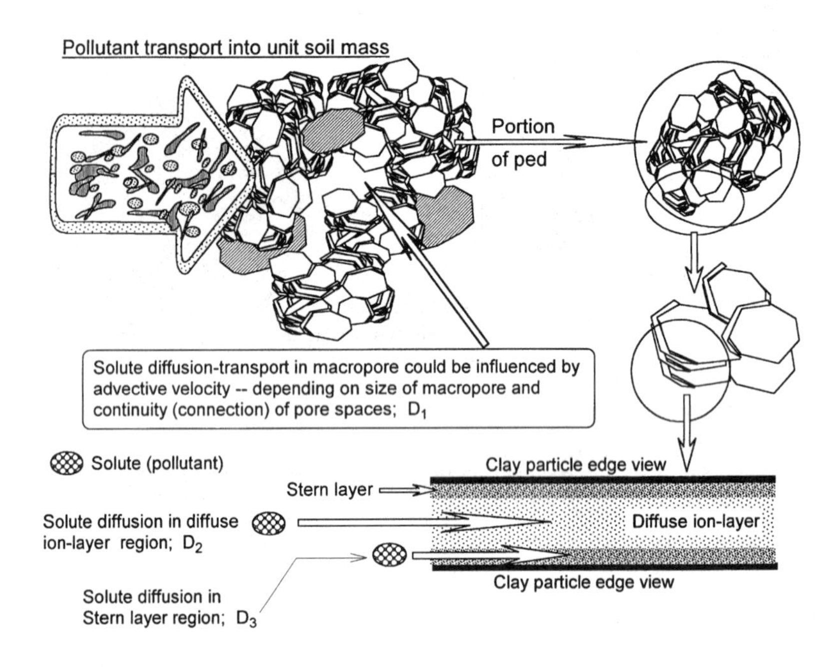

Figure 4.21 Diffusion coefficients for pollutant diffusion in macropore and micropore.

of the diffuse ion-layer may become vanishingly small. When this occurs, the values for D_2 will be very close to D_1. For that reason, and because of obvious difficulties in distinguishing between diffusive movements in the Stern layer region and the diffuse ion-layer regime, a simplification in diffusive transport in the ped unit is assumed. This relegates D_2 and D_1 into a single coefficient D_{21}, which is identified as the diffusion coefficient for transport in the ped unit.

4.9.4 Vadose Zone Transport

The irreversible thermodynamics approach taken by Elzahabi and Yong (1997) for treatment of vadose zone transport extends the original development given previously in Yong et al. (1992a). By coupling the moisture content with the contaminant solutes, and denoting the subscripts θ and *c* as moisture and concentration of contaminants, the fluxes due to the respective thermodynamic forces can be represented as follows:

$$
\begin{aligned}
J_\theta &= L_{\theta\theta}\frac{\partial\psi_\theta}{\partial x} + L_{\theta c}\frac{\partial\psi_c}{\partial x} \\
J_c &= L_{c\theta}\frac{\partial\psi_\theta}{\partial x} + L_{cc}\frac{\partial\psi_c}{\partial x}
\end{aligned}
\tag{4.23}
$$

where J_θ and J_c are the fluid and solute fluxes, respectively, $\partial\psi_\theta/\partial x$ = thermodynamic force due to the soil-water potential (i.e., soil-water potential gradient), $\partial\psi_c/\partial x$ = thermodynamic force due to the chemical potential (i.e., chemical potential gradient), and $L_{\theta\theta}$, $L_{\theta c}$, $L_{c\theta}$, L_{cc} are the phenomenological coefficients. The various diffusivity coefficients such as $D_{\theta\theta}$ (moisture), D_{cc} (solute), $D_{c\theta}$ (solute-moisture), and $D_{\theta c}$ (moisture-solute) have been obtained by Elzahabi and Yong (1997) as:

$$\begin{aligned} D_{\theta\theta} &= L_{\theta\theta}\frac{\partial\psi_\theta}{\partial x} \qquad D_{cc} = L_{cc}\frac{RT}{c} \\ D_{c\theta} &= L_{c\theta}\frac{\partial\psi_\theta}{\partial\theta} \qquad D_{\theta c} = L_{\theta c}\frac{RT}{c} \end{aligned} \tag{4.24}$$

and the final set of coupled relationships given in the following form:

$$\begin{aligned} \frac{\partial\theta}{\partial t} &= \frac{\partial}{\partial x}\left[D_{\theta\theta}\frac{\partial\theta}{\partial x} + D_{\theta c}\frac{\partial c}{\partial x}\right] \\ \frac{\partial c}{\partial t} &= \frac{\partial}{\partial x}\left[D_{c\theta}\frac{\partial\theta}{\partial x} + D_{cc}\frac{\partial c}{\partial x}\right] - \frac{\rho}{\rho_w\theta}\frac{\partial S_c}{\partial t} \end{aligned} \tag{4.25}$$

where S_c is the sorbed concentration of contaminants. Solution of the coupled flow relationships follows along lines similar to those developed previously by Yong and Xu (1988), i.e., using an identification technique for evaluation of the phenomenological coefficients. The choice of functional forms for the phenomenological coefficients has been based on experimental knowledge of the distribution of contaminants along the length of unsaturated leaching column samples, together with moisture contents associated with the distributed contaminants.

4.10 CONCLUDING REMARKS

The basic points covered in this chapter focused on processes and mechanisms of sorption of pollutants in the soil, and on the general requirements of problem conceptualization in the development and use of transport prediction models. The general term of *sorption* has been used to include the various processes that serve to remove the dissolved solutes from the porewater. Whether the removal process is through actual adsorption of the pollutants to the soil solid surfaces, or by formation of a new phase (precipitation), the significant event is the depletion of concentration of pollutants in the porewater.

The actual pollutant-soil interaction processes involved in the determination and characterization of fate of pollutants will be addressed in the next two chapters. Our concern in this chapter has been the various reactions in the porewater which affect the sorption processes. We have sought to provide a simple means for characterizing the sorption capability of soils, but need to be wary about its use, since the tests conducted for characterization (adsorption isotherms) are chemical equilibrium tests on soil solutions. Extrapolation to actual compact in situ soil cannot be confidently accomplished without supporting leaching column information.

The process of sorption of heavy metals will generally result in a drop in the pH in the immediate soil particles' environment. The amount of decrease of pH is seen to be a function of the concentration of the heavy metals coming in, and obviously the species or type of heavy metals. This is explained in terms of the release of hydrogen ions resulting from the metal-proton exchange reactions on the particle surfaces' sites and on the hydrolysis and precipitation of the metals in the porewater.

The surface functional groups such as carbonyl compounds, aldehydes, ketones, and carboxylic acids have dipole moments. The electron in their double bonds are unsymmetrically shared. Whilst they can accept protons, stability of complexes between carbonyl groups and protons is considered to be weak. Interactions occur either directly with interlayer cations or through formation of hydrogen bonds with water molecules coordinated to exchangeable cations of the soil solids.

Chemical mass transfer responsible for partitioning of contaminants constitutes a significant part of the processes involved in the transport and fate of contaminants. In the longer term consideration, the redox environment (pE) and subsequent reduction-oxidation reactions will ultimately determine the final fate of the contaminants. In the final analysis, assessment of whether retention or retardation processes are responsible for the observed partitioning and hence the attenuation of contaminants within the soil matrix constitutes the vital and critical requirement in the evaluation of the *natural attenuation capability* or *managed natural attenuation* of the soil barrier system. The dilemma facing both regulatory agencies and practitioners is obvious: If potential pollution hazards and threats to public health and the environment are to be minimized or avoided, we must ensure that the processes for contaminant attenuation in the substrate are the result of (irreversible sorption) retention mechanisms. The other alternative is to provide for circumstances that would assure dilution of contaminant and pollution concentrations to levels far below allowable levels or limits — if we are to meet safety standards designed to protect the environment and human health.

Whereas some of the defining mechanisms for sorption and desorption are generally known for many types of contaminants in interaction with specific soil fractions, and will be discussed in the next two chapters, the combined processes leading to partitioning of these contaminants in complex mixtures of contaminants and soil fractions have yet to be fully defined and understood. The preceding notwithstanding, sufficient information exists concerning chemical mass transfer and biologically mediated mass transfer that permits one to comprehend the vital differences between irreversible sorption responsible for partitioning and hence attenuation of contaminants, and temporary sorption processes and physical hindrances that lead to retardation of contaminants as the demonstrated contaminant attenuation phenomenon.

The constant k_d partition coefficient used in the retardation coefficient R for Equation 4.16 means that infinite adsorption by the soil fractions cannot be realistically accepted. So long as the concentration of contaminants keeps increasing in the contaminant leachate, sorption of the contaminants by the soil fractions will also keep increasing in the proportion given by k_d, as can be seen by the straight line in Figure 4.14. Since this cannot happen in real situations, limits must be placed on the use of this relationship, i.e., a maximum sorption capacity must be defined. Alternatively, the non-linear adsorption isotherms shown in Figure 4.14, which are more representative of field situations, should be used.

CHAPTER 5

Partitioning and Fate of Heavy Metals

5.1 INTRODUCTION

To provide a focal point to the discussions concerning inorganic pollutants, we will concentrate on the heavy metals (HMs) originating from anthropogenic activities, which find their way into the ground. Section 4.2 has stated that this group of metals comprises 39 elements. Not all of these are found in significant quantities in the soil. Those considered to be most commonly associated with anthropogenic activities include arsenic, cadmium, chromium, copper, iron, lead, mercury, nickel, silver, tin, and zinc. Most, if not all of the heavy metals are generally considered to be pollutants. They have been found to be toxic in their elemental forms or as compounds. Whilst some of these may be released into the ground from processes associated with "natural changes" in the pH and redox environments, e.g., arsenic, most of these (HMs) are found in the wastes generated from such activities and processes as steel production, electroplating and other metal processing activities, etc. A good example of this is lead (Pb). Activities that contribute to ground and groundwater pollution by lead are mining and smelting and battery production. Pb has the capability to form various complexes (chloride, sulfate, and sulfide) and hydroxide species, as has been briefly shown in Section 4.5 in connection with Figure 4.6. Depending upon the types of soil fractions in a soil mass, Pb can be sorbed onto soil fractions (a) via cation exchange reactions — typical of the reactions between metallic ions and charged surfaces, and (b) via replacement of a bound proton — as in the case of Pb bonding with hydrous oxides. Other mechanisms of Pb bonding with soil fractions also exist, e.g., through formation of inner- and outer-sphere complexes.

Interactions between HM pollutants and soil fractions leading to removal of the HMs from the porewater are of considerable interest and concern. The concern is with respect to the subsequent release of these metals from the soil solids (particles). This desorption process, which can be triggered by many events and circumstances, will permit the metals to be mobile, i.e., transported in the substrate. While the availability and mobility of desorbed pollutants falls under the category of environmental mobility, the desorption distribution coefficients will not be similar to the adsorption distribution coefficients, i.e., $k_d^{desorp} \neq k_d^{adsorp}$. This will be discussed in greater detail in the subsequent portions of this chapter.

5.2 ENVIRONMENTAL CONTROLS ON HEAVY METAL (HM) MOBILITY AND AVAILABILITY

The availability of the heavy metals is of concern in respect to uptake by plants, and ingestion by humans and other biotic receptors. The term *bioavailability* is used by professionals in many different disciplines to mean the availability of a pollutant in a form that would be toxic to the receptors under consideration. The more specific definition considers the pollutant to be available for biological actions. There are at least four possible factors that can affect the environmental mobility and bioavailability of heavy metals in soils: (a) changes in acidity of the system; (b) changes in the system ionic strength; (c) changes in the oxidation-reduction potential of the system; and (d) formation of complexes. By and large, the principal mechanisms and processes involved in heavy metal retention include precipitation as a solid phase (oxide, hydroxides, carbonates), and complexation reactions (Harter, 1979; Farrah and Pickering, 1977a, 1997b, 1978, 1979; Maguire et al., 1981; Yong et al., 1990b). The literature reports on ion-exchange adsorption as a means of "retention" should, strictly speaking, be considered as "retardation" in the present context of regulatory expectations and requirements. This is because desorption of contaminants sorbed by ion-exchange mechanisms can readily occur.

The interaction of a kaolinite soil and HM pollutants is used to illustrate some of the above points of discussion. Chapter 3 has shown that two kinds of surface charge reactions occur with kaolinites: (a) reactions in relation to the net negative charge developed from heterovalent cation substitution in the clay lattice structure, and (b) reactions at the surfaces of the edges of mineral particles — pH-dependent reactions due to hydration of broken bonds. The two types of functional groups populating the surfaces of the edges of the kaolinite particles are the hydroxyl (*OH*) groups. One type is singly coordinated to the Si in the tetrahedral lattices, whereas the other is singly coordinated to the Al in the octahedral lattices that characterize the kaolinite structure (Figures 2.9 and 3.3). Both types of edges function as Lewis acid sites, i.e., these sites can accept at least one pair of electrons from a Lewis base.

Figure 5.1 shows the pH-dependent surface charge for a kaolinite with specific surface area of 800 m^2/g at 25°C, using graphical data reported by Brady et al. (1998). Their surface complexation modelling studies indicate that the Al sites are the principal proton acceptor sites, and that these sites are more acidic for kaolinite edges than for exposed Al hydroxides. The surfaces of the kaolinite function as nucleation centres for heavy metals. Thus, in the case of sorption of heavy metal contaminants by kaolinites, if the metal concentrations in the contaminant plume are less than the cation exchange capacity (CEC) of the kaolinite, desorption occurs easily because the mechanisms controlling initial sorption are mainly non-specific. However, if the metal concentrations in the contaminant plume are greater than the CEC, desorption is more difficult because the total sorption processes will most likely include both non-specific adsorption and some specific adsorption. Release

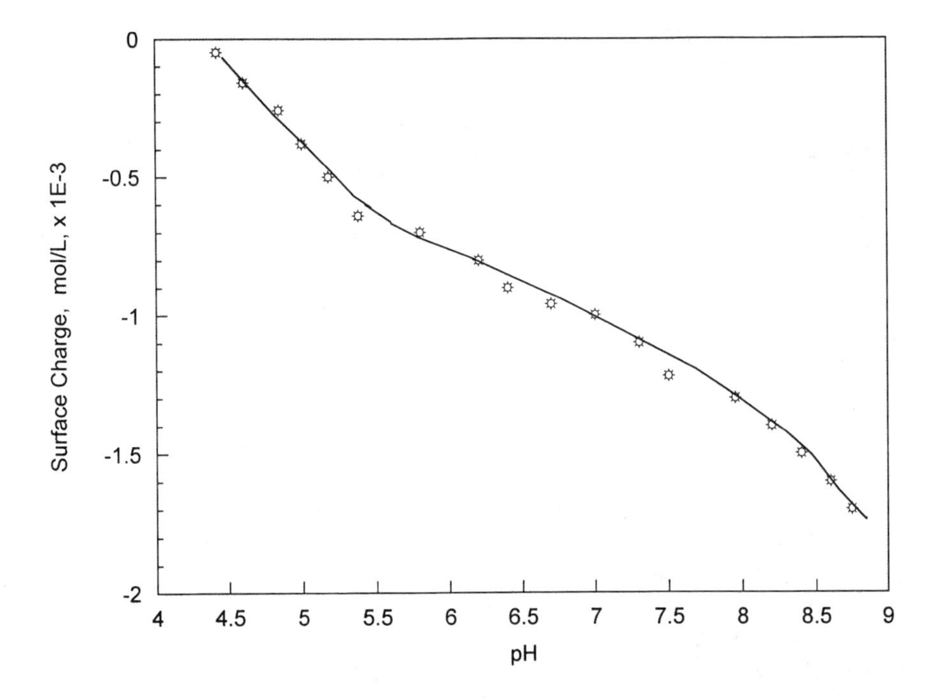

Figure 5.1 pH-dependent surface charge for kaolinite using data from Brady et al. (1998).

(desorption) of the previously sorbed metal ions can result when saturation sorption occurs and when the ions in the bulk or pore fluid are lesser in concentration than the initial sorbed ions. In addition, desorption of cations can also occur through replacement, as demonstrated in the familiar lyotropic series in Section 4.4.2:

$$Na^{+} < Li^{+} < K^{+} < Rb^{+} < Cs^{+} < Mg^{2+} < Ca^{2+} < Ba^{2+} < Cu^{2+} < Al^{+} < Fe^{3+} < Th^{4+}$$

In general, contaminant and pollutant attenuation by (sorption) retention mechanisms involve specific adsorption and other mechanisms such as chemisorption — via hydroxyl groups from broken bonds in the clay minerals, formation of metal-ion complexes, and precipitation as hydroxides or insoluble salts. Table 5.1 (using information from Bolt, 1979) shows some of the mechanisms responsible for retention of Cu, Co, Zn, Pb, and Cd in some clay minerals.

Inorganic and organic ligands in the porewater contribute significantly to the processes associated with retention and/or retardation of inorganic contaminants and pollutants such as HMs. Yong and MacDonald (1998) show that Cu and Pb retention relative to soil pH and the presence of OH, HCO_3^-, and CO_3^{2-} in the porewater are influenced by:

- Competition for metallic ions offered by the sorption sites provided by the soil fractions and the anions.
- The formation of several precipitation compounds that are dependent on the pH environment. Soluble Pb concentration is influenced by the precipitation of $PbCO_3$ (cerrusite) and $Pb(CO_3)_2(OH)_2$ (hydrocerrusite).

Table 5.1 Heavy Metal Retention by Some Clay Minerals (adapted from Bolt, 1979)

Clay Mineral	Chemisorption	Chemisorption at Edges	Complex Adsorption	Lattice Penetration *
Montmorillonite	Co, Cu, Zn		Co, Cu, Zn	Co, Zn
Kaolinite			Cu, Zn	Zn
Hectorite			Zn	Zn
Brucite			Zn	Zn
Vermiculite			Co, Zn	Zn
Illite	Zn	Zn, Cd		
Cu, Pb				
Phlogopite			Co	
Nontronite			Co	

* Lattice penetration = lattice penetration and imbedding in hexagonal cavities.

Because $PbCO_3$ precipitates at lower pH values than both calcite and dolomite, it is possible for the Pb carbonates to precipitate because of the dissolution of Mg and Ca as carbonates. In the case of soluble Cu concentration, however, its fate is controlled by the precipitation of CuO (tenorite).

Variable pH-dependent hydrolysis of metal cations such as Cu^{2+}, $CuOH^+$,$Cu(OH)_2$, Pb^{2+}, $PbOH^+$, and $Pb(OH)_2$ changes the Lewis acid strength of the aqueous species of the metals and thus affects their affinity for soil particle surfaces. This is particularly significant for borderline Lewis acids such as Pb^{2+} and Cu^{2+} since they can behave as hard or soft acids depending on the environment solution. This affects affinity relationships between metals and reactive soil surfaces, and impacts directly on sorption and desorption of the metals.

Yong and MacDonald (1998) have shown that upon apparent completion of metal sorption, the equilibrium pH of the system is reduced to values below initial pH — attributable to the many reactions in the system, including but not limited to hydrogen ions released during metal/proton exchange reactions on surface sites, hydrolysis of metals in the soil solution, and precipitation of metals. We need to distinguish between surface and solution reactions responsible for release of hydrogen ions and the corresponding change in pH. If surface complexation models are to be used, the relationship between metal adsorption and proton release needs to be established, i.e., net proton release or consumption is due to all the chemical reactions involving proton transfer.

Results from soil suspension tests indicate that sorption of Cu^{2+} by kaolinite is generally accompanied by proton release to the solution, attributable to Cu^{2+} – H_3O^+ exchange at low Cu^{2+} concentrations (McBride, 1989). At higher Cu^{2+} concentrations, enhanced hydrolysis of Cu^{2+} occurs with sorption of hydrolyzed species. Whilst the affinity of kaolinite for Cu^{2+} is normally low, this can be increased through replacement of the surface Al ions with H_3O^+ and Na^+.

5.2.1 Soil Characteristics and HM Retention

Table 5.1 shows that the mechanisms for retention of HM pollutants differ somewhat amongst the various kinds of clay minerals. Chapters 3 and 4 have provided the

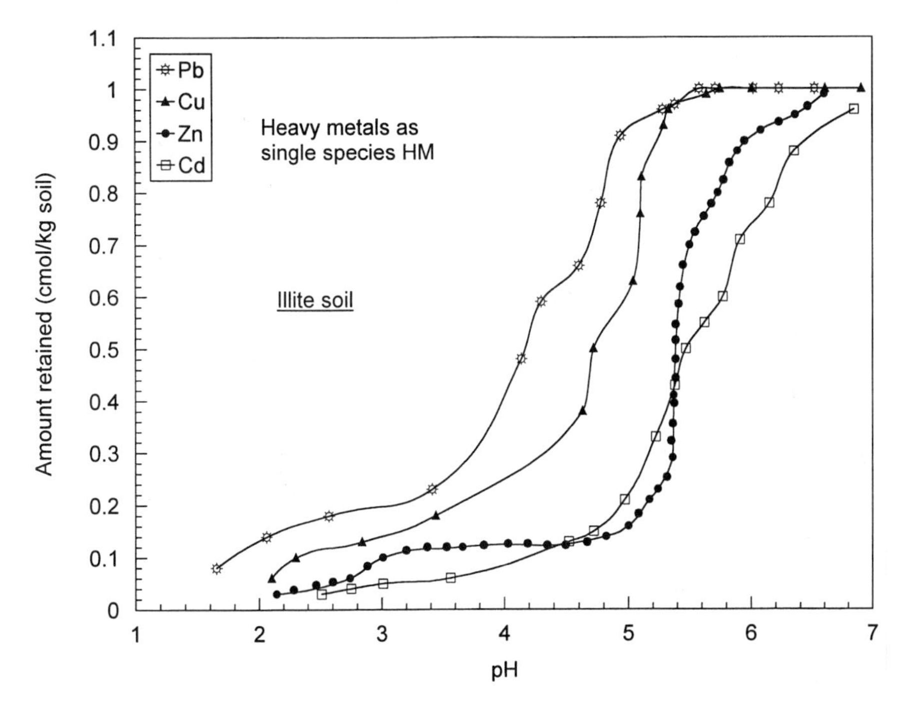

Figure 5.2 Retention of heavy metal pollutants (HMs) by illite soil. HMs introduced as single species.

details concerning the structure and surface characteristics of these kinds of clay minerals. In this section, we will use the data from Phadungchewit (1990) to illustrate the importance of competition between different kinds of heavy metals in retention by various clay minerals. Figure 5.2 shows the influence of pH on retention of Pb, Cu, Zn, and Cd by an illitic soil which contains some soil organics and carbonates. The concentration of each of the HM pollutant used for the tests conducted was maintained at 1 cmol/kg soil, either as single species pollutant or mixed species (Figure 5.3). The total HM concentration in the mixture of HM pollutants is 4 cmol/kg soil, representing the sum of the individual HM concentrations of 1 cmol/kg soil. The results shown in Figure 5.2 are for single species HM, whereas the results shown in Figure 5.3 are from tests where the soil was allowed to interact with a mixture of HM pollutants.

The results shown in both the graphs (Figures 5.2 and 5.3) indicate higher retention of Pb at all the pH values. There appears to be a retention scale (selectivity) of the order of Pb > Cu > Zn ≈ Cd for both the single species and mixed species of HM. The amount retained indicated in the graphs refers to the amount of HM removed from the aqueous phase of the soil suspensions. No attempt is made at this stage to distinguish between the various mechanisms attending sorption; neither is there any attempt at separating sorption from removal of HM solutes from the aqueous phase by precipitation mechanisms at this stage. We will address these issues later in this chapter. Several interesting observations can be made in view of the results shown in Figures 5.2 and 5.3:

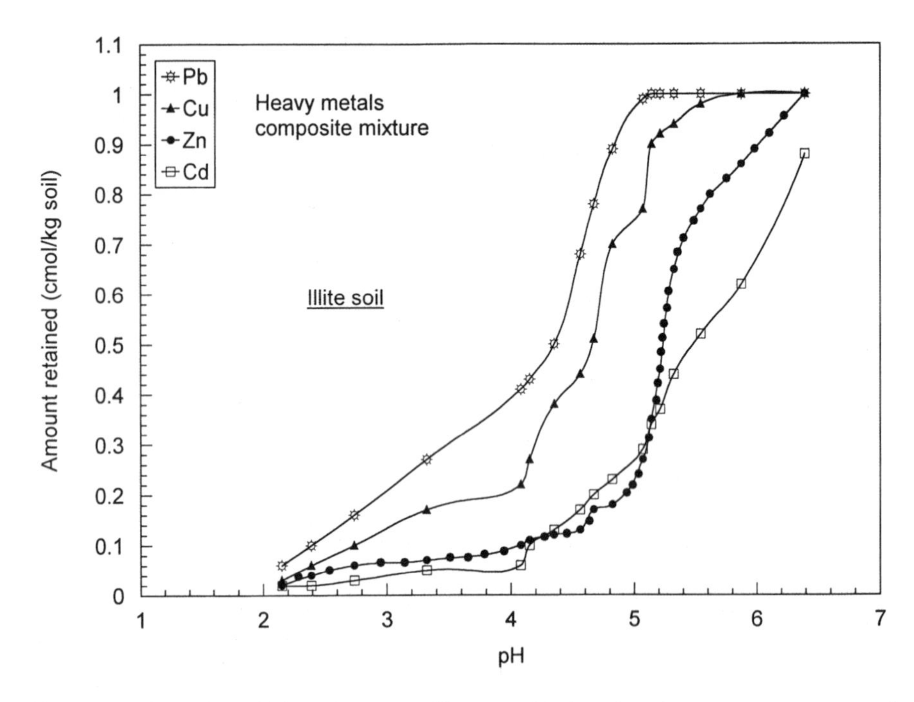

Figure 5.3 Retention of HM pollutants by illite soil. HMs introduced as composite mixture of HMs in equal proportions.

- The total retention (i.e., 100% retention) of HM at the higher pH values appears to be related to the precipitation pH of the HM. Coles et al. (2000) provide test data showing that the precipitation of Pb and Cd, forming $Pb(OH)_2$ and $Cd(OH)_2$, respectively, increases with pH, and is greater at higher metal concentrations. Furthermore, the precipitation of Pb occurs at about 2 pH units lower than that of Cd.
- The presence of other HM represented by the mixture (Figure 5.3) does not appear to change the total amount retained or the retention characteristics of the illite soil, as shown in Figure 5.4.
- Reference to Figure 4.7 shows that the precipitation pH of Pb is not significantly influenced by the presence of other HMs.
- It is not clear that the above would be maintained if the proportions of the various HMs were changed, or if the soil was different (Figures 5.5 and 5.6).
- The retention order (selectivity) for the single and mixed species suggests that we need to determine the processes which determine selectivity of HM retention.

The different soils shown in Figures 5.5 and 5.6 have reactive surfaces that are dissimilar in properties and characteristics — one from the other. These account for the differences in retention capabilities for the HM Cd. Comparison of Cd retention shown in Figures 5.5 and 5.6 in respect to competition from other HMs can be seen in Figure 5.7. The Cd-montmorillonite retention characteristics are seen to be very dependent on presence of other HMs. For comparison, the Cd-illite results from Figure 5.4 are repeated in the graph.

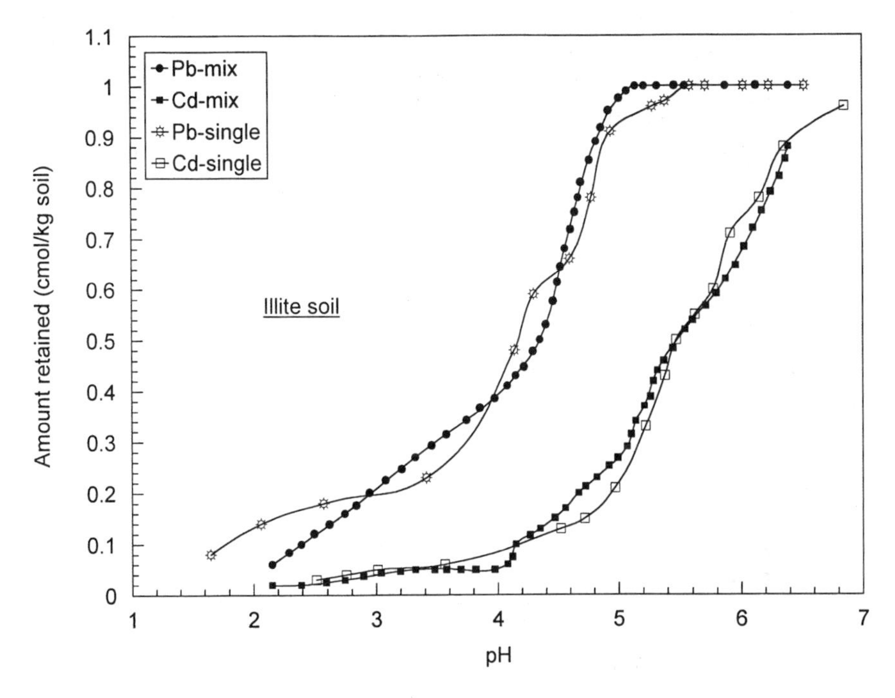

Figure 5.4 Comparison of single and mixed species of Pb and Cd retention shown in Figures 5.2 and 5.3.

The reduced Cd retention by the montmorillonite when other HMs are present in the system is because sorption of Cd is primarily via exchange mechanisms. When other HMs are present in the system, these compete for the same sorption sites. The illite soil that contains soil organics and carbonates provides for more mechanisms of HM retention. Simple generalizations on HM retention should not be made on the basis of sorption tests with limited sets of parameters and constraints. Some of the major factors that need to be considered in assessment of metal-soil interaction include: (a) mechanisms contributing to sorption of the HMs; (b) types of soil fractions involved in interaction with the HMs; (c) types and concentrations of the HMs; and (d) pH and redox environments.

5.2.2 Preferential Sorption of HMs

The results shown in Figures 5.2 through 5.7 indicate that there is a degree of selectivity in the sorption preference of heavy metals by different soils. The preferential sorption characteristics are conditioned by the types of HM pollutants and their concentrations. Additionally, the kinds of inorganic and organic ligands present in the porewater are also important factors. Preference in metal species sorption is generally called *selectivity*. This is not the same for any two soils, since this is very closely related to the nature and distribution of the reactive surfaces available in the soil. The order for selectivity remains somethat similar for the two soil types shown

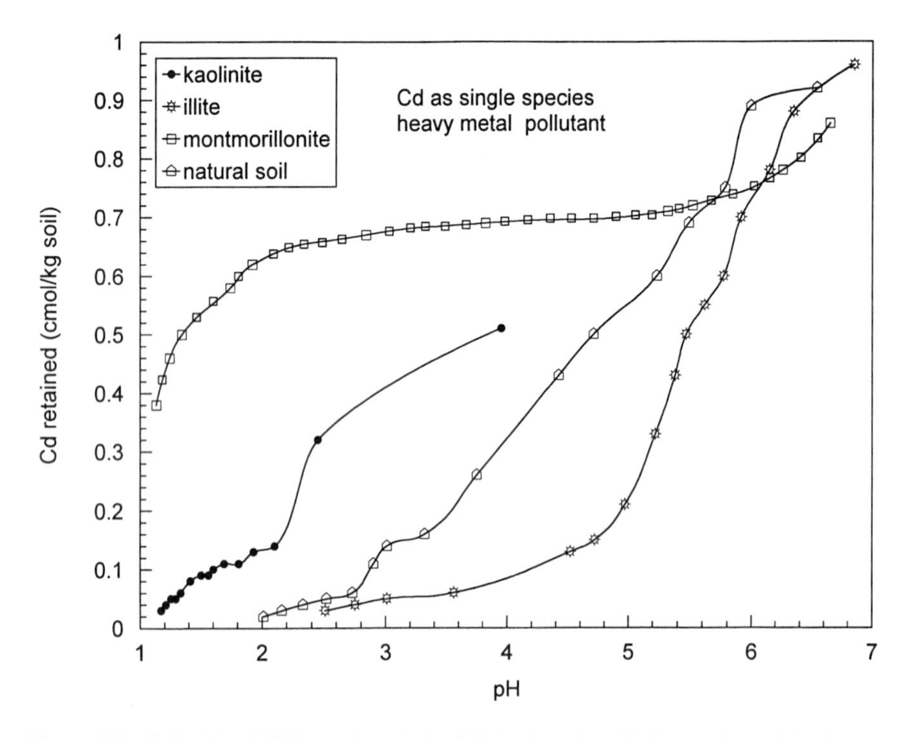

Figure 5.5 Retention of Cd by various soils. Cd introduced as single species pollutant.

in the figures, but the amounts retained and the pH influence on retention appear to be markedly affected by the presence of other metallic ions in the aqueous phase. For a constant HM pollutant presence, the greater or lesser sorption reaction kinetics will depend on the immediate pH condition established by the soil (and pollutants), and the kinds, distribution, and availability of reactive surfaces. The availability of reactive surfaces is a significant consideration in evaluation of sorption capacity and selectivity. This is discussed in the next section.

The results shown in Figures 5.2 through 5.7, which have been obtained from tests with *single species* and *composite species,* indicate that the selectivity order for the illite soil would be Pb > Cu > Zn ≈ Cd. The selectivity order for the montmorillonite soil appears to be sufficiently well defined for relatively higher pH values. For pH values below at about 4, the selectivity order appears to be Pb > Cu > Zn > Cd. As the pH values increase the selectivity order changes slightly, as seen in Figure 5.8. Results obtained from reactions at pH values below 3 are not quantitatively reliable because of dissolution processes, and should only be used for qualitative comparison purposes, i.e., dissolution processes can interfere with the HM sorption reactions. In general, selectivity is influenced by ionic size/activity, soil type, and pH of the system.

Table 5.2 shows the selectivity order reported in some representative studies in the literature. This confirms that selectivity order depends on the soil type and pH environment, conditions wherein soil-contaminant interaction is established. Elliott et al. (1986) report that for divalent heavy metals, when the concentrations applied

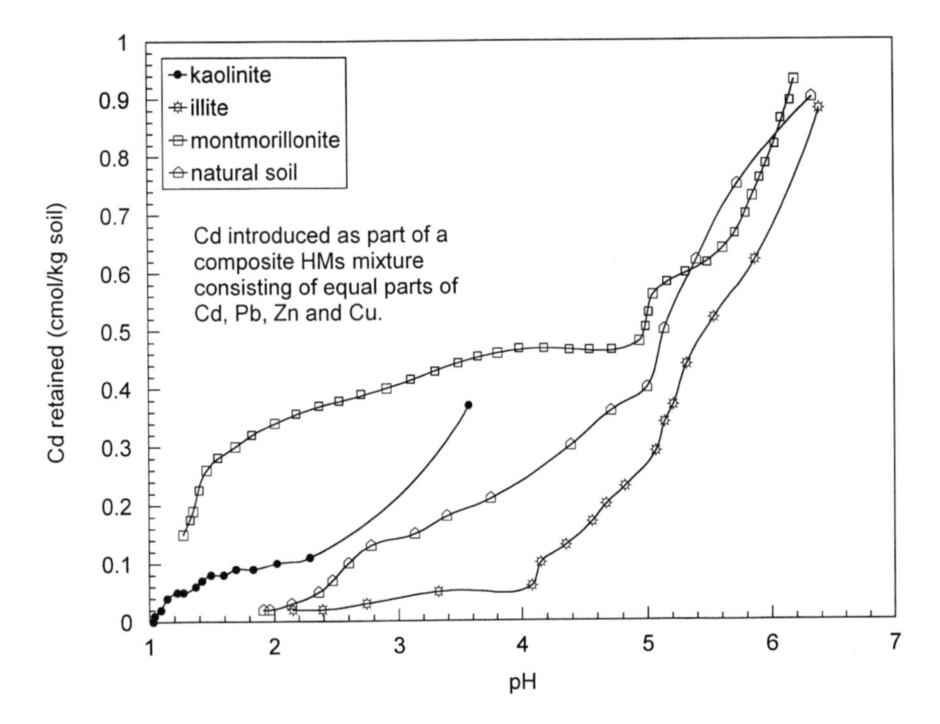

Figure 5.6 Retention of Cd by various soils. Cd introduced as part of a composite mixture of HMs consisting of equal parts of Cd, Pb, Zn, and Cu.

to soil are the same, a correlation between ionic size and selectivity order may be expected. According to Bohn (1979), the ease of exchange or the strength with which cations of equal charge are held is generally inversely proportional to the hydrated radii, or proportional to the unhydrated radii. For the heavy metals shown in the previous figures, the predicted order of selectivity based on unhydrated radii should be:

$$Pb^{2+}\ (0.120\ nm) > Cd^{2+}\ (0.097\ nm) > Zn^{2+}\ (0.0.074m) > Cu^{2+}\ (0.072\ nm)$$

Yong and Phadungchewit (1993) show a general selectivity order to be Pb > Cu > Zn > Cd.

Elliott et al. (1986) show that at high pH levels aqueous metal cations hydrolyze, resulting in a suite of soluble metal complexes according to the generalized expression for divalent metals given as:

$$M^{2+}(aq) + nH_2O \rightleftharpoons M(OH)_n^{2-n} + nH^+$$

This hydrolysis results in precipitation of metal hydroxides onto soils, which is experimentally indistinguishable from metals removed from solution by sorption mechanisms. Sorption selectivity of heavy metals may relate to the *pk* of the first hydrolysis product of the metals (Forbes et al., 1974) where *k* is the equilibrium

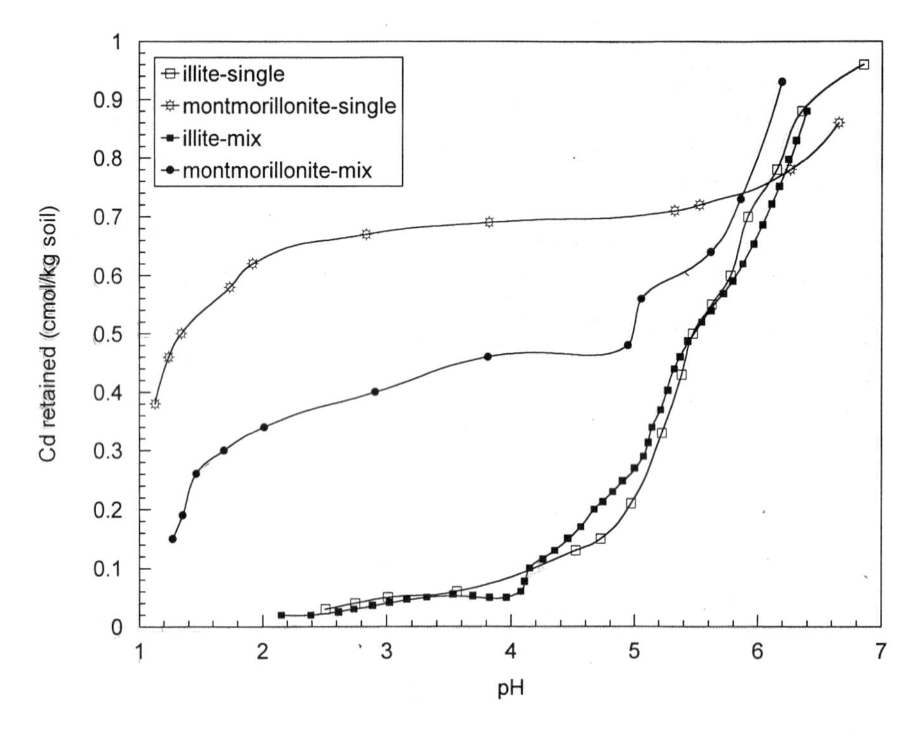

Figure 5.7 Comparison of Cd retention from single and mixed species HM pollutants for montmorillonite and illite soils using data from Figures 5.4, 5.5, and 5.6.

constant for the reaction in the above equation when n = 1. Ranking the heavy metals shown in the previous figures using the *pk* values of Pb, Cu, Zn, and Cd, we obtain a selectivity order as follows:

$$\mathrm{Pb}(6.2) > \mathrm{Cu}(8.0) > \mathrm{Zn}(9.0) > \mathrm{Cd}(10.1)$$

where the numbers in the parentheses refer to the *pk* values.

5.3 PARTITIONING OF HM POLLUTANTS

Partitioning of HM pollutants refers to the various sorption processes that result in the apportionment of HM pollutants between the soil fractions and the aqueous phase (porewater). In essence, the removal of HM pollutants from the porewater by the various sorption mechanisms results in partitioned HMs. While *partitioning* as a process is also used in conjunction with those mechanisms that result in separation of organic chemical pollutants between soil fractions and porewater, we will address the partitioning of HM pollutants in this chapter and consider partitioning of organic chemical pollutants in the next chapter.

The two main points to be considered include (a) technique for determination of partitioning and partition coefficients, and (b) technique for determination of the

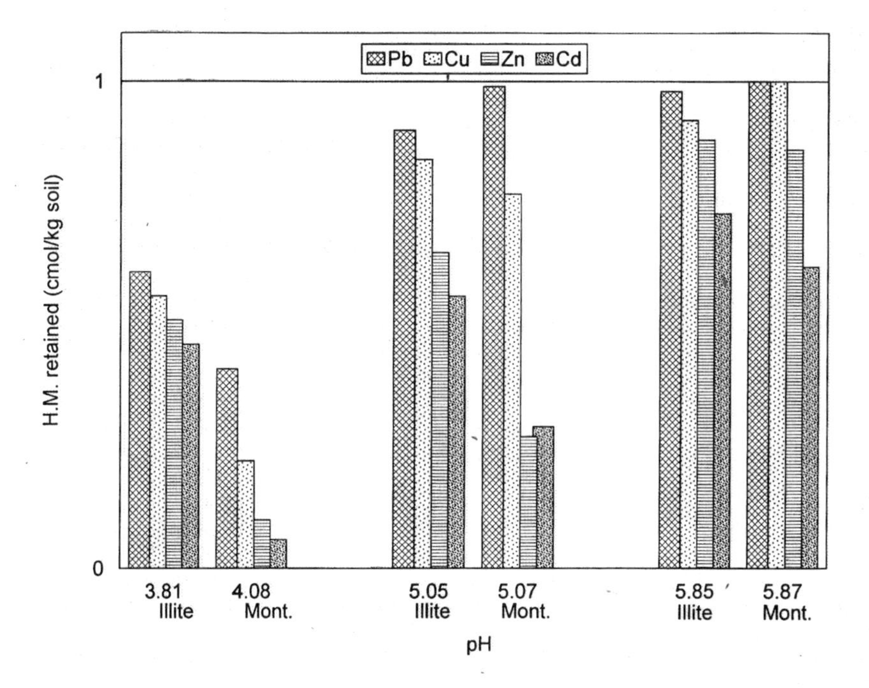

Figure 5.8 Comparison of preferential sorption of heavy metals for illite and montmorillonite soils. Equal proportions of each of the HMs used in total HM leachate.

distribution of HM pollutants amongst the soil fractions. Both points of consideration are particularly significant since they impact directly on our ability to assess and predict the fate of the pollutants.

5.3.1 Determination of Partitioning and Partition Coefficients

Section 4.7 has addressed the adsorption of HMs in terms of adsorption isotherms (continued in Section 4.8), and the distribution coefficient k_d (Section 4.8.2). We recall that these performance characteristics are obtained from soil suspension tests with specific HMs. The use of these as direct measures of partitioning of HMs is not an uncommon practice. In particular, the use of the distribution coefficient (also called the partition coefficient) k_d as a parameter in contaminant transport equations is most common.

There exists considerable controversy concerning the use of soil suspension sorption test results to represent sorption performance of compact soil in the substrate. Aside from the many variations and combinations of species and concentrations of the HMs and soil types, the main issues concern the manner in which the soils interact with the pollutants. The problems of role and effect of distribution (including fabric and structure) of the various soil fractions and availability of reactive surfaces for interaction with the HMs in the porewater need to be properly

Table 5.2 Sorption Selectivity of Heavy Metals in Different Soils

Material	Selectivity Order	References
Kaolinite clay (pH 3.5–6)	Pb > Ca > Cu > Mg > Zn > Cd	Farrah and Pickering (1977)
Kaolinite clay (pH 5.5–7.5)	Cd > Zn > Ni	Puls and Bohn (1988)
Illite clay (pH 3.5–6)	Pb > Cu > Zn > Ca > Cd > Mg	Farrah and Pickering (1977)
Illite clay (pH 4–6)	Pb > Cu > Zn > Cd	Yong and Phadungchewit (1993)
Montmorillonite clay (pH 3.5–6)	Ca > Pb > Cu > Mg > Cd > Zn	Farrah and Pickering (1977)
Montmorillonite clay (pH5.5–7.5)	Cd = Zn > Ni	Puls and Bohn (1988)
Montmorillonite clay (pH ≈ 4)	Pb > Cu > Zn > Cd	Yong and Phadungchewit (1993)
(pH ≈ 5)	Pb > Cu > Cd ≈ Zn	
(pH ≈ 6)	Pb = Cu > Zn > Cd	
Al oxides (amorphous)	Cu > Pb > Zn > Cd	Kinniburgh et al. (1976)
Mn oxides	Cu > Zn	Murray (1975)
Fe oxides (amorphous)	Pb > Cu > Zn > Cd	Benjamin and Leckie (1981)
Goethite	Cu > Pb > Zn > Cd	Forbes et al. (1974)
Fulvic acid (pH 5.0)	Cu > Pb > Zn	Schnitzer and Skinner (1967)
Humic acid (pH 4–6)	Cu > Pb > Cd > Zn	Stevenson (1977)
Japanese dominated by volcanic parent material	Pb > Cu > Zn > Cd > Ni	Biddappa et al. (1981)
Mineral soils (pH 5.0), (with no organics)	Pb > Cu > Zn > Cd	Elliot et al. (1986)
Mineral soils (containing 20 to 40 g/kg organics)	Pb > Cu > Cd > Zn	Elliot et al. (1986)

addressed. Figure 5.9 shows the differences in sorption performance between a soil suspension and compact soil samples. The concentration of sorbed pollutants determined from the column tests are identified as *sorption characteristic curves.*

The batch equilibrium adsorption isotherm curve at the top of Figure 5.9 corresponds to the type of isotherms shown previously in Figures 4.13 and 4.14. The *sorption characteristic curves* shown in the figure refer to the partitioning of the pollutants in the leaching soil column as a result of continuous input of influent leachate. The distinction in terminology is deliberate. We need to distinguish between *adsorption isotherms* determined from batch equilibrium tests on soil suspensions, and *sorption characteristic curves* determined from leaching column tests. As more influent leachate (leachant) is transported through the column, the bottom sorption characteristic curve will migrate upward toward the other characteristic curve. Because compact soil samples such as those in the leaching column do not have the same amount of exposed reactive surfaces, the top sorption characteristic curve will always remain below the batch equilibrium adsorption isotherm. Figure 5.10 (from Yong et al., 1991a) shows a typical set of results. The "scatter" in results reflects the variations in replicate testing of compact samples.

Determination of the distribution coefficient k_d has been discussed in Section 4.8.2. This distribution coefficient is also sometimes known as the *partition coefficient.* Strictly speaking, this term should be used in relation to compact soil

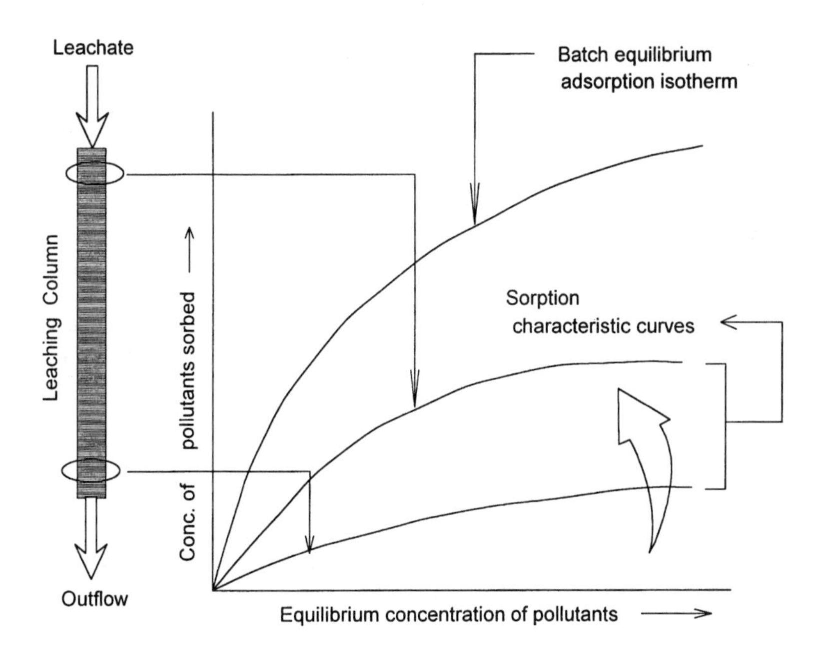

Figure 5.9 Comparison of sorption characteristics between soil solution and compact soil sample. Abscissa and ordinate values for equilibrium concentration of pollutants and sorbed concentration are obtained with respect to increasing input of influent leachate.

samples — to avoid confusing this with the distribution coefficient determined from batch equilibrium adsorption isotherms. Partition coefficients are determined from sorption characteristic curves in much the same manner as the distribution coefficients. It is important to always distinguish between the two sources of data for determination of the distribution/partition coefficients.

5.3.2 Rate-Limiting Processes

The rate of sorption of heavy metals by soil will be controlled by the sorption properties of the soil and the heavy metal pollutants themselves. Depending on the distribution of the various soil fractions, and depending on the nature of the soil fractions, sorption rates can be rapid or slow. Metal sorption kinetics related to the various oxides and soil organic matter are relatively rapid (Sparks, 1995), whereas sorption rates by clay minerals will be influenced by the nature of the interlayer characteristics. Unrestricted montmorillonites can sorb metals more rapidly than vermiculites because the absence of restriction on the montmorillonites permits expansion of the interlayer space and allows for entry of the metals. In contrast, interlayer spaces in vermiculites are restricted, and hence will impede movement of the metals in sorption processes. However, if montmorillonites are restricted, i.e., if

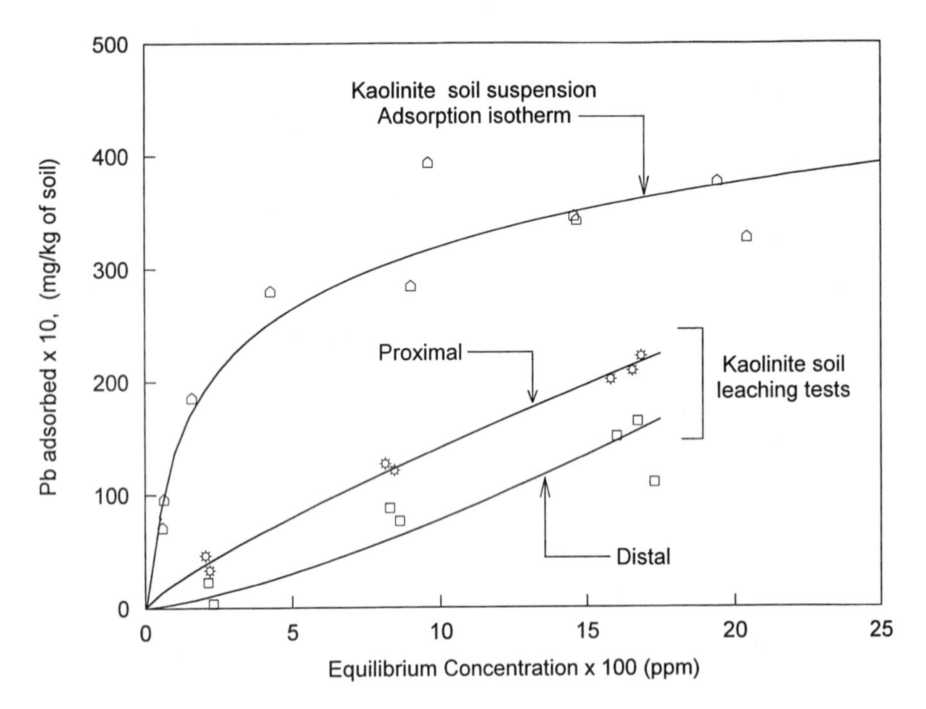

Figure 5.10 Comparison of Pb sorption curves between soil suspension and leaching column tests. The proximal and distal notations refer to locations of sampling positions in the soil column in respect to input of leachate.

montmorillonite interlayer expansion is severely constrained, sorption of the metals will become less rapid.

Interdiffusion of counterions can be considered a rate-determining step in ion exchange. This means that when a counterion *A* diffuses from its location in the DDL region (i.e., the region within the ion exchanger) into the solution, a counterion *B* from the solution must move into the space formerly occupied by counterion *A*. The ion exchanger is generally identified as the region where the ions are controlled by DDL-type forces. The process of diffusion of counterions *A* and *B* is the interdiffusion of counterions between an ion exchanger and its equilibrium solution. There are at least two rate-determining steps:

- **Particle-type diffusion** — Interdiffusion of counterions within the ion exchanger (DDL region) itself.
- **Film-associated diffusion** — Interdiffusion of counterions in the Stern layer.

The many factors and processes such as diffusion-induced electric forces, selectivity, specific interactions and non-linear boundary conditions, make it difficult to develop and specify rate laws which apply diffusion equations to ion-exchange systems. The fluxes of various ionic species are both different and coupled to one another, making it difficult to specify one characteristic constant diffusion coefficient that will describe the flux rate of the different ionic species. Stochiometry of ion

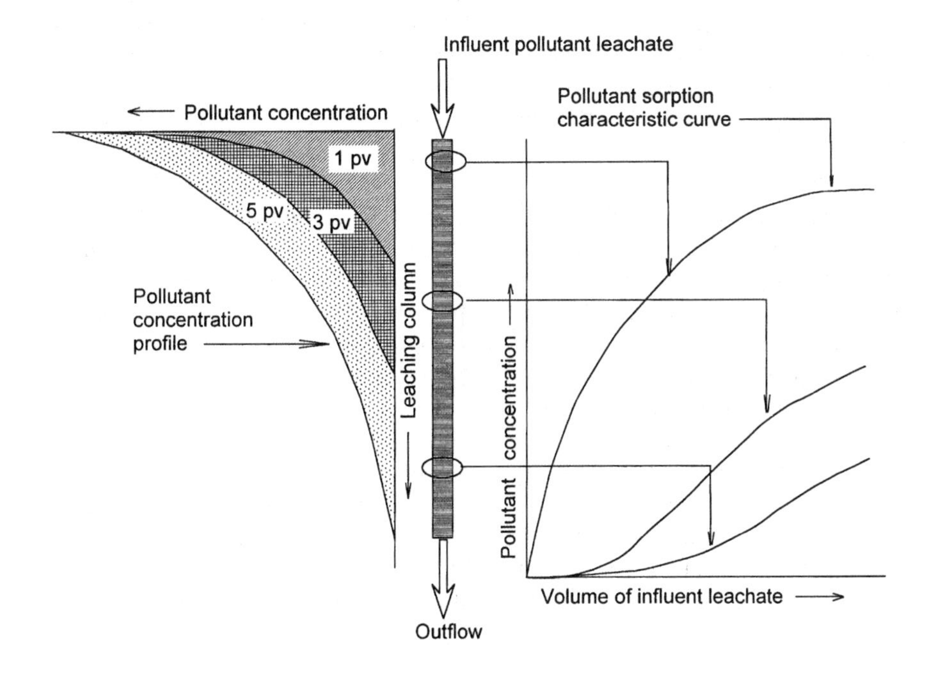

Figure 5.11 Pollutant distribution in leaching column from influent leachate transport.

exchanges requires conservation of electroneutrality between the counterions and the charged clay particle surfaces. For electroneutrality to be preserved, the different electric phenomena established must be considered in the determination of the various diffusion processes.

5.3.3 Assessment of Partitioning from Leaching Columns

The principal features of leaching column tests are shown in Figure 5.11.

The right-hand graph in Figure 5.11 shows the characteristic pollutant sorption curves that indicate sorption of the pollutants results from continuous input of the influent pollutant leachate. To avoid dealing with differing time scales when comparing the performance of different soils and different HM species, the volume of influent leachate is generally used as abscissa scale. By expressing the effluent leachate volume in terms of *pore volumes,* i.e., volume of pores in the leaching column sample (as shown in the left-hand portion of the figure), a relationship between the density of the sample and its effect on sorption can be deduced. Thus for example, 1 pv (pore volume) of leachate passing through a soil with high porosity (low density) would take less time for transport through the sample in comparison to a 1 pv leachate through a denser (low porosity) sample. We need to be careful in generalizing the pore volume-time relationship since many other factors associated with reactive surfaces and specific surface areas need to be considered. Whilst comparisons between leachate penetration and sorption performance using pvs (pore volumes) are best performed with the same soil types, we can obtain considerable benefit from comparisons between different samples so long as the proper considerations for the available reactive surfaces are made.

The left-hand graph in the same figure shows the distribution of pollutants sorbed by the soil in relation to depth. There are two components in the sorbed concentration at any one point in the soil column: (a) concentration of pollutants sorbed by soil solids at that particular point, and (b) concentration of pollutants in the porewater at that same point. The total concentration of pollutants includes both these components. As the volume of influent leachate continues to be transported through the soil column, the total sorbed pollutant concentration at any one depth will increase until the *carrying capacity* of the soil is reached. This carrying capacity is defined as the capacity for pollution sorption by the soil solids. The link between the left-hand and right-hand graphs is obvious.

We have shown previously that the pH regime has a significant influence on sorption performance of the soil solids — previously demonstrated in Figure 4.6 by the solubility-precipitation relationship for a metal hydroxide complex. Using the same type of diagram, we can show the proportions of HMs sorbed by the soil solids and the amount remaining in solution in Figure 5.12. This figure illustrates the fact that the concentration of HM pollutants in the porewater at any one point in the soil

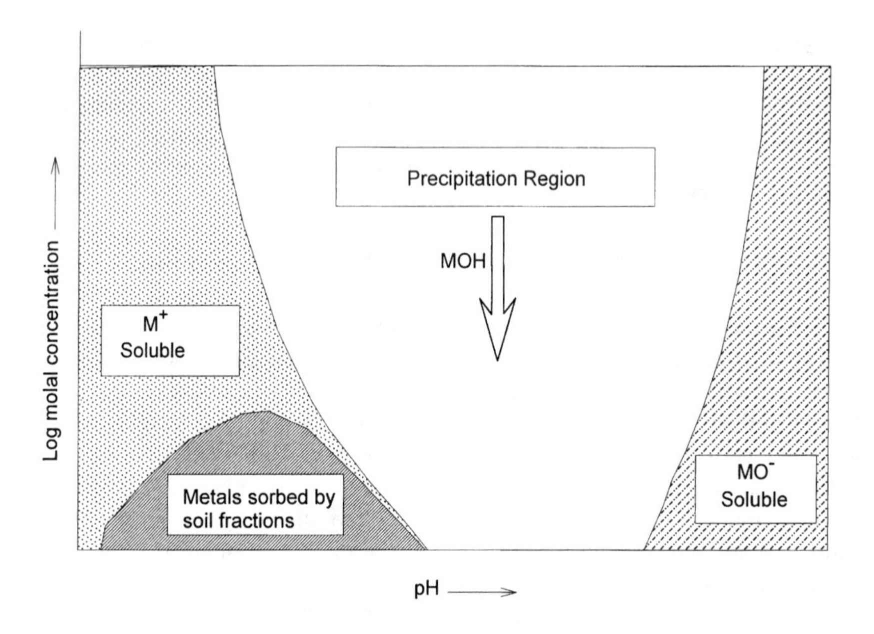

Figure 5.12 Solubility-precipitation diagram for a metal-hydroxide complex showing sorption of metals by soil fractions in relation to pH.

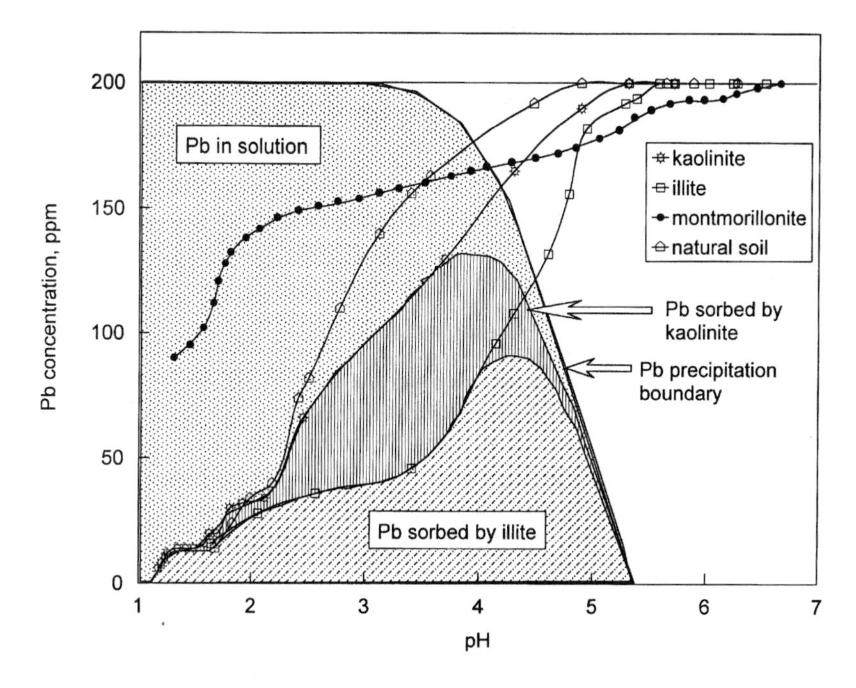

Figure 5.13 Pb removed from aqueous phase and Pb sorbed by kaolinite and illite soils.

column shown previously, and at any one time, depends not only on the availability and nature of the reactive surfaces, but also on the pH regime. The presence of inorganic and/or organic ligands in the porewater which also affect this distribution are represented by the metals remaining in solution (M^+ soluble in the diagram).

Yong (1999a) gives the example of some of the dissolved Pb series in Figure 5.13 as $PbCl^+$, $PbNO_3^+$, $PbOH^+$, and $Pb(OH)_2^o$. All the chemical reactions involved respond to the requirements of the system to seek equilibrium. We can use the results from soil-Pb interaction suspension tests such as those shown in Figure 5.13 to illustrate the sorption relationships established in the left-hand portion of the diagram shown in Figure 5.12. The relationships shown in Figure 5.13 demonstrate the need to distinguish between Pb removed from the aqueous phase and Pb sorbed by the soil fractions. The solid lines show the Pb concentration removed from the aqueous expressed in terms of ppm. This represents both the Pb sorbed by the soil fractions and Pb removal through precipitation processes. The Pb sorbed by the kaolinite and illite soils can be seen in the diagram. As noted, these two soils show significant differences in their ability to sorb Pb because of the nature of the reactive surfaces and the various soil fractions that make up the two soils.

The lessons to be learnt in regard to determination and evaluation of partitioning of HM pollutants in soil suspension tests (Figure 5.13) can be applied directly to soil leaching column studies. From Figures 5.12 and 5.13, we can see that it would be a mistake to assume that a determination of the HMs concentration remaining in the aqueous phase, together with the application of mass balance calculations, can provide a direct measure of the HMs sorbed by the soil fractions, i.e.:

$$HM_{total} - HM_{aqueous} \text{ \underline{\textit{may not be equal to}} } HM_{sorbed}$$

where:

HM_{total} = concentration of total HM applied in test;
$HM_{aqueous}$ = concentration HM remaining in aqueous phase;
HM_{sorbed} = concentration of HM sorbed by the soil fractions in the soil suspension.

Figure 5.13 shows that before a pH of about 3.6, $HM_{total} - HM_{aqueous} = HM_{sorbed}$ for the kaolinite soil. After a pH of 3.6, $HM_{total} - HM_{aqueous} \neq HM_{sorbed}$. The pH value for a similar sorption performance for the illite soil is about 3.9.

Assessment of the partitioning of HMs in soils using effluent concentrations measurements from soil column experiments and calculated sorbed HM instead of direct sorption measurements can lead to serious error, as noted from the preceding

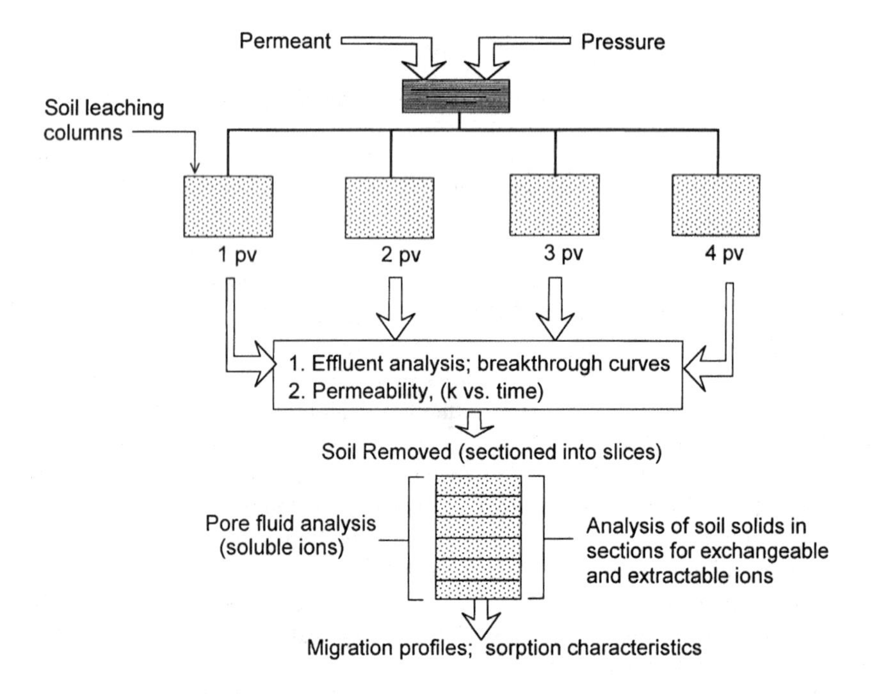

Figure 5.14 Test procedure for determination of partitioning of HM pollutants using replicate samples and various pore volumes (pvs) of HMs permeant as influent leachate.

discussion. Figure 5.14 shows the minimum required procedure for proper assessment of partitioning in soil column studies. The procedure calls for replicate sample testing with various quantities of permeant passing through the soil columns. Analyses of porewater and soil solids for soluble ions (in porewater) and exchangeable and extractable ions (from soil solids) should be conducted.

Analysis of the sorbed HM in the soil samples in the leaching columns is best performed on both the soil solids and the porewater. A mass balance consisting of the total sum of HMs sorbed by the soil solids, HMs in the porewater, and HMs in the effluent should show complete or near-complete accord. Problems associated with complete removal of HMs sorbed by soil solids generally contribute to the less-than-complete accord in mass balance calculations. The later discussion on selective sequential analysis (SSA) will demonstrate this particular problem. Figure 5.15 shows an example of the information obtained from soil sample analyses in leaching column studies. In this particular experiment, the kaolinite soil sample in the leaching column showed that it retained more Pb in the porewater that through processes associated with sorption forces. This is expected since the primary sorption mechanism for the soil is the net negative charges on the surface of the kaolinite particles. The unhatched portion of the graph represents the unaccounted Pb using mass balance calculations.

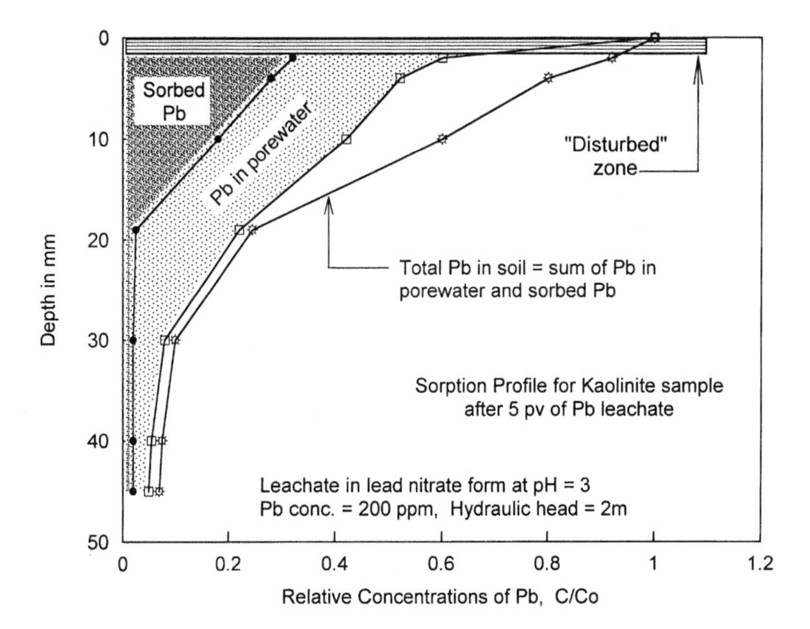

Figure 5.15 Pb concentration profile in leaching column kaolinite soil sample showing sorbed Pb and Pb in porewater (data from Darban, 1997).

Determination of partition coefficients based on column leaching tests is difficult since decisions need to be made concerning some key issues, two of the more significant ones being:

1. **Nature of pollutant loading** — Continuous and constant; sporadic and intermittent; variable concentration, etc. The development of the pollutant concentration profiles vary in accord with the amount, nature, and manner in which the leachate is transported in the leachate column.
2. **Equilibrium conditions** — The results of leaching column tests generally show a developing concentration profile as more and more leachate penetrates the sample. The question of when sorption equilibrium (total carrying capacity of the sample) is reached is an issue that must be addressed. Partition or distribution coefficients determined on the basis of leaching column tests need to recognize this.

In comparing the adsorption isotherm obtained from standard batch equilibrium tests with adsorption characteristic curves obtained from soil column leaching tests shown in Figure 5.10, we observe that the full pollutant carrying capacity of the soil column has yet to be reached. Full capacity is obtained when proximal and distal curves are identical. The choice of adsorption characteristic curve for specification of the distribution coefficient will depend on user experience and preference. The options available are:

- Standard adsorption isotherm from batch equilibrium tests.
- Different distribution coefficients for different locations distant from influent leachate, and for different elapsed times.
- A variable distribution coefficient determined according to the variation of the adsorption characteristics with both time and space.

5.3.4 Breakthrough Curves

If we use measurements from leaching column tests in the form shown in Figure 5.16, we can obtain further appreciation of the soil capability for sorption of the heavy metal pollutants. The ordinate shown in the figure expressed as "relative concentration C_i/C_o" refers to the ratio of the concentration of the target pollutant (C_i) in the outflow at the instant of time *i* to the concentration C_o of the influent target pollutant. The 50% point on the ordinate marks the point of breakthrough of the target pollutant in the candidate soil being tested. The curves in Figure 5.16 which show different sorption capability profiles can be used to provide more information on the sorption potential of candidate soil materials. Figures 5.17 and 5.18 show the breakthrough test information (from column leaching tests) for the kaolinite and illite soils (respectively). These are the same soils used for the retention tests shown previously in Figures 5.5. and 5.6. The good sorption capacity of the illite is evident from the results shown in Figure 5.18. This is not surprising if we recall the results shown in Figure 4.10, i.e., the buffering capacity curves there demonstrate clearly that the illite soil shows considerable buffering capacity in comparison to the kaolinite soil.

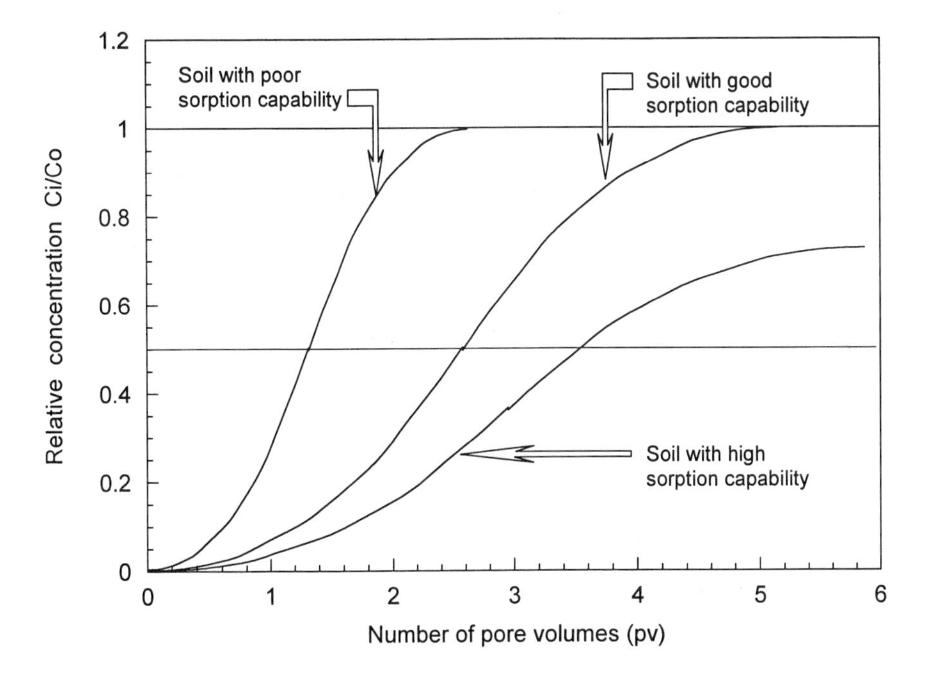

Figure 5.16 Typical breakthrough curves.

We see from Figures 5.17 and 5.18, together with the results shown previously in Figures 5.5, 5.6, and 4.10, there is a coherent pattern of sorption of heavy metals which reflects the various processes previously described as interactions between pollutants and the reactive surfaces of the soil fractions. The kaolinite, with its poor buffering capacity (Figure 4.10), shows the characteristic "soil with poor sorption capability" in Figure 5.16. The illite soil, with a well-demonstrated buffering capacity shows the characteristic "soil with good sorption capability" in that same figure. The higher illite soil pH also contributes (significantly) to the observed sorption performance because of partitioning by both sorption and precipitation mechanisms The breakthrough curves for the MR1 soil (Figure 5.19) show good buffering performance. What is interesting to observe in the figure is the pH variation in both the effluent and in the porewater. This is the same soil shown previously in Table 2.2 and Figure 2.15. Because of the initial high pH of the soil, we can deduce that the mechanisms for retention of the heavy metals in the soil column are through processes associated with precipitation.

5.4 DISTRIBUTION OF PARTITIONED HMs

The *distribution* of partitioned HMs refers to the manner in which the sorbed heavy metals are retained by each kind of soil fraction that composes the total soil under test scrutiny. In essence, the distributed HMs amongst the soil fractions reflect

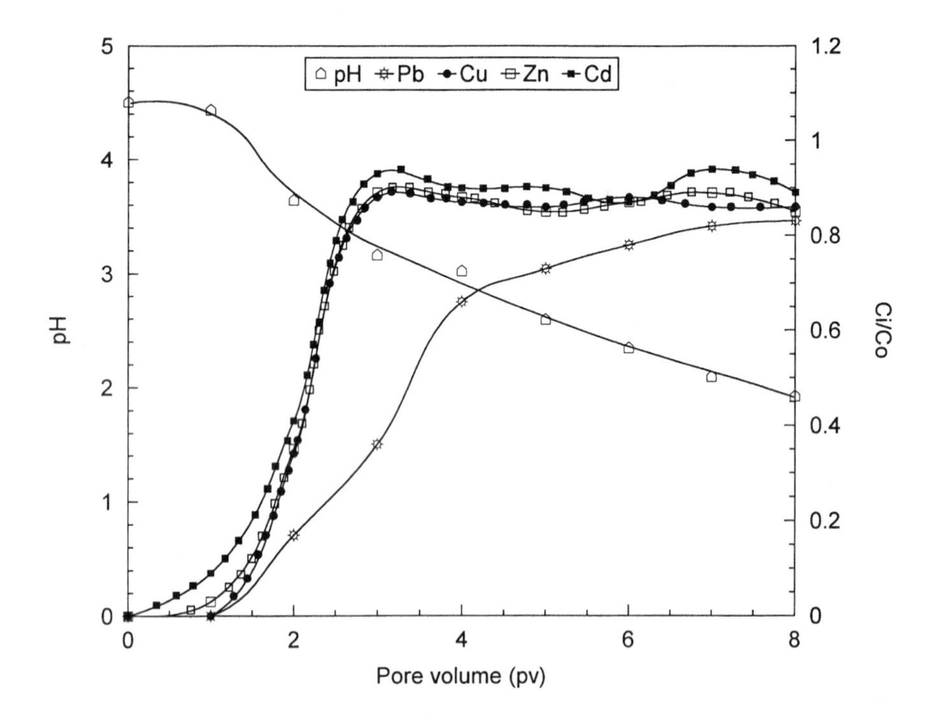

Figure 5.17 Breakthrough curves for kaolinite soil used for retention tests in Figures 5.5 and 5.6. pH of effluent is also shown. HMs are introduced as a composite mixture with equal parts of each HM.

the different sorption capabilities of the various soil fractions. We recall from Chapters 3 and 4 that the nature of the soil fractions, their reactive surfaces, and soil structure all combine to produce the resultant observed pollutant retention characteristics of a soil. Each type of soil fraction possesses different HM sorption characteristics. A knowledge of the distribution of partitioned HMs is useful since it tells us about the role of the individual soil fractions in the partitioning process, and permits us to use this information to determine the following:

- The sorption potential of candidate soils based on a knowledge of the soil composition;
- The likely fate and environmental mobility of sorbed pollutants from knowledge of the interaction or sorption processes of individual soil fractions; and
- The potential for removal of the sorbed pollutants with different remediation techniques.

Procedures for evaluation of the distribution of partitioned heavy metal pollutants, i.e., heavy metal pollutants sorbed onto soil fractions, include: (a) pollutant extraction techniques which can selectively remove the target pollutants from specific soil fractions; (b) techniques of systematic removal of soil fractions in pollutant-soil interaction studies; and (c) techniques of systematic addition of soil fractions to study pollutant retention of the laboratory-constituted soil. The last procedure, which

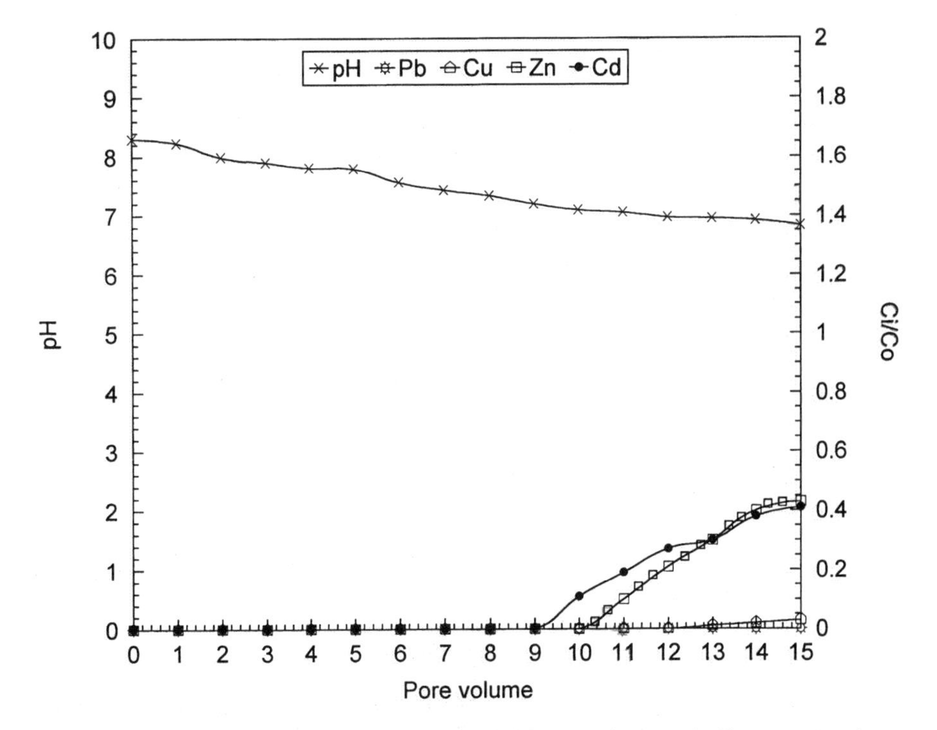

Figure 5.18 Breakthrough curves for illite soil used for retention tests in Figures 5.5 and 5.6. pH of effluent is also shown. HMs are introduced as a composite mixture with equal parts of each HM. Refer to Figure 4.10 for buffering curves for illite and kaolinite soils.

is strictly a laboratory study technique, involves the deliberate inclusion of specific soil fractions to form a laboratory-constituted soil for use in soil-column or soil-suspension tests designed to study pollutant retention.

The techniques that systematically remove target soil fractions as part of the sequential extraction test is known as the *selective soil fraction removal techniques* (SSFR) procedure. This contrasts with the *systematic soil fraction addition* (SSFA) procedure which adds specific soil fractions to produce the soil samples for pollution retention studies. The procedure of selective removal of target pollutants from various soil fractions is known as *selective sequential extraction* (SSE). The SSE and SSFR techniques have the advantage over the SSFA technique in that the SSE and SSFR techniques can be used to study or test soil samples extracted from contaminated sites. Figure 5.20 is a schematic portrayal of the SSE and SSFA techniques.

5.4.1 Selective Sequential Extraction (SSE) Procedure and Analysis

The basic idea in application of the SSE procedure centres around the removal of sorbed HMs from individual soil fractions. Chemical reagents which are chosen are designed to selectively destroy the bonds established between HM pollutants

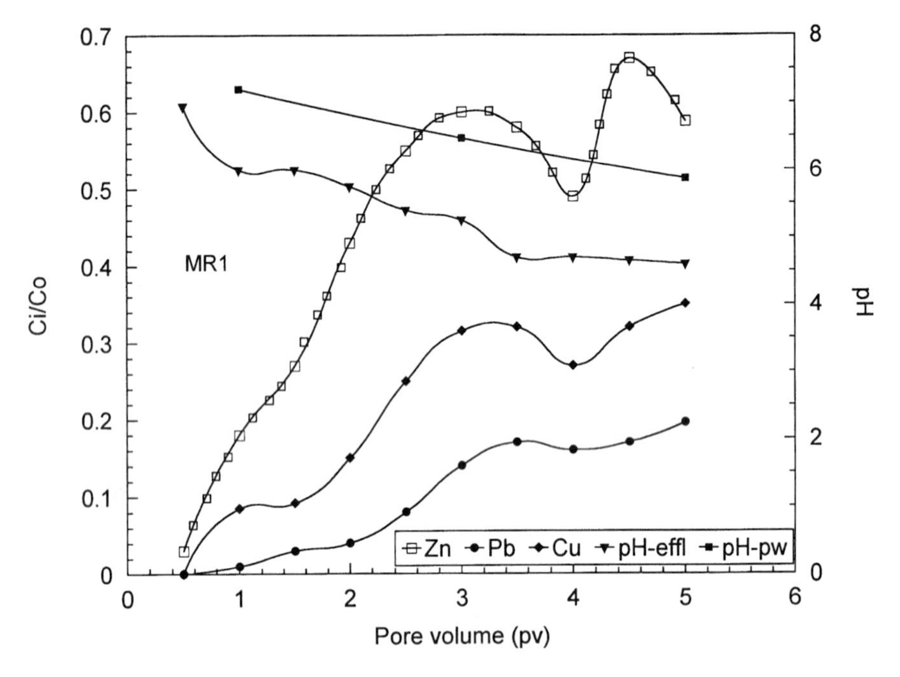

Figure 5.19 Breakthrough curves and pH curves for both effluent and porewater for MR1 soil. Refer to Figure 2.15 and Table 2.2 for MR1 soil details.

and specific individual soil fractions. Obviously, the choice of chemical reagents is the key to the success of the technique. There is no assurance that in destroying the pollutant-soil bonds, dissolution of individual soil fractions will not also occur. The published literature shows a variety of reagents used by different researchers in application of the SSE technique. Some of these are shown in Table 5.3. The results obtained in regard to apportioning (i.e., distribution) of partitioned HMs depend on the aggressiveness of the chemical reagents. Accordingly, since the quantitative results obtained are dependent on experimental techniques, the measurement of distribution of the partitioned will be *operationally defined.* In short, these measurements should be considered to be more qualitative than quantitative. They can nevertheless provide a good insight into the distribution of the partitioned HMs.

A proper application of the SSE technique requires the chemical reagents to release the HM pollutants from specific soil solids by destroying the bonds binding the HMs to the target soil fractions. The extractant reagents used have a history of application in routine soil analyses, and are classified as: concentrated inert electrolytes, weak acids, reducing agents, complexing agents, oxidising agents, and strong acids (Tessier et al., 1979). Whilst the sequence of application of the extractant reagents has not been standardized, there is general acceptance that one begins with the least aggressive extractant. The five different HM-bonding groups obtained as the operationally defined groups or phases include:

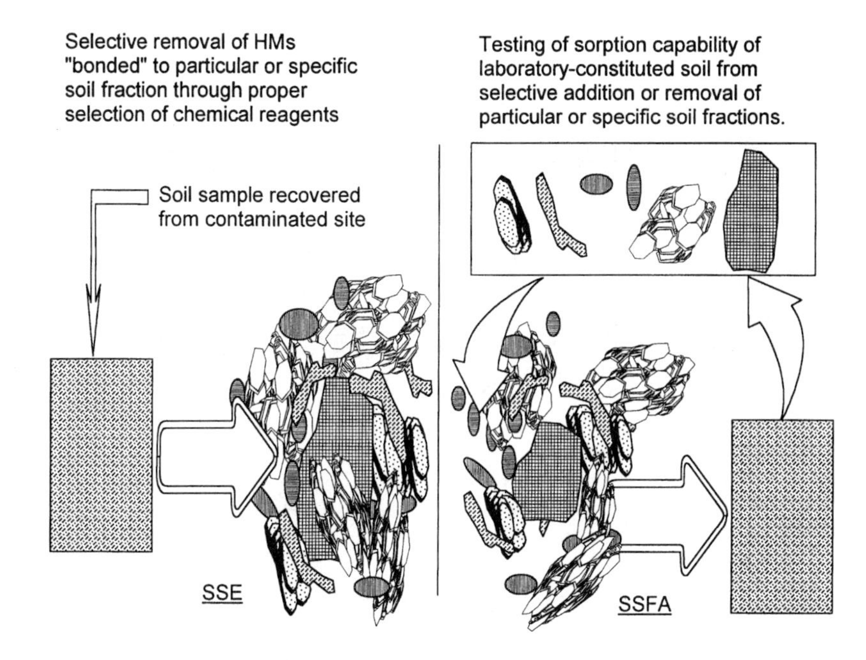

Figure 5.20 Laboratory study procedures for determination of distribution of partitioned HMs.

1. **Exchangeable metals (exchangeable phase)** — Metals extracted in this group are identified as "in the *exchangeable phase*," i.e., they are considered to be non-specifically adsorbed and ion exchangeable and can be replaced by competing cations. The soil fractions involved are mostly clay minerals, soil organics, and amorphous materials. Neutral salts such as $MgCl_2$, $CaCl_2$, KNO_3, and $NaNO_3$ are commonly used as ion-displacing extractants. These will promote release of metal ions bound by electrostatic attraction to the negatively charged sites on the soil particle surfaces.

   There is little evidence to suggest that dissolution of the soil solids occurs because of the neutral electrolytes. Pickering (1986) showed that $MgCl_2$ sediment leachate contained only low levels of Al, Si, and organic carbon, confirming the weakness of the neutral salts in interaction with the clay surfaces, sulfides, and organic matter. If the neutral salt solutions are applied at neutral pH, dissolution of Fe or Mn oxides is not expected, and we would only expect minimal dissolution of carbonates. Other types of salts such as NH_4Cl and NH_4OAc, may dissolve considerable amounts of compounds such as $CaCO_3$, $MgCO_3$, $BaCO_3$, and $MgSO_4$. If $CaSO_4$, and then NH_4OAc are used, they can cause some dissolution of Mn oxyhydrates and metal oxide coatings.
2. **Metals associated with carbonates (carbonate phase)** — Metals precipitated or co-precipitated as natural carbonates can be released by application of acidified

Table 5.3 Summary of Some Extraction Procedures

Authors	Exchangeable	Bound to carbonates	Bound to Fe-Mn oxides	Bound to Org.Mat.	Residual
Tessier et al. (1979)	1-$MgCl_2$	2-NaOH/HOAc	3-$NH_2OH.HCl$ in 25% HOAc	4-H_2O_2/HNO_3 + NH_4OAc	5-HF + $HClO_4$
Chester and Hughes (1967)	1-NH_3OHCl + CH_3COOH	2-NH_3OHCl + CH_3COOH	3-NH_3OHCl + CH_3COOH		
Chang et al. (1984)	1-KNO_3	4-Na_2EDTA		3-NaOH	5-HNO_3 (70-80°C)
Emmerich et al. (1982)	1-KNO_3	4-Na_2EDTA		3-NaOH	5-HNO_3
Gibson and Farmer (1986)	1-CH_3COONH_4 pH 7	2-CH_3COONa pH 5	3-4-Hydroxyl-ammonium + HNO_3/Acetic Acid	5-H_2O_2 + HNO_3 85°C	6-Aqua regia + HF + Boric acid
Yanful et al. (1988)	1-$MgCl_2$ + Ag thiourea	2-CH_3COONa + CH_3COOH	3-$NH_2OH.HCl$	4- + sulphides H_2O_2 + HNO_3	5-HNO_3 + $HClO_4$ + HF
Clevenger (1990)	1-$MgCl_2$	2-NaOAc/HOAc		3-HNO_3/H_2O_2	4-HNO_3 (boiled)
Belzile et al. (1989)	1-$MgCl_2$	2-CH_3COONa/ $NH_2OH.HCl/HNO_3$ Room temp.	3-Mn Oxide $NH_2OH.HCl$/ HNO_3, NH_4OAC/HNO_3	4- +Sulf H_2O_2/HNO_3, NH_4OAc/HNO_3	
Guy et al. (1978)	1-(exch.+ adsor. + organic) $CaCl_2$ + CH_3COOH + K-pyrophosphate	4- (carb. + adsor. + fe-Mn nodules) NH_3OHCl + CH_3COOH	2-(metal oxides + org.) H_2O_2 + Diothinite + Bromoethanol		
Engler et al. (1977)	1-(exch.+ adsorb.) NH_4OAc	2-$NH_2OH.HCl$		3-H_2O_2/HNO_3	4-$Na_2S_2O_4/HF/HNO_3$
Yong et al. (1993)	1-KNO_3	2-NaOAc pH 5	3-$NH_2OH.HCl$	4-H_2O_2 (3 steps)	5-$HF/HClO_4$ + HCl

1, 2, 3, 4, 5 indicates the sequence of the extraction.

acetate as the extractant. A solution of 1 *M* HOAc-NaOAc (pH 5) is generally sufficient to dissolve calcite and dolomite to release the metals bound to them without dissolving organic matter, oxides, or clay mineral particle surfaces.

3. **Metal contaminants associated with metal oxides (hydroxide/oxide phase)** — The contaminant metals released in this sequence of extractant treatment are those metals which are attached to amorphous or poorly crystallized Fe, Al, and Mn oxides. The metal oxides include ferromanganese nodules, ranging from completely crystalline to completely amorphous, which occur as coatings on detrital particles, and as pure concretions. Their varying degree of crystallization results in several types of association with the heavy metals: exchangeable forms via surface complexation with functional groups (e.g., hydroxyls, carbonyls, carboxyls, amines, etc.) and interface solutes (electrolytes), moderately fixed via precipitation and co-precipitation (amorphous) and relatively strongly bound. (A more complete recounting of the various interacting/retentive mechanisms can be found in Yong et al., 1992a).

   The extractant selected for oxyhydrates should not attack either the silicate minerals or the organic matter. A good example is the one used by Chester and Hughes (1967), i.e., a combination of an acid reducing agent (1 *M* hydroxylamine hydrochloride) with 25% (v/v) acetic acid for the extraction of ferromanganese oxides.
4. **Metals associated with organic matter (organic phase)** — The binding mechanisms for metals in association with organic matter include complexation, adsorption, and chelation. Because of the different types of binding mechanisms, some overlapping effects will be obtained with those methods designed to release exchangeable cations. However, the general technique used with respect to metal binding to organic matter is to obtain release of the metals as a result of oxidation of the organic matter. The oxidants are generally used at levels well below their (organic matter) solubilities.
5. **Metals contained in the residual fraction** — This metal fraction is generally not considered to be significantly large. The metals are thought to be contained within the lattice of silicate minerals, and can become available only after digestion with strong acids at elevated temperatures. Determination of the metal associated with this fraction is important in completing mass balance calculations.

The information given in Table 5.3 and demonstrated in Figure 5.21 indicates that as we progress from step 1 to step 5, the extraction procedures become very aggressive. As the level of aggression increases (in respect to extraction), we can expect a corresponding increase in the retention capability of the soil (for the HM pollutants). This can be used as an indicator of the degree of retention capability, i.e., a qualitative measure of the irreversibility of sorption by the soil.

The SSE results shown in Figure 5.21 refer to the "Pb sorbed by illite" shown in Figure 5.13. This can be compared with the results for Cd sorbed by the same illite soil (Figure 5.22). We have shown previously in the discussion following Figure 5.13 that as the pH increases beyond pH 4, onset of precipitation and co-precipitation of Pb occurs. The distribution shown in Figure 5.21 indicates that these are in the form of (or associated with) carbonates and hydrous oxides. The influence of pH is pronounced when the precipitation pH of Pb is reached (and beyond). It is not easy to distinguish between sorption-retention (physisorption and chemisorption)

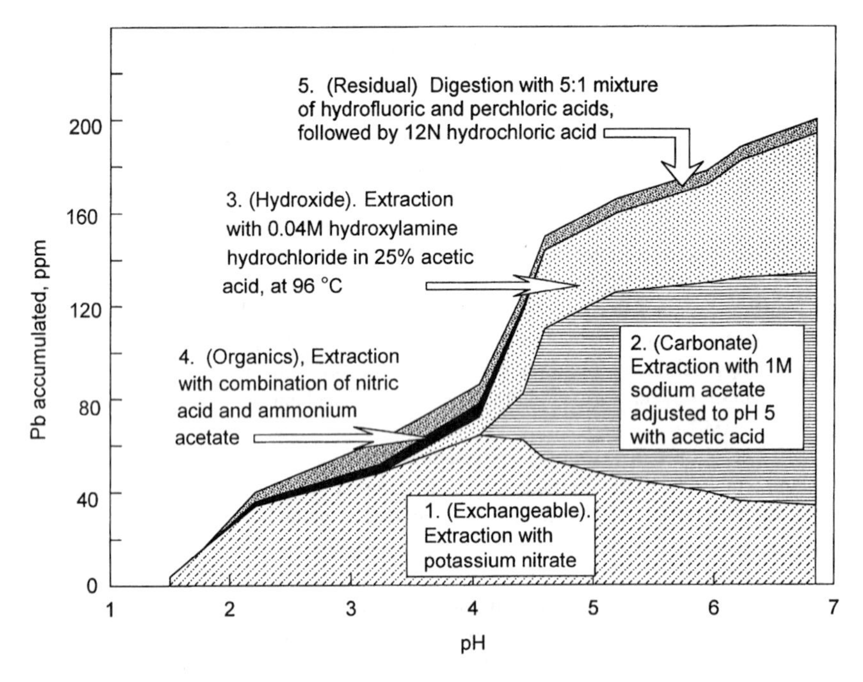

Figure 5.21 Example of application of SSE procedure on illite soil with Pb pollutant. The distribution of partitioned Pb should be compared with the results shown in Figure 5.13.

and precipitation-retention (i.e., retention by precipitation mechanisms). The exchangeable ions can be considered to be non-specifically adsorbed and ion exchangeable, i.e., they can be replaced by competing cations. These ions are lesser in quantity when the precipitation pH of Pb is exceeded because other competing mechanisms begin to dominate. At pH values below 4, Pb is present in the solution as a free cation (Pb^{2+}), and the dominant mechanism for Pb retention is by cation exchange. The amounts of Pb retained increases as pH increases. When the soil solution pH increases to a certain level, Pb begins to form hydroxy species, the beginning of Pb retention by the hydroxide fractions. The Pb precipitated or co-precipitated as natural carbonates can be released if the immediate environment is acidified.

In contrast, the distribution of the partitioned Cd for the same illite soil shows significantly lesser partitioned Cd throughout the entire pH spectrum. The previous discussions on selectivity or preferential sorption of HMs tell us that this is not unexpected. The value of SSE studies lies in the information provided in respect to the distribution of partitioned different species of HM pollutants amongst the various soil fractions, as shown, for example, by comparing Figure 5.22 (Cd distribution) with Figure 5.21. There are at least two significant reasons for the distinct differences in Cd associated with the different phases: (a) ionic activity of the HM in question, and (b) precipitation pH difference between Cd and Pb.

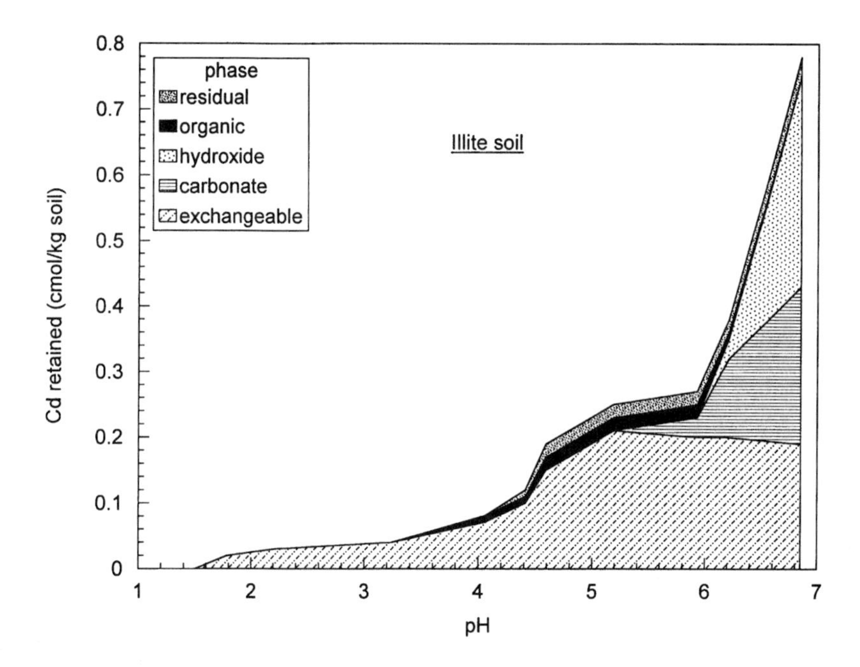

Figure 5.22 Distribution of partitioned Cd in illite soil using SSE procedure.

5.4.2 Selective Soil Fraction Addition (SSFA) Procedures and Analysis

Application of the selective soil fraction addition (SSFA) technique is necessarily limited to laboratory-prepared soil samples. The common procedure generally begins with a basic soil fraction such as a clay mineral (e.g., kaolinite) as the first soil sample. In this case, it will be a total clay mineral laboratory-constituted sample. Interaction with HM pollutants can either be conducted as soil-suspension studies which will generate adsorption isotherm results, or as soil leaching column tests where adsorption characteristics will be obtained.

Subsequent additions of other kinds of soil fractions such as carbonates, hydrous oxides, and soil organics are undertaken sequentially. With the addition of each specific soil fraction, a new combined laboratory-constituted sample is obtained. The proportions of each type of added soil fraction and the sequence of addition are experimental design procedures determined by the project or study objectives. The total number and types of soil fractions used to constitute the final soil sample are factors determined (again) by experiment objectives. Pollutant interactions are conducted after each soil-fraction addition, and evaluation of the adsorption isotherms or adsorption characteristics will provide for qualitative assessments of influence of soil fractions on distribution of partitioned HM. It bears repeating that the results

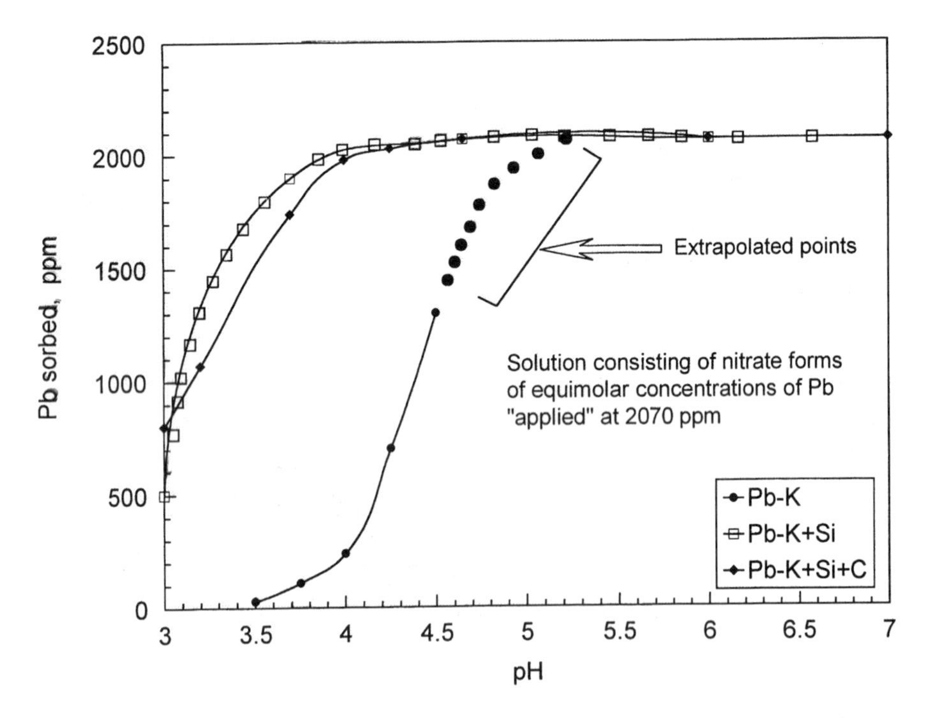

Figure 5.23 Effect of selective addition of soil fractions on sorption of Pb. Note that K, Si, and C refer to kaolinite, silica gel, and carbonates, respectively (data from Darban, 1997).

obtained are directly conditioned by the experimental design factors. Nevertheless, they provide useful clues into the sorption performance of soil fractions.

We will illustrate the preceding with an example of a series of experiments conducted and reported by Darban (1997). Figure 5.23 shows the sorption curves obtained using the addition option of the SSFA technique, using some of the data reported by Darban. The basic soil fraction used was a kaolinite, identified as K in the graph. To this, 10% (by weight) silica gel was added to provide for the K + Si samples. The K + Si + C samples used to provide the resultant curve in Figure 5.23 consisted of kaolinite plus 5% (by weight) of silica gel and 5% (by weight) calcium carbonate.

To gain an insight into the reasons for the performance of the retention curves shown in the figure, it is necessary to obtain some information on the reactive properties of the soil materials. The significant pieces of information such as soil pH, CEC (meq/100 g soil), and specific surface area SSA (m^2/g) are given in the following order:

> kaolinite (K), kaolinite and silica gel soil (K + Si), kaolinite, silica gel, and calcium carbonate soil (K + Si + C).

The properties of interest for the various soil mixtures K, K + Si, and K + Si + C are:

> Soil pH ≈ 4.5, 5.0, NA, and 7.0; CEC ≈ 8, 67, NA meq/100 g soil, and 55; SSA ≈ 12, 118, and 97 m^2/g soil, respectively (NA = not available).

The information presented in Figure 5.23 tells us that:

1. The retention capability for kaolinite by itself is limited. This is evident from the CEC and SSA information, and is well understood from previous discussions on the reactive surface characteristics of the mineral.
2. The addition of silica gel and calcium carbonates to the kaolinite mineral as the basic soil material changes the sorption characteristics of the total soil. Because separate samples of K + S and K + S + C were used for evaluations, we can observe that insofar as sorption of Pb is concerned, there is very little difference in the characteristics of sorption of Pb by both K + S and K + S + C.
3. The information regarding pertinent surface properties (CEC and SSA) indicates that even though half of the silica gel was used in the K + Si + C soil, the decrease in CEC and SSA values was not in proportion to the decrease in amount of silica gel used. This tells us that whilst the contribution to retention of the metals made by the carbonates is lesser than the silica gel, it is sufficient to promote greater retention — most likely through co-precipitation of the metals at the pH levels above metal precipitation pH.

5.4.3 Selective Soil Fraction Removal (SSFR) Procedure and Analysis

To describe the selective soil fraction removal (SSFR) procedure, we use the results reported by Xing et al. (1995). Figure 5.24 provides the bar-chart representation of the results of SSFA soil suspension sorption tests on a black soil containing montmorillonite. The parent material for the black soil was granite, and its pH was 6.16. Organic material content was 53.5 g/kg of soil, whilst the clay content was 216 g/kg of soil. The reported CEC was given as 26.1 cmol/kg.

In the general procedure for selective soil fraction removal from a natural soil (soil containing many soil fractions), the reagents used for soil fraction removal are generally similar to those used in selective sequential analyses. One begins with a natural soil that is allowed to interact with the pollutants of interest, either through leaching column experiments or as soil suspension tests similar to those used in batch equilibrium studies. Companion soil samples are needed since each sample will undergo removal of one or more soil fractions. The procedure used by Xing et al. (1995) for removal of the various soil fractions was somewhat similar to that used by Yong and MacDonald (1998). Removal of carbonates was achieved by using 1 *M* NH_4Ac-HAc at pH 5. To distinguish between the amorphous Fe and the structural Fe (in the layer lattice), 0.2 *M* $(NH_4)_2C_2O_4$ + 0.2 *M* $H_2C_2O_4$ was used to extract the amorphous Fe, whilst structural Fe was removed by the same extractant but with 0.1 *M* ascorbic acid added to the reagent. Extraction of the Fe was conducted in the dark. Determination of partitioned HMs was conducted in the same manner as the regular SSE procedure.

As in the case of SSE experiments, one begins with the removal of the exchangeable phase for one SSFR sample, followed by additional removal of carbonates in the next, etc. Figure 5.24 shows initial removal of E (exchangeable) from one sample. The general procedure will be to subject this sample to interaction tests with a specified HM leachate. A second sample will have both E and C (carbonates)

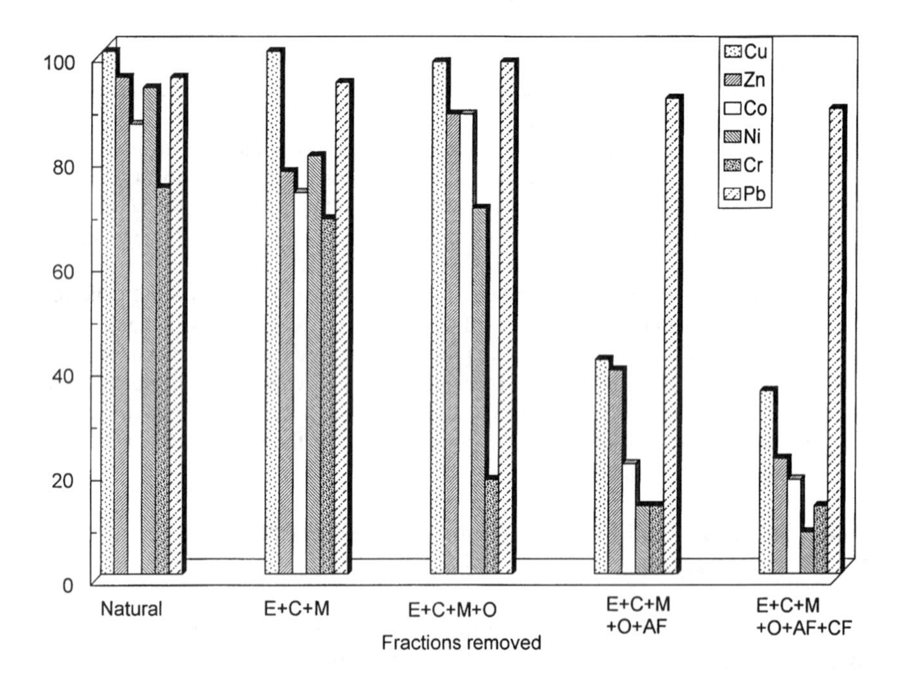

Figure 5.24 Selective soil fraction removal option in SSFA procedure. Sorption results are shown in terms of percentage of HM species introduced in the soil suspension test. Note that E, C, M, O, AF, and CF refer to: exchangeable, carbonates, Mn oxides, amorphous Fe oxide, and crystal Fe oxide, respectively. (Data from Xing et al., 1995.)

removed before being subject to interactions. The procedure continues until all the soil fractions of interest are removed.

The results shown in Figure 5.24 are total interaction results obtained in soil suspension-type tests. These procedures allow for maximum interaction between the reactive surfaces of the various soil fractions and the HM pollutants, and allows us to obtain some insight into the role played by the individual soil fractions in sorption of HM, as in the SSE and SSFA types of tests. In describing the evident features of the results shown in the figure, it is necessary to bear in mind that unless information concerning the various proportions of the soil fractions are available, it can be difficult to fully appreciate the significance of the results. However, at first glance, we can ask, "What do the results in Figure 5.24 show?"

1. The removal of the E (exchangeable), C (carbonates), and M (manganese oxides) from the natural soil does appear to result in significant changes in the sorption of Cu and Pb. The other HMs show some measurable decrease in sorbed amounts.
2. The above points to selectivity in sorption of different HM species.
3. Removal of the organics appear to decrease the sorption capability for Ni and Cr on the one hand, and slightly increases the sorption capability for Zn and Co.
4. Removal of amorphous iron oxide (AF) seems to have a significant effect on the capability of the soil to sorb HM.

5.5 SOIL COMPOSITION, STRUCTURE, AND HM PARTITIONING

We need to recall that the intent of the sorbed (or partitioned) HM extraction tests identified as SSE, SSFR, and SSFA tests is to determine the sorption of HMs by the different soil fractions that constitute the soil sample under study. No direct accountability for the role of natural soil structure on the retention properties of each of the soil fractions of the soil is included in the test results obtained. This is because the SSE test is generally conducted using procedures similar to those for batch equilibrium studies. The SSFR procedure, which selectively removes the target soil fraction with chemical reagents, also determines retention of the HMs by the remaining soil fractions using techniques similar to the SSE procedure. Both these test procedures can be conducted on actual contaminated soil samples recovered from the field and on laboratory-contaminated samples. The SSFA test on the other hand is strictly a laboratory test procedure since the soil sample used for study is obtained from systematic construction by addition of target soil fractions. Contamination of the SSFA soil sample by heavy metal pollutants can be conducted either as soil-solution contamination, or as contaminant leaching column procedures. Assessment of the retention characteristics of the individual soil fractions can be made in a fashion similar to the SSE procedure. A schematic of the essence of the procedures is given in Figure 5.25.

It is the source of the original sample that distinguishes between the SSE, SSFR, and SSFA types of tests. The common thread throughout the extraction process begins with the soil solution (shown in the middle of the diagram) and ends with the reuse of the cleaned residue for the next extraction step. The intent of the SSE technique is to determine the retention characteristics of individual soil fractions from selective destruction of the bonds between the heavy metals and the target soil fraction. The SSFR technique is basically similar to the SSE, except that selective soil fractions are removed prior to conduct of the extraction tests. All tests are essentially designed to measure the optimum retention capacity of the various soil fractions, a condition obtained by default because of the use of soil solutions. By and large, dispersants are not usually used to disperse the soil solids during the extraction procedure. Hence, it is not entirely clear as to whether the full reactive surfaces of all the soil solids are individually available for reaction with the extractants.

5.5.1 Comparison of Results Obtained

The studies reported by Yong and MacDonald (1998) provide a direct comparison between the SSE and the SSFR types of results (Figure 5.26). The sequential removal of carbonates and amorphous oxides from the illitic soil tested by Yong and MacDonald (1998) was somewhat similar to that used by Xing et al. (1995) in the results reported in Figure 5.24. In this case the SSFR procedure consisted of the following 2-step approach: (a) extraction of soil carbonates using 1 *M* NaOAc at pH 5 (adjusted with HOAc) at a soil:solution ratio of 40:1, and (b) extraction of amorphous oxides using 0.1 mol/L oxalic acid, buffered to pH 3 by ammonium oxalate and mixed in the dark at the same soil:solution ratio.

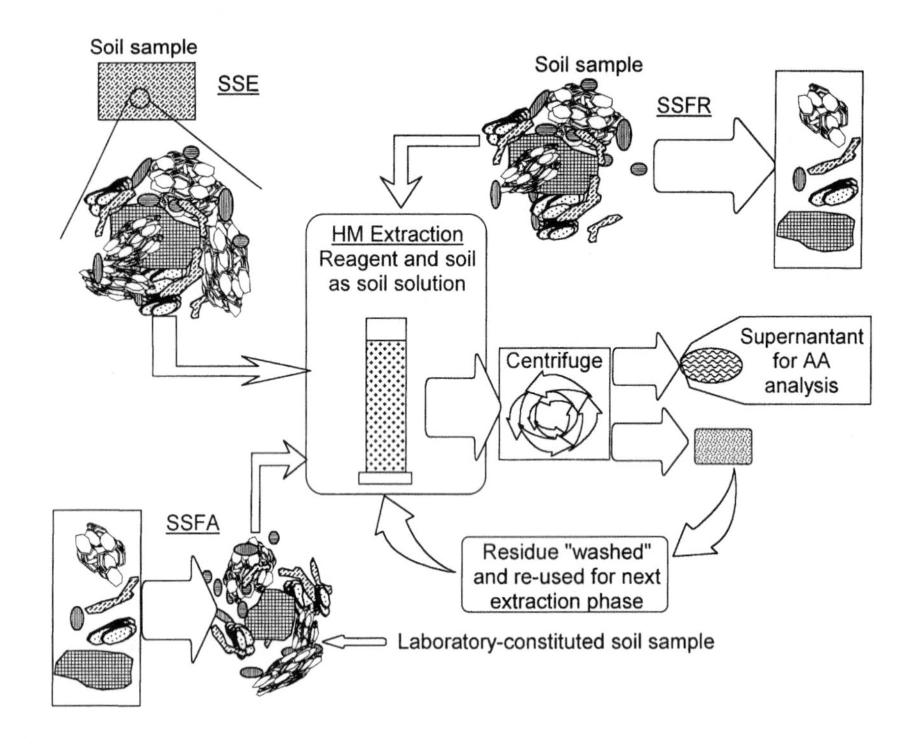

Figure 5.25 General scheme for study of partitioned HM pollutants using SSE, SSFR, and SSFA techniques.

The "Ret" curve shown in Figure 5.26 refers to the Pb retained (or Pb that can be extracted) using the SSFR procedure. In this SSFR sample, both the carbonates and amorphous oxides were removed using the methods described above. The Pb retained, or conversely, the Pb that can be extracted (if such were the case) is seen to be considerably larger at the lower pH values. The standard SSE type of test results shown in the same graph indicate that the amount of exchangeable ions extracted (identified as "Exc" in the graph) decreases from a high value at pH 3 to their lowest value at about pH 5.5. The largest amount of Pb is held by the carbonate soil fractions. The significant point of note is the large difference in Pb retained by the SSFR sample between the pH 3 and pH 5.5 range which is not accounted for in the SSE test. The large difference seen in Pb retention in the SSFR results or extracted (SSE results), especially at the lower pH range where Pb exists in solution below its precipitation pH, raises some very interesting questions. Both these methods of data gathering (SSFR and SSE) depend very highly on operational procedures and reagents used. Quantitative comparisons cannot be made with any accuracy.

The SSFR results obtained by Yong and MacDonald (1998), as shown in Figure 5.26, indicate that the differences in Pb retention with both carbonates and amorphous oxides removed are very small. However, the results shown by Xing

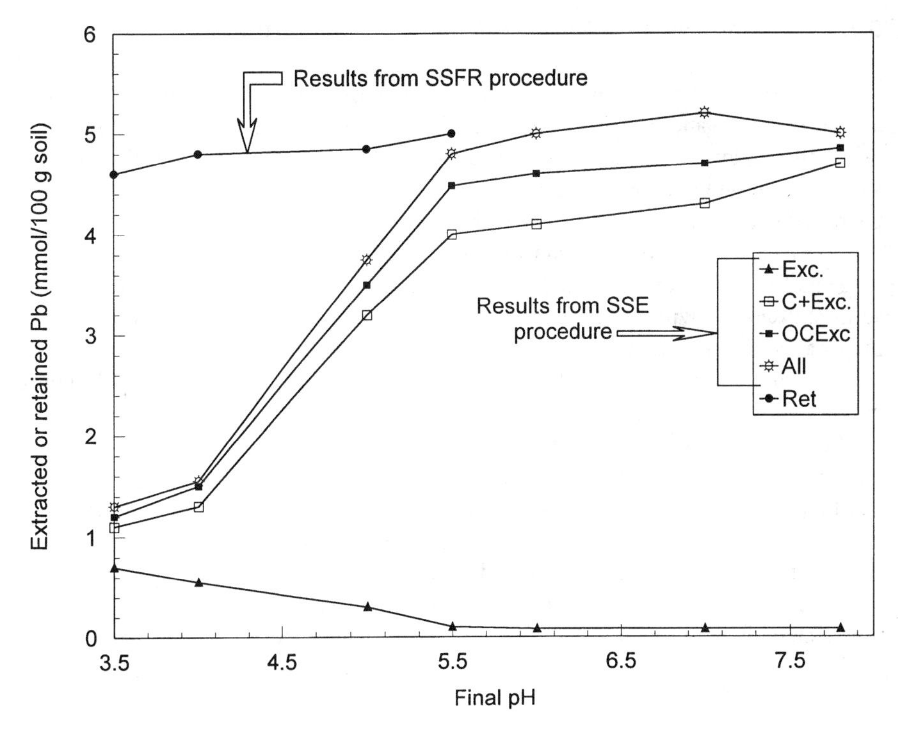

Figure 5.26 Comparison between SSFR and SSE assessment of Pb partitioning in an illite soil. Note Exc = exchangeable, C = carbonates, OC = amorphous oxides and carbonates, "Ret" = retained by SSFR sample. (Adapted from Yong and MacDonald, 1998.)

et al. (1995) in Figure 5.24, in respect to Cu and Cr and some of the other HMs, show distinct differences when carbonates and oxides were removed. Several hypotheses can be advanced to explain the differences obtained. To a very large extent, it is assumed that in reconstructing the soil after removal of a specific soil fraction, the resultant SSFR structure will be conditioned by what is left in the soil. Amorphous oxides and carbonates are known to provide coatings on soil particles and bridges between particles — dependent on the concentration of these soil fractions and on the pH of the immediate environment. We can assume that in the absence of dispersants used to fully disperse the soil solids in the soil solution, the different soil structures obtained in the SSFR constituted compact soil samples will be somewhat reflected in the soil solutions produced for the extraction procedure. The chances of some residual soil structural units (peds and such) acting as individual soil solids in the soil solution cannot be discounted. We need to recognize that both SSE and SSFR types of extraction tests can be conducted on actual soil samples retrieved from contaminated sites. Hence, whilst some residual soil structural units could participate in the production of the extraction test results, there are no means available at present for direct determination of such participation and their effects on the measured results.

5.5.2 Column Studies for Soil Structure and Partitioning

The preceding discussions show that the use of treatments involving soil solutions at one stage or another in the SSE, SSFR, and SSFA types of test is designed primarily to gain insight into the role of soil composition on retention of pollutants such as heavy metals. For determination of the role of soil structure on retention characteristics, we need to use leaching tests with soil columns and intact soil samples. This means that instead of using soil solutions and interactions between reagents for extraction of the pollutants, extraction of the heavy metal pollutants in soil column tests will be through leaching of extractants through the samples. The importance of soil structure in retention of pollutants can be illustrated by comparing the retention characteristics determined by batch equilibrium tests and by leaching column tests.

Figure 5.27 shows the basic elements of the proposition which says that if equilibrium sorption of the contaminant is obtained in a leaching column test, this can be plotted on the same diagram with the adsorption isotherm of that same soil obtained from batch equilibrium. At any one position in the leaching soil column, equilibrium sorption can be obtained when the sorption capacity of the soil is reached. When this capacity is reached, no further partitioning of the pollutants in the leachate being transported through the sample will occur, i.e., the contaminants

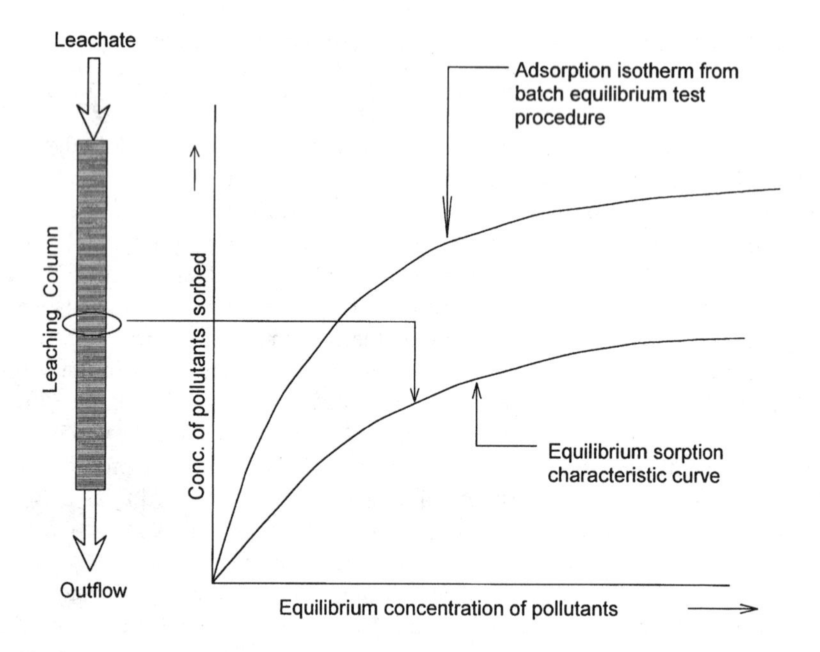

Figure 5.27 Comparison between adsorption isotherm from batch equilibrium test procedure and equilibrium sorption characteristic curve from leaching column test.

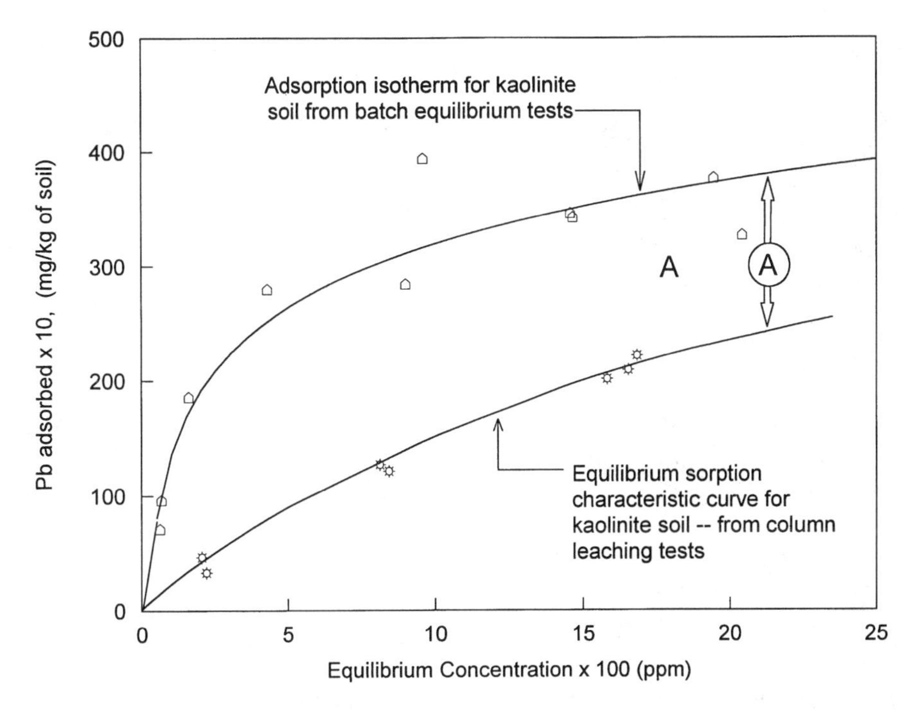

Figure 5.28 Kaolinite soil, adsorption isotherm and equilibrium sorption characteristic curve.

or pollutants will remain in the leachate as it is transported through the soil. At that time, the equilibrium sorption characteristic curve shown in Figure 5.27 will be obtained.

The adsorption isotherm obtained for the same soil and the same set of contaminants using batch equilibrium tests shown in the diagram represents the sorption performance of a totally dispersed soil in a soil solution. This provides us with maximum sorption of the pollutant under study. In contrast, since intact soil samples are used in the leaching column test, not all the reactive soil surfaces are available. Masking and changes of surface reactive forces due to coatings by oxides, and formations of peds and other microstructural features contribute to the significant changes in available reactive surfaces. This will result in lower sorption capacities, e.g., the equilibrium sorption characteristic curve shown in Figure 5.28. The test results for the kaolinite soil in the diagram include the double-headed arrow "A" drawn at the beginning of the region where the adsorption isotherm curve appears to be closely parallel to the leaching column sorption characteristic curve. This value "A" appears to be constant as one proceeds further to the right of the graph. When the sorption capacity of the soil is reached, both for batch equilibrium and leaching column tests, we would expect the sorption curves to exhibit asmyptotic behaviour. It is reasonable to conclude that if "A" is sensibly constant, or if the test results are extended so that the two curves would show a reasonably constant "A" value, this would represent the Pb retention modification due to the influence of soil structure. Thus, if different soil structures for the same soil are obtained, as for example

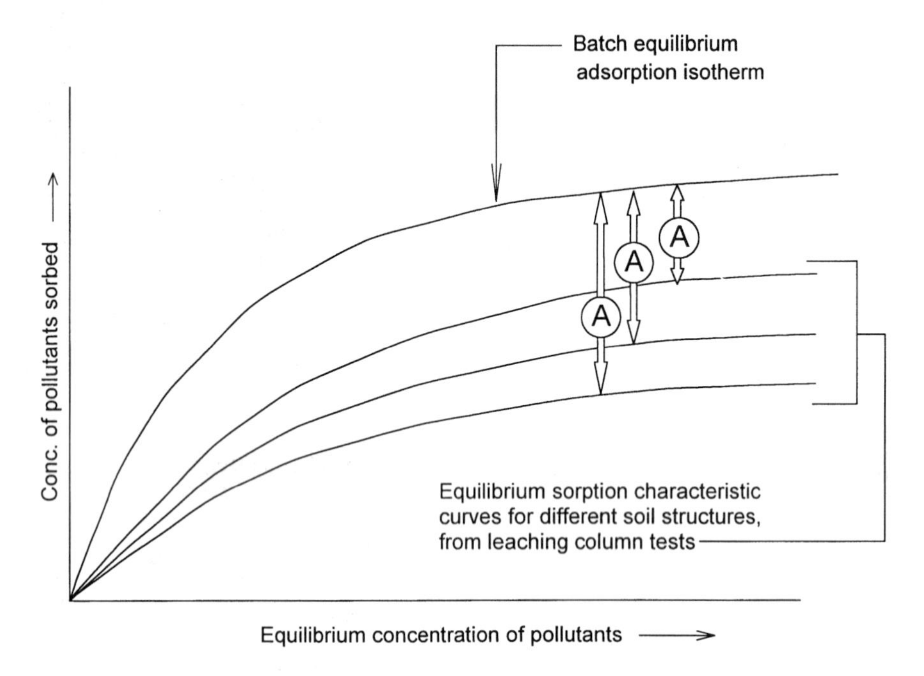

Figure 5.29 Soil structure influence on equilibrium sorption characteristic curves.

demonstrated as different densities, the results obtained would be shown as those represented in Figure 5.29.

The results shown in Figure 5.28 suggest that when equilibrium sorption is reached in column leaching tests, the reduction in sorption of Pb appears to be a constant proportion of the Pb sorbed in the batch equilibrium sorption tests. A strong case can be made for interpretation of the constant "A" value as a proportional reduction in the surface area presented to (i.e., interacting with) the Pb. In this particular instance, the results shown in Figure 5.28 appear to suggest a reduction of about 25% of the surface area presented by the batch equilibrium tests. If we assume that the batch equilibrium test procedure provides total particle dispersion and therefore all surfaces are exposed to interaction with Pb, this can be construed to mean that the surface area presented might perhaps be represented by the specific surface area SSA measured, for example, by the EGME method prescribed by Carter et al. (1986) and Eltantawy and Arnold (1973). For the case of sorption of Pb by the kaolinite sample, we can argue that the leaching sample used to provide the equilibrium sorption results shown in Figure 5.28 would have a specific surface area roughly equal to 75% of the SSA.

The effect of different soil structures obtained as a result of different soil densities (for the same soil), as represented in Figure 5.29 can thus be evaluated in terms of the changes in the surface areas presented to an intruding contaminant or pollutant carried in a leachate stream. Test results on sorption of different HMs have shown that preferential sorption occurs. This suggests that while we have seen a reduction of about 25% in the SSA presented by the kaolinite soil to Pb interaction, we need to conduct specific tests with different HM pollutants to produce the effective interaction SSA. The evidence suggests that because of the various mechanisms of interaction and retention, the effective interaction SSA presented to pollutants is a direct function of the soil structure, concentration of the particular pollutant, presence and distribution of other pollutants, and the pH and pE of the immediate environment.

5.6 CONCLUDING REMARKS

The information obtained in respect to the contribution made by each type of soil fraction (clay minerals, hydrous oxides, organics, carbonates, etc.) to overall retention or sorption of HM pollutants provides us with the ability to: (a) assess and discriminate between various soil types in regard to contaminant attenuation capabilities; (b) produce "designer" soil materials that could be more effective soil liners in engineered barrier systems; and (c) evaluate the fate of HMs in relation to the distribution and manner in which the HMs are held within the soil. We realize that soil structure must somehow have some influence on the pollution-retention (i.e., contaminant-retention) characteristics of a soil.

The use of extraction tests to provide information on retention of contaminants by the various soil fractions do not completely satisfy the requirements for analysis of the contributions from soil structure. Soil column leaching tests using intact soil samples have the potential to provide information on soil structure influences. Recognizing that all test results in batch equilibrium tests and in column leaching tests are operationally influenced and conditioned, it is still possible to use the results in a qualitative sense. The ramifications arising therefrom can be articulated as follows: (a) a means for rough determination of the reduction in SSA can be obtained, and (b) the role of soil structure in control of partitioning of contaminants and pollutants in leachate streams being transported through soil can be assessed.

It is not possible to provide quantitative information on the contribution made by each soil fraction in HM sorption in an intact soil unless the soil structure is well defined and understood. Figure 5.30 shows the change in iep of the ferrihydrite-kaolinite mixture discussed previously in Section 3.5 in Chapter 3. We recall, in the sketch given in Figure 3.18 that the distribution of the ferrihydrite in the mixture is sensitive to the pH of mixture-formation. The example shown in the previous figure demonstrated how the microenvironment (pH change) can affect the distribution of iron oxide in a soil composed of iron oxide and kaolinite particles. This difference in distribution is reflected in the change in the pH_{iep} — as witness the pH_{iep} of 5.3 and 6.6, depending on whether the mixture was formed at pH 9.5 or 3, respectively. If the mixing of the iron oxide with the kaolinite particles occurs at a pH of 3, the oxides would form coatings around the mineral particles, because of electrostatic attraction between the positive surface charges on the oxides and the net negative planar surface charges on the mineral particles. Changing the pH of the mixture from pH 3 to pH 9.5 does not change this coating arrangement.

The Pb concentration profiles shown in Figure 5.31 should be compared with the concentration profiles shown previously in Figure 5.15. By introducing silica gel as the amorphous material, we have enhanced the sorption characteristics of the

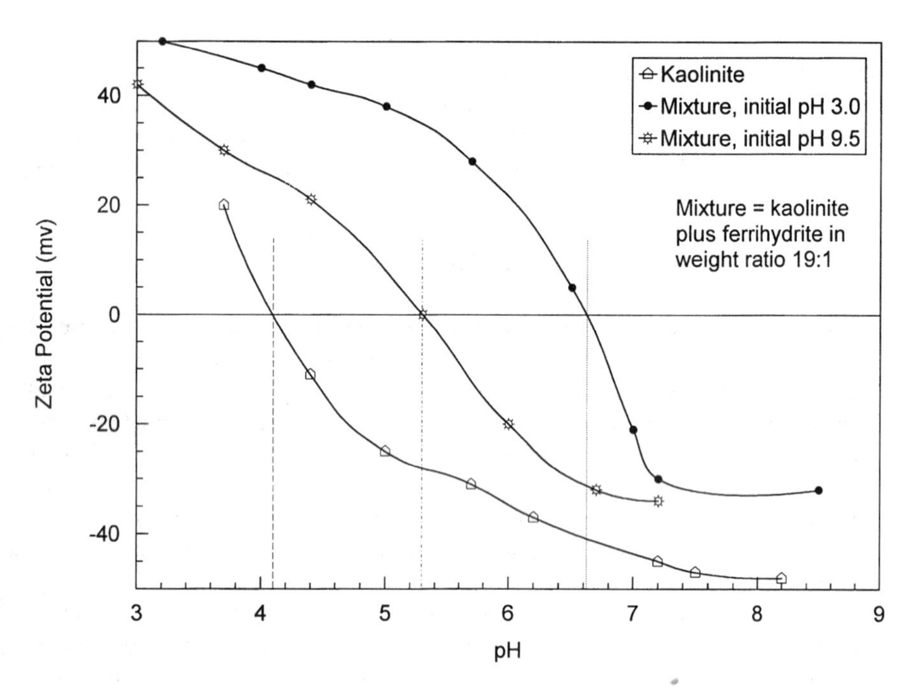

Figure 5.30 Influence of pH and amorphous material (ferrihydrite) on iep of kaolinite-ferrihydrite mixture.

same kaolinite shown in Figure 5.15. The amount of Pb sorbed by the mixture solids is considerably greater than the Pb in the porewater. The opposite is true for the kaolinite only sample shown in Figure 5.15. The various surface properties of the mixture of silica gel and kaolinite shown in the figure itself will tell us why the sorption characteristics of the mixture have been improved. As is immediately obvious from the figure and from the typical test results shown for determination of distribution of partitioned HMs, by conducting SSE, SSFR, and SSFA types of studies it is possible to obtain some insight into manner of distribution of the HMs. This type of information is useful not only in determining the kinds of soil materials to use in soil-engineered barriers, but also in assessing the environmental mobility of these kinds of inorganic pollutants. In addition, we can develop strategies for remediation, based on the knowledge of the distribution of the heavy metal pollutants (Mulligan et al., 2001).

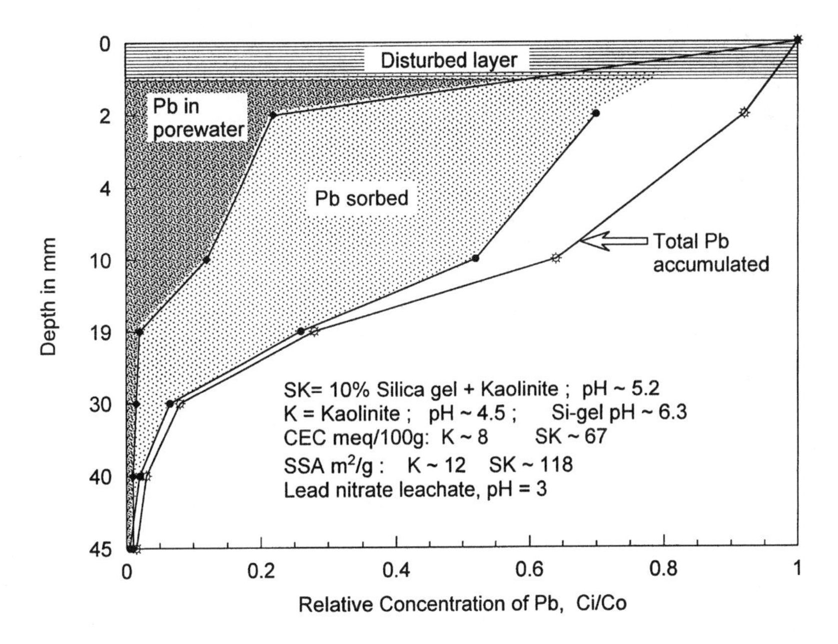

Figure 5.31 Influence of amorphous silica on structure of kaolinite-silica gel mixture and on retention of Pb.

CHAPTER 6

Persistence and Fate of Organic Chemical Pollutants

6.1 INTRODUCTION

Various kinds and forms of interactions occurring between organic chemicals (as pollutants) and the various soil fractions will participate in the determination of the fate of these pollutants. These interactions can be more complex than those previously described in interactions between inorganic pollutants and soil fractions. In soils contaminated by organic chemicals, the additional factor of microbial presence needs to be considered. Biotic redox plays a significant role in the determination of the persistence and fate of organic chemical pollutants. Since these chemicals are generally susceptible to degradation by biotic processes, determination of the fate of the pollutant chemicals is most often considered in terms of the resistance to degradation of the pollutants and/or their products. When evidence shows that a particular organic pollutant resists biodegradation, the pollutant is identified as a recalcitrant (organic chemical) pollutant, and the study of the fate of the pollutant includes determination of the persistence of the pollutant — see Section 6.4 for the definitions of *recalcitrance* and *persistence*.

The difficulties in seeking to determine the various abiotic and biotic processes responsible for pollutant fate and persistence lies not only with the means and methods for analyses, but also with the various dynamics of the problem. Whilst the records of numerous field studies show the presence of both organic and inorganic pollutants co-existing in a contaminated site, determination of the fate of these pollutants has generally focused on inorganic and organic chemicals as separate pollutants in the site. It is only recently that more detailed consideration has been given to the influence of one (e.g., inorganic) on the other (organic chemicals) in respect to control of the fate of these pollutants.

In the strictest sense, the persistence and fate of organic chemical pollutants in the soil substrate is controlled by, or is dependent on, such processes as: (a) chemical reactions between the chemicals themselves; (b) reactions with the various soil fractions; and (c) hydrolysis, photolysis, and biodegradation. However, for the purpose of this book, we will be considering the persistence and fate of organic chemical pollutants in respect to controls exercised in the soil through interactions with the soil constituents. Some attention to microbial activities will be paid as the occasion arises. The focus of this chapter will be on the fate of organic chemical pollutants as influenced by microenvironmental factors such as pH, ligands present, redox potential, nature of the soil fractions and their reactive surfaces, and the synergistic-antagonistic relationships established by the presence of the myriad of inorganic and organic contaminants.

In general, the results of interactions between soil fractions and pollutants include both organic and inorganic-driven processes such as:

1. **Sorption**, occurring principally as a result of ion-exchange reactions and van der Waals forces, and chemical adsorption (chemisorption), which involves short-range chemical valence bonds;
2. **Complexation** with inorganic and organic ligands;
3. **Precipitation**, i.e., accumulation of material (solutes, substances) on the interface of the soil solids to form new (insoluble) bulk solid phases; and
4. **Redox** reactions.

In addition to the characteristics and properties of the soil fractions and pollutants, microenvironmental conditions will dictate which of the processes may be more dominant than the others. Distinguishing between physical (electrostatic and electromagnetic) and chemical adsorption, and the results of the various processes contributing to the binding of organic chemical pollutants to soil fractions is not easy. The various processes and mechanisms will be examined in the next few sections.

6.2 ADSORPTION AND BONDING MECHANISMS

As in the case of the inorganic pollutants discussed in Chapter 5, adsorption reactions or processes involving organic chemicals and soil fractions are governed by: (a) the surface properties of the soil fractions; (b) the chemistry of the porewater; and (c) the chemistry and physical-chemistry of the pollutants. We recall that in the case of inorganic pollutants, the net energy of interaction due to adsorption of a solute ion or molecule onto the surfaces of the soil fractions is the result of both short-range chemical forces such as covalent bonding, and long-range forces such as electrostatic forces. Adsorption of inorganic contaminant cations is related to their valencies, crystalinities, and hydrated radii.

By and large, organic chemical compounds develop mechanisms of interactions that are somewhat different from those given previously in Table 5.1. Consider the transport of PHCs (petroleum hydrocarbons) in soils as a case in point. Interaction between oil and soil surfaces is important in predicting the oil retention capacity of the soil and the bioavailability of the oil. (We define *bioavailability* as the degree to which a pollutant is available for biologically mediated transformations.) The interaction mechanisms are influenced by soil fractions, the type of oil, and the presence of water. As in the case of inorganic contaminant-soil interaction, the existence of surface active fractions in the soil such as soil organic matter (SOM), amorphous materials, and clays can significantly enhance oil retention in soils — to a very large extent because of large surface areas, high surface charges, and surface characteristics.

The problem of first wetting is most important in the case of organic chemical penetration into the soil substrate. The nature of the liquid that surrounds or is made available to the dry surfaces of the soil fractions is critical for subsequent bonding of contaminants — inorganic or organic. Alcohols, for example, which have *OH* functional groups, are directly coordinated to the exchangeable cations on soil mineral particle surfaces when these particles are dry. However, with the presence of water (i.e., when the soil is wet), since the cations are hydrated, the attachment of the alcohols to the soil particle surfaces is through water bridges.

We have seen from the previous chapters that for the inorganic contaminants and pollutants, diffuse ion-layers and Stern layers can be well developed, and evaluations of transport and fate of the contaminants can be made with the aid of the DDL models. If the surfaces of the soil solids are first wetted with water, the development of the Stern layer will influence and affect soil-oil bonding relationships, and the amount of oil associated with the soil fractions will decrease in proportion to the amount of first wetting, i.e., in proportion to the extent of Stern layer development (amount of water layers surrounding the soil particle surfaces). Because of their low aqueous solubilities and large molecular size, penetration into the Stern layers is not easily achieved by many organic chemicals, e.g., the effective diameter of various hydrocarbon molecules varies from 1 to 3 nm for a complex hydrocarbon type in contrast to a water molecule which has a diameter of approximately 0.3 nm. Thus, it is very important that determination of retention of hydrocarbons (HCs) and most NAPLs (non aqueous-phase liquids) must consider first wetting and residual wetting of the soil-engineered barriers and soil substrate.

Research results from tests with organic chemical pollutants in leaching and fluid conductivity experiments have often shown significant shrinkage in the soil samples tested. Suggestions have been made concerning the inability of the diffuse double layers (DDL) to fully develop. Interaction of clay minerals with organic chemicals with dielectric constants lower than water will result in the development of thinner interlayer spacing because of the contraction of the soil-water system. We can consider the transport of organic molecules through the soil substrate as being by diffusion and advection through the macropores, with partitioning between the pore-aqueous phase and soil fractions occurring throughout the flow region. The weakly adsorbed molecules will tend to move more quickly through the connected aqueous channels. Hydrophobic substances such as heptane, xylene, and aniline, which are well partitioned, will develop resultant soil-organic chemical permeabilities that will be much lower than the corresponding soil-water permeability. By and large, organic fluid transport in soil is conditioned not only by the hydrophobicity or hydrophilic nature of the fluid, but also by other properties such as the dielectricity of the substance. This will be further evident from the examination of the partitioning of organic chemicals during, and as a result of, transport in the soil.

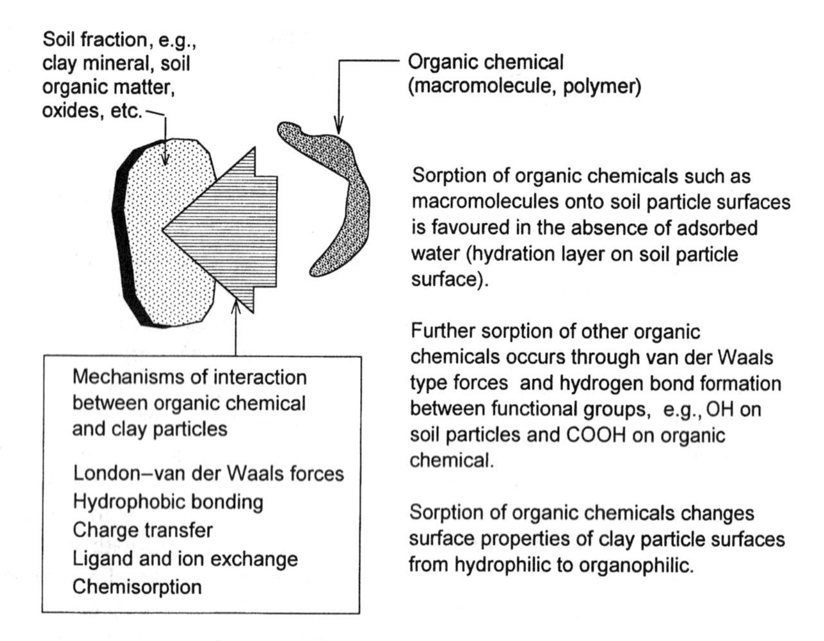

Figure 6.1 Examples of some mechanisms of interactions between organic chemical pollutants and clay particles.

6.2.1 Intermolecular Interactions

The interactions occurring at the intermolecular level that contribute directly to the mechanisms for "binding" organic chemicals to soil fractions can be physically motivated, chemically motivated, or exchange motivated. These processes are shown in simple schematic form in Figure 6.1. Whilst not all of these are included in the sketch, the basic sets of forces, reactions, and processes that constitute the major sets of interactions include:

- London-van der Waals forces;
- Hydrophobic reactions;
- Hydrogen bonding and charge transfer;
- Ligand and ion exchanges; and
- Chemisorption.

The London-van der Waals forces consist of three types: (a) Keesom forces developed as a result of instantaneous dipoles resulting from fluctuations in the electron distributions in the atoms and molecules; (b) Debye forces developed as a result of induction; and (c) London dispersion forces. Whilst the London-van der Waals influence decreases in proportion to the inverse of the sixth power of the separation distance R between molecules, i.e., proportional to $1/R^6$, the result of their interactions can lead directly to disruption of the liquid water structure immediately next to the soil solids. This leads to the development of entropy-generation hydrophobic bonding. Larger-sized organic molecules tend to be more favourably adsorbed because of the greater availability of London-van der Waals forces.

Hydrophobic reactions contribute significantly to the bonding process between these chemicals and soil fractions — particularly soil organic matter. The tendency for organic chemical molecules to bond onto hydrophobic soil particle surfaces, such as soil organic matter, is in part because this will result in the least restructuring of the pre-existing water structure in the soil pores. This phenomenon allows for water in the vicinity of the organic chemical to continue its preference for association with itself (i.e., water-to-water attachment) as opposed to being in close proximity with the hydrophobic moiety of the organic chemical. This type of interaction results in the development of organic-soil particle bonding, which is referred to as *hydrophobic bonding.*

Charge transfers, or more specifically charge transfer complex formation (of which hydrogen bonding is a special case), are complexes formed between electron-donor and electron-acceptor molecules where some overlapping of molecular orbitals occurs together with some exchange of electron densities (Hamaker and Thomson, 1972). These transfer mechanisms appear to be involved in bonding between chemicals and soil organic matter because of the presence of aromatic groups in humic acids and humins. In the case of hydrogen bonding, the hydrogen atom provides the bridging between two electronegative atoms (Dragun, 1988) via covalent bonding to one and electrostatic bonding to the other (atom).

For ligand exchange to occur as a sorption (binding) process, it is necessary for the organic chemical to have a higher chelating capacity than the replaced ligand. Humic acids, fulvic acids, and humins are important soil fractions in such exchanges and also in ion exchange phenomena. Because organic ions can be hydrophobic structure makers or breakers, the structure of water becomes an important factor in establishing the extent and rate of ion exchange sorption phenomena. As in the case of electrostatic interactions and chemical sorption between inorganic pollutants and soil fractions, the ionic properties of the organic ion are significant features that require proper characterization. This will be considered further when the influence of functional groups is examined.

Ion exchange mechanisms involving organic ions are essentially similar to those that participate in the interaction between inorganic pollutants and soil fractions. Because molecular size is a factor, the structure of water immediately adjacent to the soil particle surfaces becomes an important issue in the determination of the rate and extent of sorption — similar to the processes associated with ligand exchange. Fulvic acids are generally hydrophilic and thus produce the least influence on the structuring of water. This contrasts considerably with humins which are highly hydrophobic, i.e., these play a high restructuring role in the water structure.

It is a mistake to assume or expect that bonding relationships between organic pollutants and soil fractions at the intermolecular level are the result of any one process. Because of the different types of reactive surfaces represented by the various soil fractions, and because of the variety in functional groups for both the organic chemical pollutants and the soil fractions, it is reasonable to expect that bonding between the pollutants and soil will comprise more than one type of process, e.g., ion exchange and hydrophobic bonding.

6.2.2 Functional Groups and Bonding

A simple initial characterization of organic chemical pollutants distinguishes between organic acids/bases and non-aqueous phase liquids. The latter (i.e., NAPLs) are liquids that exist as a separate fluid phase in an aqueous environment, and are not readily miscible with water. They are generally categorized as NAPL densities greater than (DNAPLs) or less than water (LNAPLs). Because DNAPLs are heavier than water, they have a tendency to plunge all the way downward in the substrate until progress is impeded by an impermeable boundary (see Figure 4.3). The major constituents in the DNAPL family in soils include those associated with anthropogenic sources, e.g., chlorinated hydrocarbons such as PCBs, carbon tetrachloride, 1,1,1-trichloroethane, chlorophenols, chlorobenzenes, and tetrachloroethylene. The chemistry of the soil porewater is influential in the partitioning processes, i.e., processes that remove the solutes from the porewater phase to the surfaces of the soil fractions. The bonding relationships between organic chemical pollutants and soil fractions are controlled not only by the constituents in the porewater (inorganic and organic ligands), but also by the chemically reactive groups of the pollutants and the soil fractions.

The functional groups for soil fractions and organic chemical compounds (pollutants), which are chemically reactive atoms or groups of atoms bound into the structure of a compound, are either acidic or basic. As noted in Chapter 4, the nature of organic compounds is considerably different from the soil fractions — except for the soil organic matter. In the case of organic chemicals, the nature of the functional groups in the (organic) molecule, shape, size, configuration, polarity, polarizability, and water solubility are important in the adsorption of the organic chemicals by the soil fractions. Since many organic molecules (amine, alcohol, and carbonyl groups) are positively charged by protonation (adding a proton or hydrogen), surface acidity of the soil fractions becomes very important in the adsorption of these ionizable organic molecules. The adsorption of the organic cations is related to the molecular weight of the organic cations. Large organic cations are adsorbed more strongly than inorganic cations by clays because they are longer and have higher molecular weights. Depending on how they are placed, and depending on the pH and chemistry of the soil-water system, the functional groups will influence the characteristics of organic compounds, and will thus contribute greatly in the development of the mechanisms which control accumulation, persistence, and fate of these compounds in soil.

Whilst the hydroxyl functional group is the dominant reactive surface functional group for most of the soil fractions (clay minerals, amorphous silicate minerals, metal oxides, oxyhydroxides, and hydroxides), the soil organic matter (SOM) will contain many of the same functional groups identified with organic chemicals, e.g., hydroxyls, carboxyls, carbonyls, amines, and phenols, as shown previously in

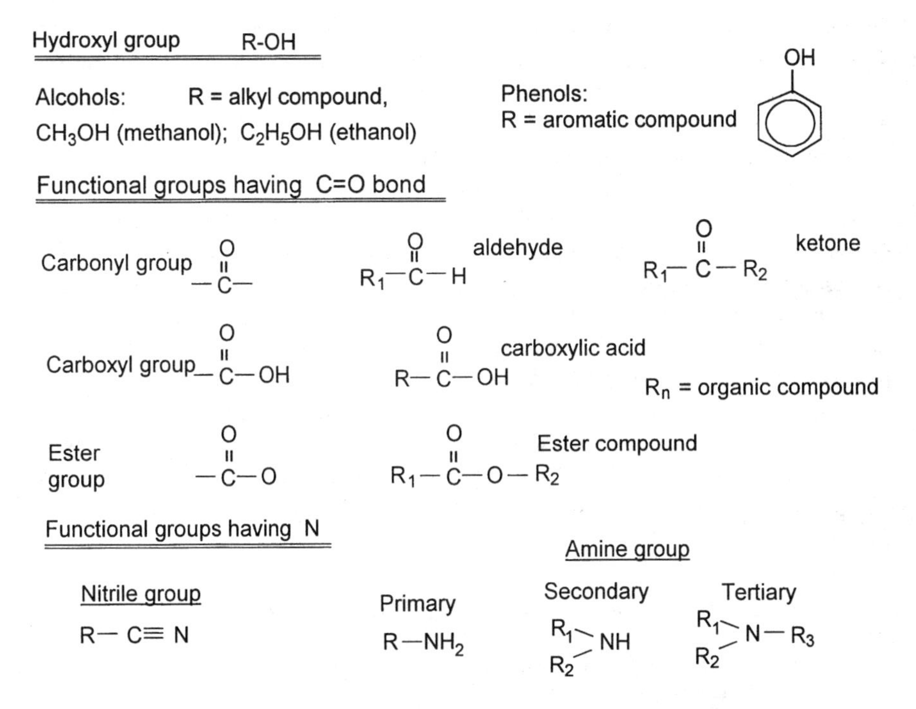

Figure 6.2 Some common functional groups for organic chemical pollutants.

Figure 3.2 and Table 3.2. For organic chemical pollutants, the hydroxyl functional group is present in two broad classes of compounds:

1. Alcohols, e.g., methyl (CH_3–), ethyl (C_2H_5–), propyl (C_3H_7–), and butyl (C_4H_9–);
2. Phenols, e.g., monohydric (aerosols) and polyhydric (obtained by oxidation of acclimatised activated sludge, i.e., pyrocatechol, trihydroxybenzene.

Alcohols are hydroxyl alkyl compounds (R– OH), with a carbon atom bonded to the hydroxyl group. The more familiar ones are CH_3OH (methanol) and C_2H_5OH (ethanol), as seen in Figure 6.2. Phenols, on the other hand, are compounds which possess a hydroxyl group attached directly to an aromatic ring.

Alcohols are considered to be neutral in reaction since the *OH* group does not ionize. Adsorption of the hydroxyl groups of alcohol can be obtained through hydrogen bonding and cation-dipole interactions. Most primary aliphatic alcohols form single-layer complexes on the negatively charged surfaces of the soil fractions, with their alkyl chain lying parallel to the surfaces of the soil fractions. Double-layer complexes are also possible with some short-chain alcohols such as ethanol. Alcohols acts as acids when they lose their *OH* proton and will act as bases when their oxygen atom accepts a proton.

In the group of organic chemicals with carbon-oxygen double bonds (*C=O* carbonyl functional group), we should note that the *C=O* bonds are polarized due to the high electro-negativity of the oxygen *O* relative to the carbon *C*. This is because of the greater electron density over the more electronegative oxygen atom. The *C* functions as an electrophilic site and the *O* is in essence a nucleophilic site. We could say that the electrophilic site is a Lewis acid and the nucleophilic site is a Lewis base.

Organic chemical pollutants with: (a) functional groups having a *C*=*O* bond, e.g., carboxyl, carbonyl, methoxyl, and ester groups, and (b) nitrogen-bonding functional groups, e.g., amine and nitrile groups, are fixed or variable-charged organic chemical compounds. They can acquire a positive or negative charge through dissociation of H^+ from or onto the functional groups, dependent on the dissociation constant of each functional group and the pH of the soil-water system. The fate of organic chemical pollutants can be significantly affected when a high pH regime replaces an original lower pH regime in the soil. As with the case of organic compounds with *OH* functional groups, a high pH regime will cause these functional groups (i.e., groups having a *C*=*O* bond) to dissociate. The release of H^+ (dissociation) would result in the development of negative charges for the organic chemical compounds, as shown for example by a carboxyl compound and an alcohol as follows:

$$R - COOH \rightleftharpoons R - COO^- + H^+$$

$$R - OH \rightleftharpoons R - O^- + H^+$$

where R represents any chemical structure (e.g., hydrocarbon moiety) and COOH is the carboxyl functional group. If cation bonding was initially responsible for sorption between organic chemicals and the soil fractions, charge reversal (i.e., to negative charges) will result in the possible release of the organic chemical pollutant. When this happens, the released organic chemical pollutant could be sorbed by those soil fractions which possess positive-charged surfaces, e.g., edges of kaolinites, oxides, and soil organics. If such soil fractions are unavailable, the pollutants will be free to move. This situation is not desirable since it represents a classic case of environmental mobility of pollutants.

Carbonyl compounds (aldehydes, ketones, esters, amides, and carboxylic acids) are often obtained as products of photochemical oxidation of hydrocarbons. They most often possess dipole moments because the electrons in the double bond are unsymmetrically shared. Aldehydes have one hydrocarbon moiety (R) and a hydrogen atom (H) attached to the carbonyl (*C*=*O*) group as shown in Figure 6.2. They can be oxidized to form carboxylic acids. Ketones, on the other hand, have two hydrocarbon moieties (R and R_1) attached to the carbonyl group. Whilst they can accept protons, the stability of complexes between carbonyl groups and protons is considered to be very weak. The carboxyl group of organic acids (benzoic and acetic acids) can interact either directly with the interlayer cation or by forming a hydrogen bond with the water molecules coordinated to the exchangeable cation associated with the soil fractions. Adsorption of organic acids depends on the polarizing power of the cation. Because of their ability to donate hydrogen ions to form basic substances, most carboxyl compounds are acidic, weak acids, as compared to inorganic acids.

The amino functional group NH_2 is found in primary amines. Much in common with alcohols, amines are highly polar and are more likely to be water-soluble. Their chemistry is dominated by the lone-pair electrons on the nitrogen, rendering them nucleophilic. As shown in Figure 6.2, the amino group consists of primary, secondary, and tertiary amines depending on the nature of the organic compound R_n. They can be adsorbed with the hydrocarbon chain perpendicular or parallel to the reactive surfaces of the soil fractions, depending on their concentration. The phenolic functional group, which consists of a hydroxyl attached directly to a carbon atom of an aromatic ring, can combine with other components such as pesticides, alcohol, and hydrocarbons to form new compounds, e.g., anthranilic acid, cinnamic acid, ferulic acids, gallic acid, and *p*-hydroxy benzoic acid.

The various petroleum fractions in petroleum hydrocarbons (PHCs) are primarily constituted by non-polar organics with low dipole moments (generally less than one), and dielectric constants less than three. Adsorption of nonionic organic compounds by soil fractions is governed by the CH activity of the molecule; the CH activity arises from electrostatic activation of the methylene groups by neighbouring electron-withdrawing structures, such as $C{=}0$ and $C{=}N$. Molecules possessing many $C{=}0$ or $C{=}N$ groups adjacent to methylene groups would be more polar and hence more strongly adsorbed than those compounds in which such groups are few or absent.

The chemical structures of petroleum hydrocarbons such as monocyclic aromatic hydrocarbons (MAHs) and polycyclic aromatic hydrocarbons (PAHs), shown in Figure 6.3 for example, indicate that there are no electron-withdrawing units such as $C{=}0$ and $C{=}N$ associated with the molecules. Accordingly, the PHC molecules would be weakly adsorbed (mainly by van der Waals adsorption) by the soil functional groups, and do not involve any strong ionic interaction with the various soil fractions.

Weakly polar (resin) to non-polar compounds (saturates and aromatic hydrocarbons) of PHCs develop different reactions and bonding relationships with the surfaces of soil fractions. Weakly polar compounds are more readily adsorbed onto soil surfaces in contrast to non-polar compounds. The adsorption of non-polar compounds onto soil surfaces is dominated by weak bonding (van der Waals attraction) and is generally restricted to external soil surfaces, primarily because of their low dipole moments (less than 1) and their low dielectric constants (less than 3) (Yong and Rao, 1991). Aqueous solubility and partition coefficients are important factors which control the interactions of organic compounds. Most hydrocarbon molecules are hydrophobic and have low aqueous solubilities. As shown in the next section, partitioning of PHCs onto soil surfaces occurs to a greater extent than in the aqueous phase. This results in lower environmental mobility and higher retention of the PHCs.

Studies on the desorption of PHCs using soil column leaching tests show that these can be desorbed as an aqueous phase or as a separate liquid phase (i.e., non-aqueous phase liquid — NAPL). Figure 6.4 shows the results of a leaching cell experiment with a clayey silt contaminated with 4% (by weight) PHC. The water solubility of the PHC is a significant controlling factor in determination of whether the PHC is desorbed as an aqueous phase or as a NAPL. As can be seen in Figure 6.3, the water solubility (ws) of the different PHC types varies considerably. When the

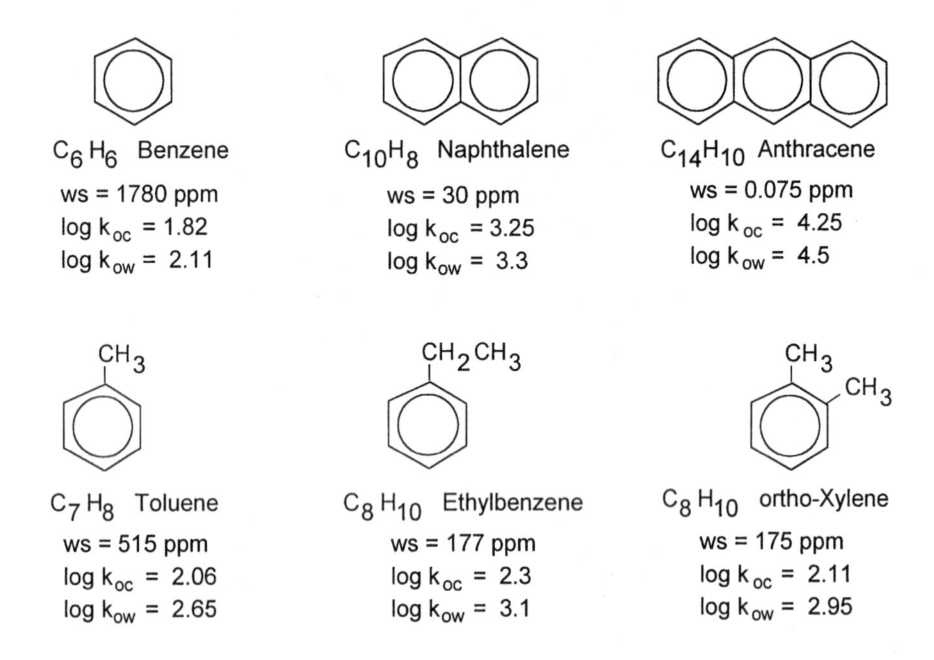

Figure 6.3 Typical petroleum hydrocarbon (PHC) compounds and their log k_{oc}, log k_{ow}, and water solubility (ws) values.

desorbed PHC remains as a NAPL, viscosity and surface wetting properties are critical. Light hydrocarbons are more likely to volatilize and be leached, whereas heavier constituents will tend to be retained in the soil fractions.

6.3 PARTITIONING OF ORGANIC CHEMICAL POLLUTANTS

The distribution of organic chemical pollutants between soil fractions and porewater is generally known as *partitioning.* By this, we mean that the chemical pollutants are partitioned such that a portion of the pollutants in the porewater (aqueous phase) is removed from the aqueous phase. We have seen from the study of partitioning of heavy metals that this assumption of sorption by the soil fractions may not be totally valid. This is because precipitation of the heavy metals will also serve to remove the heavy metals from solution. Since we do not have equivalent precipitation mechanisms for organic chemical pollutants, it is generally assumed that the total partitioned organic chemicals are sorbed or attached to the soil solids. The partitioning or distribution of the organic chemical pollutants is described by a coefficient identified as k_d, much similar to that used in the description of partitioning of HM pollutants in the previous chapter. As defined previously, this coefficient refers to the ratio of the concentration of pollutants held by the soil fractions to the concentration of pollutants remaining in the porewater (aqueous phase), i.e., $C_s = k_d C_w$, where C_s

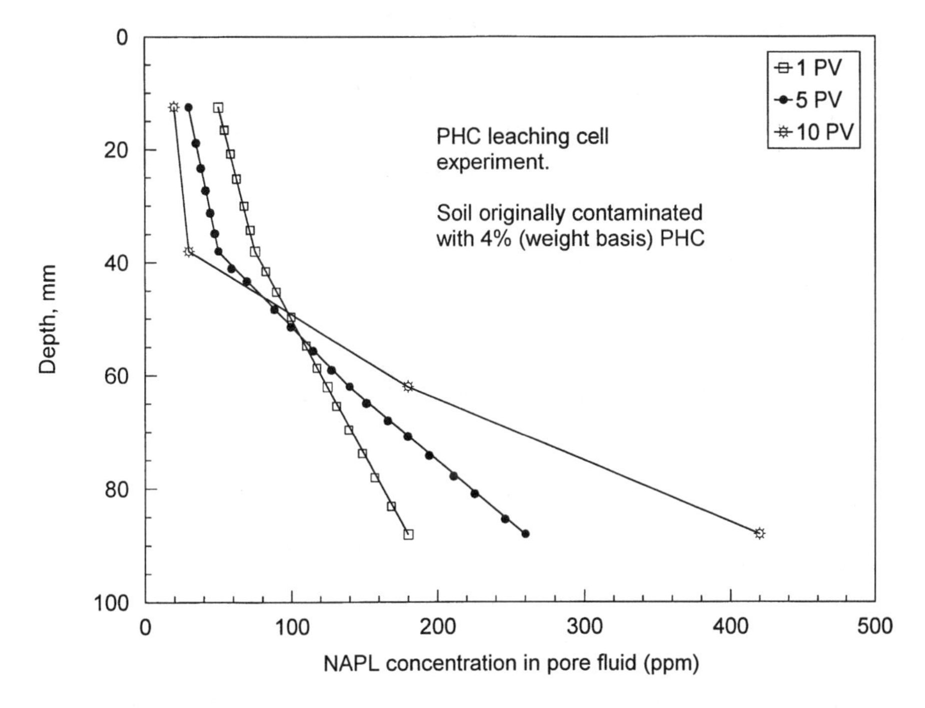

Figure 6.4 Results from leaching cell experiments on a clayey silt contaminated with PHC.

refers to the concentration of organic pollutants sorbed by the soil fractions, and C_w refers to the concentration remaining in the aqueous phase (porewater), respectively.

6.3.1 Adsorption Isotherms

The partitioning of organic chemical pollutants in the soil is not the result of a single interaction mechanism or one type of process between pollutants and soil fractions. Many processes contribute to the partitioning of the pollutants. The partitioning coefficient k_d is generally obtained using procedures similar to those described in Chapter 5 in respect to adsorption isotherms. The soil-suspension tests utilize target pollutants and specified (or actual) soil fractions. Figure 6.5 shows three classes of adsorption isotherms describing the partitioning behaviour of organic chemicals.

The general Freundlich isotherm given previously as Equation 4.11 is used to characterize the three classes.

$$C_s = k_1 C_w^n \qquad (6.1)$$

To avoid confusion with the isotherms used previously in the inorganic pollutant sorption tests, we will use the relationship shown in the graph depicted in Figure 6.5. As before, denoting C_s and C_w as the organic chemical sorbed by the soil fractions and remaining in the aqueous phase, respectively, the k_1 and n terms are better known

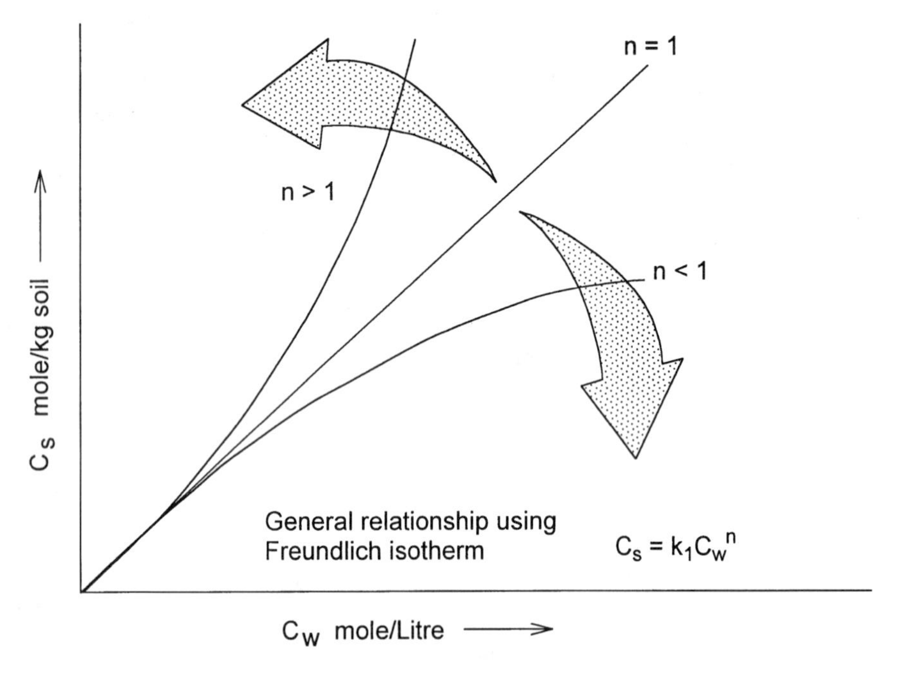

Figure 6.5 Categories of adsorption isotherm for organic chemical sorbed onto soil fractions. The shape of the curves are essentially defined by n.

as Freundlich constants. Previously, in Section 4.8.1 these were identified as k_1 and k_2, respectively. The relationship shown in Equation 6.1 is identical to Equation 4.11. The parameter n is associated with the nature of the slope of any of the curves shown in Figure 6.5. When $n = 1$, linearity is obtained, and one concludes therefrom that the sorption of the chemical pollutant by the soil fractions is a constant proportion of the available pollutant. When $n < 1$, the sorbed chemical pollutant decreases proportionately as the available pollutant increases, suggesting therefore that all the available mechanisms for sorption are being exhausted. However, when $n > 1$ we obtain the reverse situation. For such a situation to exist, enhancement of the sorption capacity of the soil must result from sorption of the chemical pollutant, i.e., sorption of the chemical pollutant increases the capacity of the soil to proportionately sorb more pollutants. These are shown in the adsorption isotherm test data from Hibbeln (1996) for a PAH and substituted PAHs such as naphthalene, 2-methyl naphthalene, and 2-naphthol (Figure 6.6).

We should recognize, as we did in Chapter 4, that the case of $n > 1$ in the Freundlich relationship has a limiting condition, i.e., it is not reasonable to expect that organic pollutants will be sorbed in ever increasing amounts without limit. Because the properties of both organic chemicals and soil fractions participate in this sorption process, and because the distribution of soil fractions and organic chemicals are also participants in this total process, it is difficult to establish where and what these limits are without systematic characterization experiments.

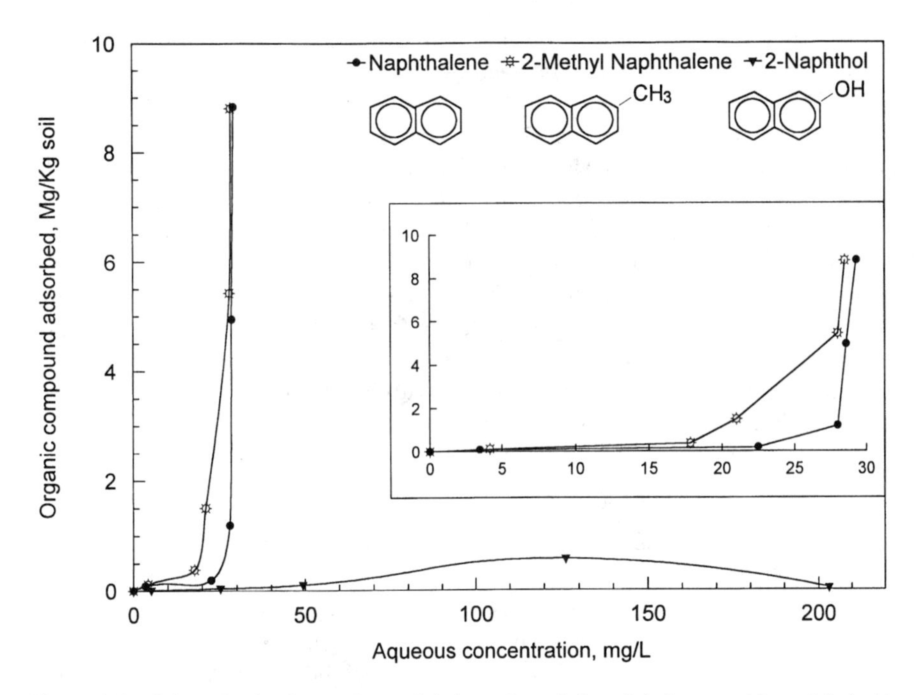

Figure 6.6 Adsorption isotherms for naphthalene, 2-methyl naphthalene, and 2-naphthol with kaolinite as the soil medium. Inset in Figure is the "enlarged" view of the isotherms for naphthalene and 2-methyl naphthalene. (Data from Hibbeln, 1996.)

The water solubility of an organic chemical pollutant is of significant importance in the control of the fate of the pollutant. Organic molecules, by and large, demonstrate less polar characteristics than water, and their varied nature (size, shape, molecular weight, etc.) render them as being considerably different than water. The water solubility of organic molecules will influence or control the partitioning of the organic pollutant, and the transformations occurring as a result of various processes associated with oxidation/reduction, hydrolysis, and biodegradation. The results shown in Figure 6.6 are a case in point. Both the naphthalene ($C_{10}H_8$) and 2-methyl naphthalene ($C_{11}H_{10}$) have water solubilities that are closely similar, e.g., 30 mg/L and 25 mg/L, respectively. In contrast, the water solubility of the 2-naphthol ($C_{10}H_8O$) is about between 25 to 30 times larger than the naphthalene and 2-methyl naphthalene, respectively. As might be intuitively expected, the higher water solubility allows for a greater amount of chemical pollutant to be retained in the aqueous phase. This will result in lower sorption by the soil solids (curves for naphthelene and 2-methyl naphthalene shown in Figure 6.6).

6.3.2 Equilibrium Partition Coefficient

The *equilibrium partition coefficient*, i.e., coefficient pertaining to the ratio of the concentration of a specific organic pollutant in other solvents to that in water, is considered to be well correlated to water solubilities of most organic chemicals.

Chiou et al. (1977), for example, reported good correlations between solubilities of organic compounds and their *n*-octanol-water partition coefficient k_{ow}. Because *n*-octanol is part lipophilic and part hydrophilic (i.e., it is amphiphilic), it has the ability to accommodate organic chemicals with the various kinds of functional groups shown in Figure 6.2. The dissolution of *n*-octanol in water is roughly eight octanol molecules to 100,000 water molecules in an aqueous phase, a ratio of about 1 to 12,000 (Schwarzenbach et al., 1993). Water-saturated *n*-octanol has a molar volume of 0.121 L/mol as compared to 0.16 L/mol for pure *n*-octanol. This close similarity allows us to ignore the effect of the water volume on the molar volume of the organic phase in experiments conducted to determine the octanol-water equilibrium partition coefficient.

The relationship for the *n*-octanol-water partition coefficient k_{ow} given in terms of the solubility S (Chiou et al., 1982) is seen in Equation 6.2:

$$\begin{aligned} \log k_{ow} &= 4.5 - 0.75 \log S \qquad (ppm) \\ \log k_{ow} &= 7.5 - 0.75 \log S \qquad (ppb) \end{aligned} \tag{6.2}$$

The k_{ow} octanol-water partition coefficient has been widely adopted as a significant parameter in studies of the environmental fate of organic chemicals. It has been found to be sufficiently correlated not only to water solubility, but also to soil sorption coefficients. In the experimental measurements reported, the octanol is considered to be the surrogate for soil organic matter. Organic chemicals with low k_{ow} (e.g., less than 10) may be considered to be relatively hydrophilic. They tend to have high water solubilities and small soil adsorption coefficients. Conversely, chemicals with a high k_{ow} value (e.g., greater than 10^4) are very hydrophobic and are not very water-soluble (i.e., they have low water solubilities). Solvent systems that are almost completely immiscible (e.g., alkanes-water) are fairly well behaved, and if the departures from ideal behaviour exhibited by the more polar solvent systems are not too large, a thermodynamic treatment of partitioning can be applied to determine the distribution of the organic chemical without serious loss of accuracy.

Aqueous concentrations of hydrophobic organics such as polyaromatic hydrocarbons (PAHs), compounds such as nitrogen and sulphur heterocyclic PAHs, and some substituted aromatic compounds indicate that the accumulation of the hydrophobic chemical compounds is directly correlated to the organic content (soil organic matter SOM) of a soil. A large proportion (by weight) of SOM is carbon, and as we have noted in Table 3.2 and Figure 3.2, the SOM functional groups are similar to most of the organic chemicals. They (SOM) occupy a position inbetween water and hydrocarbons insofar as polarity is concerned. Because of their composition and structure, they are well suited for hydrophobic bonding with organic chemical pollutants.

Studies have shown that whereas the variability in sorption coefficients between different soils may be due to characteristics of soil fractions (surface area, cation exchange capacity, pH, etc.), and the amount and nature of the organic matter present, a good correlation of sorption can be obtained with the proportion of organic carbon in the soil. The partition coefficient k_{ow} can be related to the organic content coefficient k_{oc}. The organic carbon content in soil organic matter can be used to characterize the k_d performance. Amongst the relationships commonly used are (Olsen and Davis, 1990):

$$k_{oc} = \frac{k_d}{f_{oc}} = 1.724\ k_{om} \tag{6.3}$$

where f_{oc} refers to the organic carbon content (dimensionless) in the SOM, and k_{om} refers to the partition coefficient expression using the (soil) organic matter content. Values for k_d and k_{oc} for numerous organic chemicals can be found in the various handbooks detailing the environmental data for such chemicals. A sampling of these (log k_{oc}) can be seen in Figure 6.3. We should note that soils with very low organic matter content (less than 1% by weight) will tend to give high values for k_{oc} because of the competing sorption processes offered by the other soil fractions in the soil. Because of that, Equation 6.3 should not be used when $f_{oc} < 1$. McCarty et al. (1981) give a critical minimum level for the organic carbon content $f_{oc\text{-}cr}$ as:

$$f_{oc-cr} = \frac{SSA}{200(k_{ow})^{0.84}} \tag{6.4}$$

where *SSA* denotes the specific surface area of the soil.

The graphical relationship shown in Figure 6.7 uses some representative values reported in the various handbooks (e.g., Verscheuren, 1983, Montgomery and Welkom, 1991) for log k_{ow} and log k_{oc}. The PHC compounds shown in Figure 6.3 and the chlorinated benzenes shown in Figure 6.8 are identified in the chart. The black squares which do not have individual names attached include such organic chemicals as fluorene, arachlor 1248, arachlor 1254, benzyl alcohol, dibenzofuran, pyrene, endrin, lindane, methoxychlor, chloroethane, trichloroethylene, dichloroethylene, and vinyl chloride. The values used for log k_{ow} are considered to be the mid-range results reported from many studies. Not all log k_{oc} values are obtained as measured values. Many of these have been obtained through application of the various log k_{oc}-log k_{ow} relationships reported in the literature, e.g., Kenaga and Goring (1980) and Karickhoff et al. (1979). The approximate relationship shown by the solid line Figure 6.7 is given as:

$$\log k_{oc} = 1.06 \log k_{ow} - 0.68 \tag{6.5}$$

Equation 6.5 covers chemicals ranging from PAHs to pesticides and PCBs. We can compare this to other relationships shown in Equation 6.6, which were obtained for certain classes of chemical compounds.

$$\begin{aligned}
\log k_{oc} &= \log k_{ow} - 0.21 && (PAHs) \\
\log k_{oc} &= 1.029 \log k_{ow} - 0.18 && (Pesticides) \\
\log k_{oc} &= 0.72 \log k_{ow} + 0.49 && (Chlorinated\ benzenes)
\end{aligned} \tag{6.6}$$

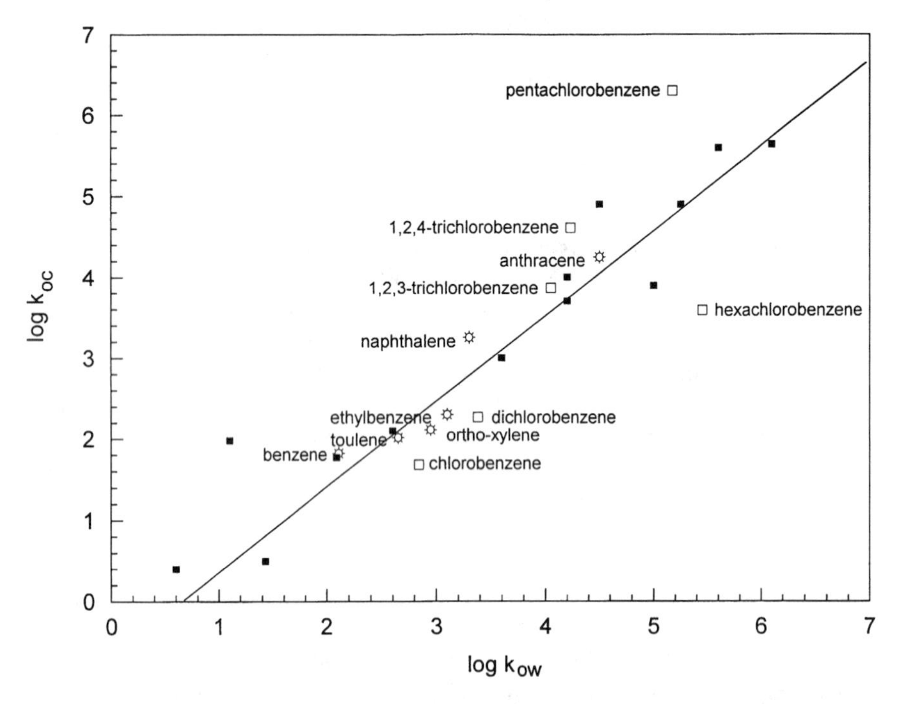

Figure 6.7 Relationship between log k_{oc} and log k_{ow} for some organic chemicals. Names on graph refer to "open" symbols. Black squares (un-named) include: fluorene, arachlor 1248, arachlor 1254, benzyl alcohol, dibenzofuran, pyrene, endrin, lindane, methoxychlor, chloroethane, trichloroethylene, dichloroethylene, and vinyl chloride.

The first relationship shown in Equation 6.6 was reported by Karickhoff et al. (1979) in respect to 10 PAHs, whilst the second one referring to pesticides was reported by Rao and Davidson (1980). The relationship describing the group containing chlorinated benzenes, which also includes methylated benzenes, has been reported by Schwarzenbach and Westall (1981).

Studies on adsorption of the hydrocarbons by the active soil fractions' surfaces show that adsorption occurs only when the water solubility of the PHCs is exceeded and the hydrocarbons are accommodated in the micellar form. Instead of using the k_{ow} and k_{oc} partition coefficients, the accommodation concentration of hydrocarbons in water is sometimes used to reflect the partitioning tendency of organic substances between the aqueous and soil solids. Hydrocarbon molecules with lower accommodation concentrations in water (i.e., higher k_{oc} values) would be partitioned to a greater extent onto the soil fractions than in the aqueous phase. From the results of Meyers and his co-workers (1973, 1978), it is shown that one can expect to obtain a general inverse relationship between the accommodation concentration of the hydrocarbons and the proportion (percent) adsorbed; i.e., the lower the accommodation concentration of the hydrocarbon in water, the greater the tendency of the organic compound to be associated with the reactive surfaces of the sediment fractions. The important consequence of such a relationship is that the aromatic fraction of petroleum products, which are the most toxic, would have the least affinity for

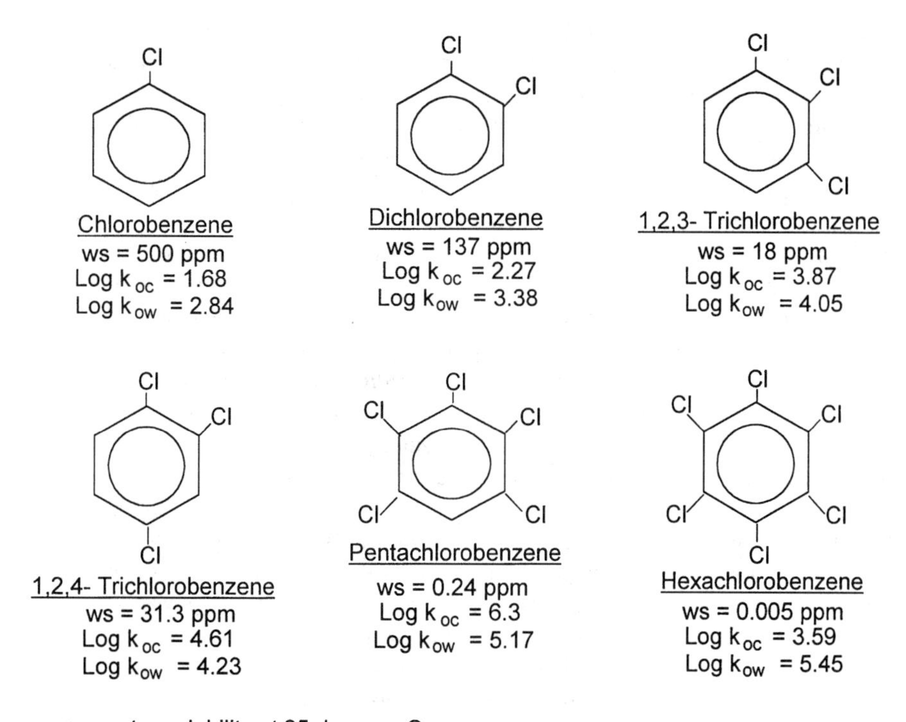

Figure 6.8 Some chlorinated benzenes and their log k_{oc}, log k_{ow}, and water solubility (ws) values.

the reactive surfaces associated with the soil fractions. As might be expected, a study of adsorption data of hydrocarbons shows that anthracene is substantially adsorbed, as can be confirmed by the high k_{oc} value and the very low solubility of the organic compound in water (Figure 6.3). The higher accommodation concentrations of the aromatic hydrocarbons inhibit their association with the clay particles.

6.4 INTERACTIONS AND FATE

6.4.1 Persistence and Recalcitrance

The term *persistence* has been defined generally in the previous chapters. At that time, we referred to persistence as "the continued presence of a pollutant in the substrate." The persistence of inorganic and organic pollutants differ in respect to meaning and application. Chapter 5 defines the persistence of heavy metals (representative of the major inorganic pollutants) in the same spirit as the general definition, i.e., continued presence of the inorganic HM pollutant in the soil in any of its oxidation states. In this section, we need to recognize that organic chemical pollutants can undergo considerable transformations because of microenvironmental factors.

By *transformations* we mean the conversion of the original organic chemical pollutant into one or more resultant products by processes which can be abiotic, biotic, or a combination of these. Whether or not the transformed products can be identified as degraded products is, to a very large extent, dependent on how one categorizes or defines *degradation*. Converted organic chemical compounds resulting from biotic processes, defined as intermediate products along the pathway toward complete mineralization can be safely classified as degraded products. Transformed products resulting from abiotic processes in general do not classify as being intermediate products along the path to mineralization. However, this is not easily distinguished because some of the transformed products themselves may become more amenable to biotic transformations, i.e., combination transformation processes.

We define *persistent organic chemical pollutants* as those organic chemical pollutants that are resistant to conversion by abiotic and/or biotic transformation processes. The continued presence of the original pollutant or its various transformed states is testimony to the persistence of the original pollutant. *Recalcitrant organic chemical pollutants* are those persistent organic chemical pollutants that are totally resistant to conversion by abiotic and/or biotic transformation processes. The persistence of organic chemical pollutants in soils depends on at least three factors: (a) the physico-chemical properties of the pollutant itself; (b) the physico-chemical properties of the soil (i.e., soil fractions comprising the soil); and (c) the microbial forms present in the soil, which can degrade or assimilate the organic chemical pollutants. The abiotic reactions and transformations resulting therefrom are sensitive to factors (a) and (b). All of the factors are important participants in the dynamic processes associated with the activities of the microorganisms in the biologically mediated chemical reactions and transformation processes.

6.4.2 Abiotic and Biotic Transformation Processes

Abiotic transformation processes occur without the mediation of microorganisms. These kinds of (abiotic) processes include chemical reactions such as hydrolysis and oxidation-reduction. Whilst photochemical reactions classify under abiotic processes, because these form a minor part of the processes that occur in the contaminated ground, these will not be addressed in this book. *Biotic* transformation processes are biologically mediated transformation reactions, and include associated chemical reactions arising from microbial activities. The principal distinction between the transformation products from these two processes (abiotic and biotic) is the fact that abiotic transformation products are generally other kinds of organic compounds. This contrasts with mineralization of organic chemical compounds as the transformation product for biotic processes. Biologically mediated transformation processes are the only types that can lead to mineralization of the subject organic chemical compound. Whilst complete conversion to CO_2 and H_2O (i.e., mineralization) may not be achieved, the intermediate products obtained during this process point toward complete mineralization. The conversion products obtained from abiotic and biotic transformation processes can themselves become recalcitrant. A good example of this can be found in the PCE (CCl_2CCl_2) example shown in Section 4.1.1, discussed further in a later section.

Transformations from biotic processes occur under aerobic or anaerobic conditions. The transformation products obtained from each will be different. Complete mineralization of the organic compound can occur if the compound is a primary substrate, as opposed to transformation resulting from partial degradation of the compound due to biological processes. As might be expected, biotic transformation processes under aerobic conditions are oxidative. The various processes include hydroxylation, epoxidation, and substitution of OH groups on molecules. Anaerobic biotic transformation processes are most likely reductive. These could include hydrogenolysis, H^+ substitution for Cl^- on molecules, and dihaloelimination (McCarty and Semprini, 1994).

6.4.3 Nucleophilic Displacement Reactions

Abiotic transformation processes can occur with or without net electron transfer. We refer to *non-reductive chemical reactions*, which involve attacks by nucleophiles on electrophiles. A *nucleophile* is an electron-rich reagent (nucleus-liking species) containing an unshared pair of electrons, whilst an *electrophile* has an electron-deficient (electron-liking species) reaction site and forms a bond by accepting an electron pair from a nucleophile. Nucleophiles are generally negatively charged and because of their "nucleus-liking" nature they are "positive charge-liking." Electrophiles, on the other hand, are generally positively charged and because they are "electron-liking," this means that they are also "negative charge-liking." Oxidation-reduction reactions classify under the latter category of processes which include electron transfer. Figure 6.9 shows a schematic of chemical transformation reactions with and without electron transfer.

Some common inorganic nucleophiles include $HCO_3^-, ClO_4^-, NO_3^-, SO_4^{2-}, Cl^-,$ HS^-, OH^-, and H_2O. As seen in Figure 6.9, hydrolysis is a specific instance of nucleophilic attack on an electrophile. We define *hydrolysis reaction* as that chemical reaction between an organic chemical and water. In this reaction, the water molecule or OH^- ion replaces groups of atoms (or another atom) in the organic chemical. A new covalent bond with the OH^- ion is formed, with cleavage of the bond and the "leaving group X" in the reacting organic molecule. No change in the oxidation state of the organic molecule is involved in the transformation.

The term *neutral hydrolysis* is often used to refer to nucleophilic attack by H_2O. This is to distinguish it from acid-catalyzed and base-catalyzed hydrolysis where catalytic activity is accomplished by the H^+ and OH^- ions, respectively. This distinction is necessary since both acid-catalysis and base-catalysis impact directly on the kinetics of hydrolysis, i.e., pathway and rate of hydrolysis kinetics. The products of hydrolysis reactions are generally compounds, which are more polar in comparison to the original chemical compound, and will therefore have different properties. The same cannot be said for transformation products obtained as a result of nucleophilic attack on electrophiles, i.e., reactions that do not include water (Schwarzenbach et al., 1993). Detailed types, situations, and examples of nucleophilic-electrophilic reactions can be found in the reference texts on organic chemistry and environmental organic chemistry.

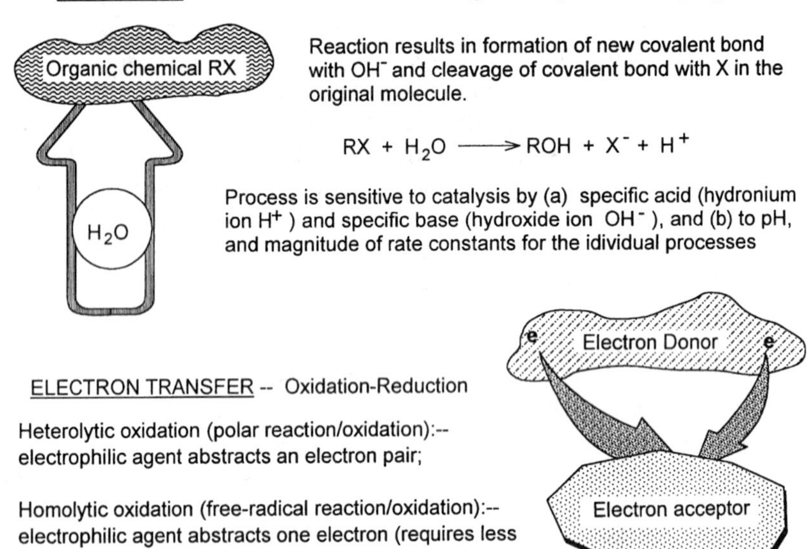

Figure 6.9 Schematic portrayal of chemical transformation reactions with electron transfer (oxidation-reduction), and without net electron transfer (hydrolysis).

6.4.4 Soil Catalysis

Soil-catalyzed hydrolysis reactions associated with the surface acidity of clay minerals can be significant because they can affect the hydrolysis half lives of the reacting organic chemicals, i.e., they affect the kinetics of hydrolysis. Measurements on surface acidity of many clay minerals have shown that these can be at least anywhere from 2 to 4 units lower than that of bulk water (Mortland, 1970; Frenkel, 1974). The surface acidity of kaolinite minerals, for example, derives from the surface hydroxyls on the octahedral layer of the mineral particles. Figure 6.10 shows the effect of moisture content on the acidity of a kaolinite, using data reported by Solomon and Murray (1972). As can be seen from the figure, the surface acidity is reduced dramatically as the moisture content of the soil is increased. Accordingly, the catalytic activity will be correspondingly decreased.

Surface acidity in the case of montmorillonites is due to isomorphous substitution and to interlamellar cations. The layer of water molecules next to the charged lamellar sheet are strongly polarized, resulting in the loss of protons. The charge and nature of the cations affect the degree of catalytic activity since these cations impact directly on the polarizing power and the degree of dissociation of the water in the inner Helmholtz plane (adsorbed water). We would expect the surface acidity

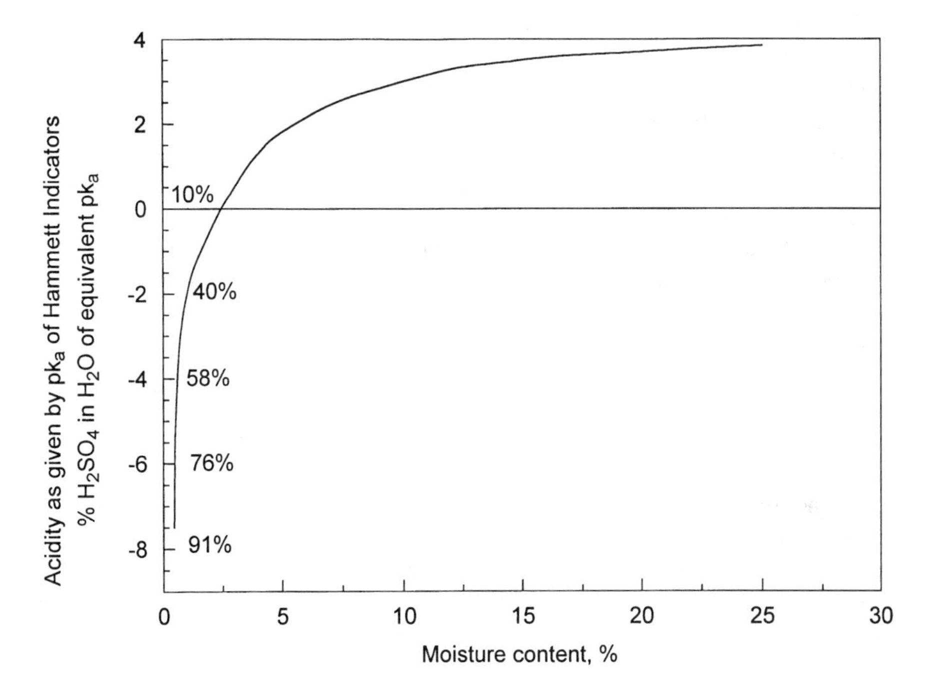

Figure 6.10 Effect of moisture content on surface acidity of kaolinite. (Adapted from Solomon and Murray, 1972.)

of montmorillonites to increase as we increase the valency of the exchangeable cations.

In heavy metal-contaminated soils, metal-ion catalysis of hydrolysis occurs through the heavy metals sorbed by the soil fractions. At least two mechanisms might be involved (Larson and Weber, 1994; Stone, 1989):

- The sorbed heavy metals that function as Lewis acids can coordinate the hydrolyzable functional groups of the subject organic chemical, thus making it more electrophilic.
- Nucleophilic attack by the metal hydroxo groups associated with the clay mineral surfaces.

In the metal ion catalyzed hydrolysis reactions, using esters as an example, coordination of the lone pair electrons (of the oxygen) results in polarization of the carbonyl functional group which in turn will make it more susceptible to nucleophilic attack by H_2O or OH^- (Larson and Weber, 1994). Such direct polarization processes can accelerate hydrolysis rates by four orders of magnitude (Buckingham, 1977). Formation of a metal-coordinated nucleophile, which is more reactive than a corresponding free nucleophile, is also possible with metal ion-catalyzed hydrolysis (Plastourgou and Hoffmann, 1984). The increased acidity of water molecules results in production of OH^-.

6.4.5 Oxidation-Reduction Reactions

Oxidation-reduction reactions can occur in interactions between organic chemical pollutants and soil fractions under abiotic and biotic conditions. In contrast to transformations occurring through nucleophilic replacement reactions where no net transfer of electrons occurs, electron transfer occurs in oxidation-reduction (redox) reactions. A general brief discussion of redox reactions and the redox potential has been given in Chapter 4. At this time, we need to understand how these affect the fate of organic chemical pollutants in soil. The significant points that require attention are the nature and result of electron transfer between the interacting participants (pollutants, microorganisms, and soil fractions). We recall that: (a) the chemical reaction process defined as *oxidation* refers to a removal of electrons from the subject of interest, and (b) reduction refers to the process where the "subject" (electron acceptor or *oxidant*) gains electrons from an electron donor (*reductant*). By gaining electrons, a loss in positive valence by the subject of interest results and the process is called a reduction.

It is not often easy to distinguish between redox reactions that occur abiotically and those that occur under biotic conditions since direct involvement of any (or some) microbial activity cannot be readily ruled out. It is not clear that insofar as organic chemical pollutants are concerned whether there is a critical requirement to distinguish between the two — since redox conditions are more likely than not to be the product of factors which include microbiological processes. The number of functional groups of organic chemical pollutants that can be oxidized or reduced under abiotic conditions is considerably smaller than those under biotic conditions (Schwarzenbach et al., 1993). Quantification of reaction rates is difficult because the interactions between the pollutants occur with both microorganisms and the many different soil fractions, thus making determination of reaction pathways almost impossible. Whilst the scarcity of kinetic data makes it difficult to provide for quantitative calculations of redox reactions, it is nevertheless instructive and informative to obtain a qualitative or descriptive appreciation of these reactions.

Abiotic redox reactions of organic chemical pollutants in soil systems occur when electron acceptors such as those described above are present. Clay soils function well as electron acceptors (oxidizing agents or oxidants), i.e., they are electrophiles. The structural elements of clay minerals such as Al, Fe, Zn, and Cu can transfer electrons to the surface-adsorbed oxygen of the clay minerals. These can be released as hydroperoxyl radicals (–OOH), which can function as electron acceptors, i.e., these radicals can abstract electrons from the organic chemical pollutants. An example of this type of reaction is shown in Figure 6.11 using the results reported by Yong et al. (1997). The possible mechanisms for oxidizing the two kinds of phenols shown in the figure include (a) the structural elements of the montmorillonite clay (Fe, and Al), and (b) the partially coordinated aluminium on the edges of the clay minerals. These function as Lewis acids which can accept electrons from aromatic compounds. In addition, the exchangeable cations such as Fe(III) and Cu(III) contribute to phenol polymerization through coupling of radical cations with phenols. We see from the results given in Figure 6.12 that more effective oxidation of the 2,6-dimethylphenol is obtained by the Fe-clay — presumably because of the greater oxidizing capability of the Fe(III).The intermediate product formed is a 2,6-dimethylphenol dimer of mass

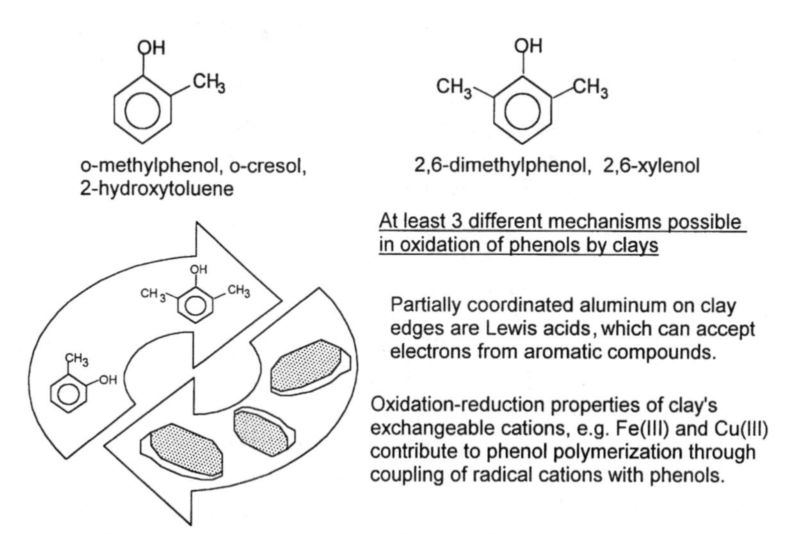

Figure 6.11 Mechanisms involved in oxidation of phenols by clay minerals.

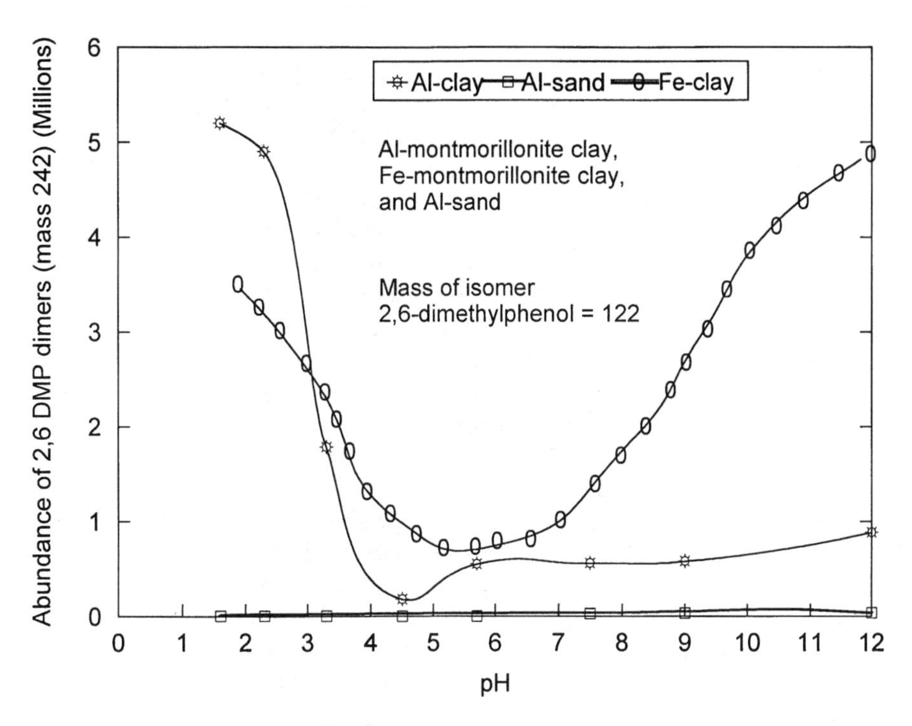

Figure 6.12 Oxidation of 2,6-dimethylphenol by Al-clay, Fe-clay and Al-sand (Data from Yong et al., 1997.)

242, as shown by the degree of abundance on the ordinate of the graph in Figure 6.12. Other intermediate products such as trimers and traces of oligomers of the 2,6-dimethylphenol have also been obtained (Desjardins, 1996).

In a *biologically mediated redox* reaction, the metabolic process is generally *catabolic* (i.e., energy releasing) and the result is a transfer of electrons from the organic carbon, resulting thereby in the oxidation of the pollutant. Common electron acceptors in the soil system are oxygen, nitrates, sulphates, Fe^{3+}, Mn^{4+} and other trace metals. The activities of microorganisms, which result in transformation of the original organic chemical, can also alter the physical and chemical nature of soils. These will directly change the interactions between soil fractions and pollutants. The biogenic processes that are of importance are biodeposition, fluid transport, stabilizing mechanisms, and macrofaunal-microbial interactions. These processes impact directly on the nature and distribution of pollutants within the soil. Biogeochemical processes influence the distribution of hydrocarbons in soils through selective removal and/or selective production. Microbial degradation can slowly but preferentially remove *n*-alkanes from a petroleum-contaminated soil, leaving behind the more resistant isoprenoids, cycloalkanes and cycloalkenes, and aromatics. Relative rates of microbial degradation proceed as *n*-alkanes > branched alkanes > cyclic alkanes. A combination of diffusion, water solubilization and transport, evaporation, and microbial degradation can be responsible for observed changes in aromatic hydrocarbon concentrations and composition.

The low water solubilities of organochlorine compounds such as PCBs and DDT, combined with their very slow rate of microbial degradation, make these compounds recalcitrant. Because of their low solubilities, they tend to persist in the soil. Since the lower chlorinated isomers of PCBs are more readily degraded, the higher chlorinated compounds will dominate as the persistent compounds of PCBs found in the soils. In addition to the MAHs representative of the PHCs shown in Figure 6.3, the chlorinated hydrocarbons which also are considered as MAHs, e.g., chloro-, dichloro-, trichloro-, pentachloro-, and hexachlorobenzene shown in Figure 6.8 have been found to be quite persistent, i.e., their presence in soils and particularly in lake and river sediments have been well established (Oliver and Nicol, 1982, 1984; Oliver 1984; Oliver and Pugsley, 1986). Analysis of (soil) sediment cores from Lake Ontario indicate that these MAHs have been accumulating in the lake's soil sediments since the early 1900s. There appears to be little evidence of either microbial oxidation or anaerobic dehalogenation of chlorobenzenes (C_6H_5Cl).

Since lake and river sediments are composed primarily of soil fractions, information obtained from studies on sediments provide useful direct clues to soil-pollutant interactions, particularly in respect to persistence and fate of the pollutants. Bosma et al. (1988) suggest that trichlorobenzenes ($C_6H_3Cl_3$) can be transformed to dichlorobenezes ($C_6H_4Cl_2$) in some sediments under anaerobic conditions with half-lives ranging from a few days to over 200 days. Dichlorobenzene, also known as ortho-dichlorobenzene, is used primarily as solvent for carbon removal and degreasing of engines. With the k_{oc} value as shown in Figure 6.8, the dichlorobenzene partitions well to sediments, and particularly the organic fractions (SOM, soil organic matter), and because of its low anaerobic degradation, it is very persistent. Although there exist three isomers of trichlorobenzene ($C_6H_3Cl_3$), 1,2,4-, 1,2,3- and 1,3,5-, the isomer 1,2,4- is most common. The low water solubilities and high log k_{ow} and log k_{oc} values indicate that 1,2,4-trichlorobenzene partitions well to the soil fractions. As in the case of the dichlorobenzene, the trichlorobenzene is well adsorbed by the SOM and will persist and accumulate under anaerobic conditions. The similarly high values of k_{oc} for pentachlorobenzene (C_6HCl_5) and hexachlorobenzene (C_6Cl_6) are also indicative of the ability to partition to soil fractions, in common with the trichlorobenzenes.

The pentachlorobenzene that has been identified in waste streams from pulp and paper mills, iron and steel mills, inorganic and organic chemical plants, petroleum refineries, and activated sludge waste water treatment plants (Meyers and Quinn, 1973; Laflamme and Hites, 1978) appears to have the highest k_{oc} value of the various chlorobenzenes. The low water solubilities of the dichloro-, trichloro-, pentachloro- and hexachlorobenzenes combined with their respective high k_{oc} values indicate that they can be well adsorbed by the soil fractions. Desorption of chlorobenzenes from soil fractions can occur (Oliver, 1984, Oliver et al., 1989).

The effects of biodegradation, or the resistance to biodegradation as an indication of the persistence of the organic chemicals in polluted sediment, have been recorded in many instances. Sediment soil contamination by pentachlorophenol (PCP) which is relatively soluble in water at pH 6 can be degraded microbially (Crosby, 1972). On first glance, we would associate the relative solubility of the chemical with a low potential for sorption (of the PCP) by soil fractions. However, there is evidence (Munakata and Kuwahara, 1969) showing substantial amounts of PCP associated with soil fractions. This suggests that PCP may not be readily degradable in the presence of particle bonding. Results from Pierce et al. (1980) over a two-year period study of PCP spill into a creek show a reduced presence of PCP from an original maximum concentration of about 1.35 mg/kg air dry sediment to about 0.2 mg/kg, in the contaminated creek. The degradation products detected included pentachloroanisole (PCA) and 2,3,4,5-, 2,3,4,6,- and 2,3,5,6-tetrachlorophenol (TCP).

Anaerobic dehalogenation of organic chemicals has been briefly shown in Chapter 4 in the case of degradation of tetrachloroethylene or perchloroethylene (PCE, C_2Cl_4) to trichloroethylene (TCE, C_2HCl_3), to 1,2-dichloroethylene (DCE, $C_2H_2Cl_2$) and to vinyl chloride (VC, C_2H_3Cl). The structural changes and the changes in the properties of the intermediate products are shown in Figure 6.13. Beginning with PCE, where the log k_{oc} value indicates good partitioning to the soil fractions, degradation of the PCE to TCE and onward to VC, show that the log k_{oc} values diminish considerably to a very low value for the vinyl chloride. As the PCE continues to degrade, more of the chemical substance is released into the aqueous phase (porewater). This is particularly true for VC, where the low values of log k_{oc} and high water solubility values suggest that this chemical can be environmentally mobile.

6.5 CONCLUDING REMARKS

The various sorption processes that contribute to bonding between organic chemical pollutants and soil fractions include partitioning (hydrophobic bonding) and accumulation — through adsorption mechanisms involving the clay minerals and

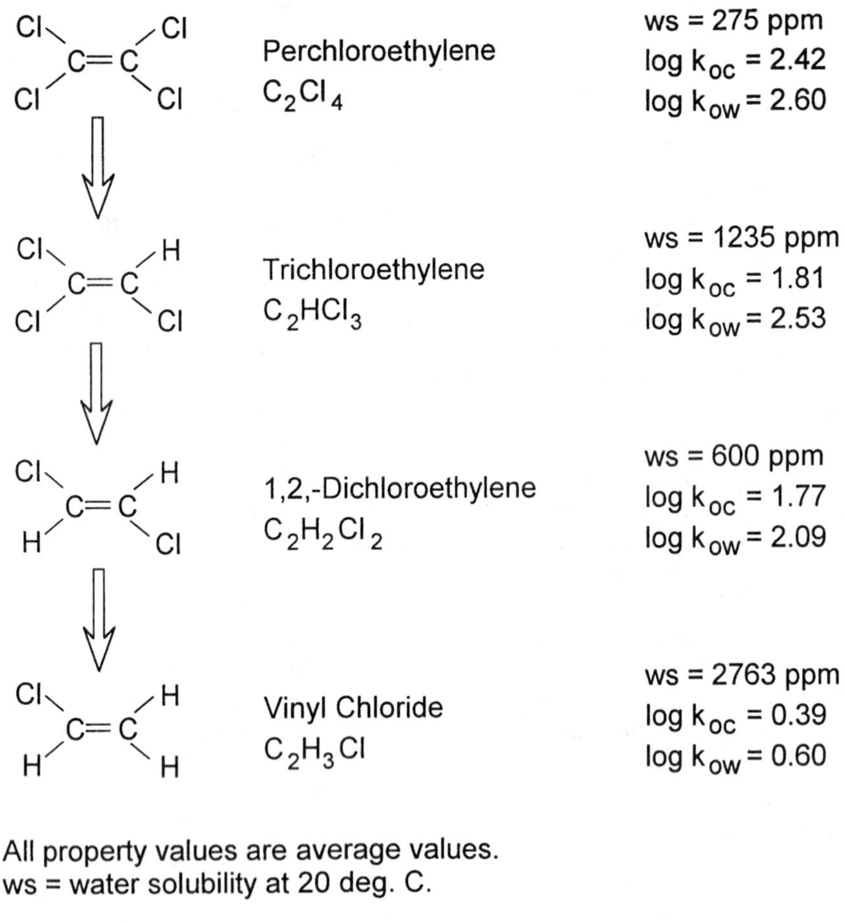

Perchloroethylene C_2Cl_4 — ws = 275 ppm; log k_{oc} = 2.42; log k_{ow} = 2.60

Trichloroethylene C_2HCl_3 — ws = 1235 ppm; log k_{oc} = 1.81; log k_{ow} = 2.53

1,2,-Dichloroethylene $C_2H_2Cl_2$ — ws = 600 ppm; log k_{oc} = 1.77; log k_{ow} = 2.09

Vinyl Chloride C_2H_3Cl — ws = 2763 ppm; log k_{oc} = 0.39; log k_{ow} = 0.60

All property values are average values.
ws = water solubility at 20 deg. C.

Figure 6.13 Degradation of PCE and associated changes in log k_{ow} and log k_{oc}.

other soil particulates such as carbonates and amorphous materials. The more prominent properties affecting the fate of organic molecules by soil fractions include the following:

- **Soil fractions** — Surface area, nature of surfaces (composition of surface fractions), surface charge (density, distribution and origin), surface acidity, CEC, exchangeable ions on the reactive surfaces, configuration of the reactive surfaces.
- **Organic chemical molecules** — Functional groups, structure, charge, size, shape, flexibility, polarity, water solubility, polarizability, partitioning and equilibrium constants.
- **Soil system (soil environment)** — Microbial community, energy sources, temperature, inorganic/organic ligands available, pH, pE, salinity, physical gradients (fluxes).

The various transformation reactions catalyzed by microorganisms have been listed under the general category of soil system to a very large extent because the attention in this book is concerned more with the soil aspect of the problem, and on the biological aspects of pollutant fate. Nevertheless, as we have seen, it is not possible to discuss the persistence and fate of organic chemical pollutants without paying some required attention to the role of microorganisms. Much remains to be researched, studied, and learnt about this aspect of the problem.

It is not possible to discuss the fate of organic pollutants without paying attention to the desorption potential of such pollutants. Given the complex distribution and nature of the various soil fractions, the organic chemicals and the soil system (listed previously), and particularly the myriad of intermediate products that are formed with time, desorption phenomena can be most difficult to quantify. It is argued that desorption of organic chemical pollutants occur continuously with time, and that the state of any chemical in the ground is the result of various chemical and physical transformations — where the physical component includes transport, sorption, and desorption. All these various processes occur simultaneously or in some fashion that we have yet to fully characterize and quantify.

CHAPTER 7

Interactions and Pollutant Removal

7.1 INTRODUCTION

Remediation and rehabilitation of contaminated land require several critical decisions. Various factors, problems, and conditions need to be considered. However, these are not always easy to define or characterize. The various issues and problems that need consideration are all tied into the following sets of concern:

- Nature and distribution of the pollutants in the contaminated site;
- Threat posed by the pollutants;
- Land use requirements and intentions;
- Zoning laws and regulations;
- Economics of decontamination or pollutant removal;
- Technology available and efficiency (best available technology?), timing, etc.; and
- Risks.

For the purpose of treatment of the subject of *pollutant removal* we will define *decontamination* to mean the removal of contaminants (pollutants and non-pollutants) — with no particular reference to whether this means partial or complete removal of the contaminants. Where necessary, the term *complete decontamination* will be used to indicate complete removal of contaminants.

We have shown the overall nature of the problem previously in Figure 1.12. All the factors stated above and the specific points noted in Figure 1.12 need to be considered. They do not, however, necessarily need to be considered in equal terms or proportions before reaching the decision point that (a) dictates or limits the type of level of land use attainable as the rehabilitation process, or (b) specifies the land use and thus establishes the requirements and technologies for site decontamination. The basic points identified in Figure 7.1 integrate the major items shown in Figure 1.12. These points refer to the requirements for a knowledge of the various bonding mechanisms and processes which bond the pollutants to the soil fractions. A knowledge of the nature of the bonding mechanisms established between the

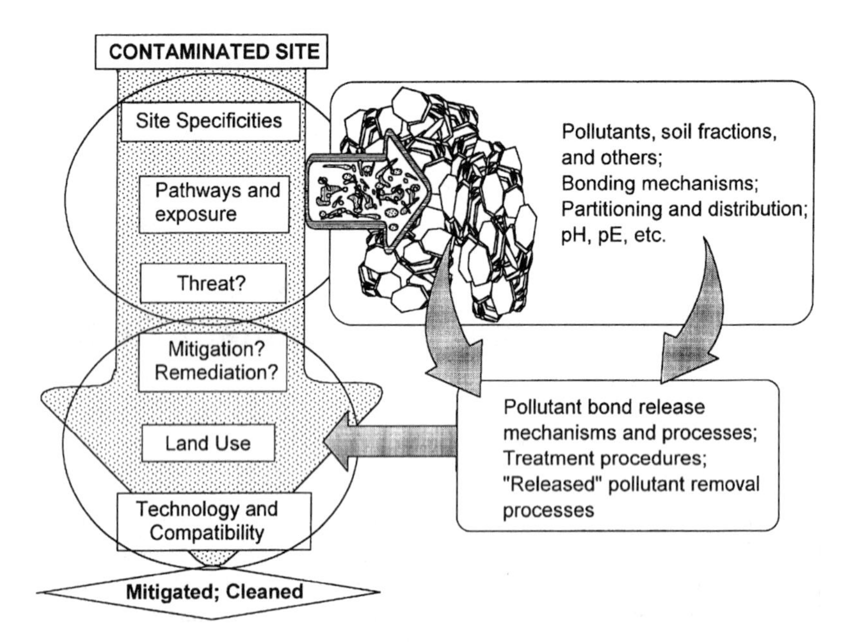

Figure 7.1 Basic approach for development of effective decontamination processes.

contaminants and the soil fractions is considered to be essential if we are to determine the various parameters prominent in the control of the persistence and fate of the contaminants, as seen in the previous chapters. This knowledge is also considered to be necessary for development or structuring of the decontamination (remediation) process. A good working knowledge of these mechanisms and processes would permit us to determine the most effective treatment procedure for release of the pollutants. This will allow for structuring of the required technology for removal of the released pollutants from the contaminated site.

We recognise that decontamination requirements and procedures involve not only technical/scientific issues relating to the site and contaminants, but also land use requirements and capabilities. For consideration of land use capabilities and/or requirements, we need to be concerned with the required level of clean-up, time frame allowed, costs or budget, processes, and impact on immediate surroundings. The focus of the material in this book does not include the land utilization aspects of the problem at hand. These are serious considerations. The interested reader is advised to consult the proper publications dealing with land use.

The interactions between pollutants and reactive surfaces of the various types of soil fractions that have been examined in the previous chapters are now revisited (in a general sense) with a focus that seeks to provide the basis for evaluation of the basic requirements for effective removal of the pollutants. The general types of procedures and methodologies for pollutant removal as remedial treatment of contaminated ground are addressed in the next chapter.

7.2 BASIC DECONTAMINATION CONSIDERATIONS

The principal issues identified in the top right-hand box in Figure 7.1 relate to the nature of pollutants and the various processes that define their fate in the contaminated site. Until a proper appreciation of the nature of the pollutants and their bonding mechanisms is established or determined, the structuring of effective procedures for pollutant removal would be difficult. The various methods of treatment of contaminated soil for removal of contaminants and pollutants from contaminated ground can be categorized into three generic categories as follows:

1. Physico-chemical and chemical;
2. Biological;
3. Thermal, electrical, infrared, acoustic, etc.

All these methods of treatment seek to reduce the pollutant-holding capability of the soil fractions as a means for release of the pollutants. Reduction of pollutant-holding capability is achieved in the case of heavy metal pollutants by changes in the microenvironment that surrounds the soil mass, e.g., changes in pH, pE, and ionic strength. For organic chemical pollutants, destruction of the functional groups associated with the pollutants is amongst the group of simple procedures.

7.2.1 Pollutant–Soil Interactions and Pollutant Removal

As discussed in the previous chapters, the interactions occurring between pollutants and soil fractions which classify under the broad category of adsorption or bonding of pollutants are essentially physical and chemical adsorption processes. To remove the sorbed or bonded pollutants, the forces binding the pollutants to the soil fractions need to be destroyed or weakened. Coulombic or ionic forces established between positively charged and negatively charged atoms are amongst the strongest of the forces of attraction between atoms and molecules. Since these decrease as the square of the distance separating the atoms, the removal of pollutants that depend on these forces for bonding to the soil fractions will be facilitated if the separation distance between them can be increased. This can be achieved by introducing external agents or treatment procedures that can:

- Overcome the energies of interaction developed between the pollutants and soil fractions;
- Weaken the energies of interaction; and
- Compete with the soil fractions for sorption of the pollutants.

A good case of competition for sorption of heavy metal (HM) pollutants, for example, is the use of organic ligands which have a high affinity for HMs. If ligand-facilitated dissolution of minerals and other soil fractions can be initiated, release of sorbed metal ions would result, thus allowing for competitive sorption to occur. The use of ethylenediaminetetraacetic acid (EDTA) is a good example of this procedure.

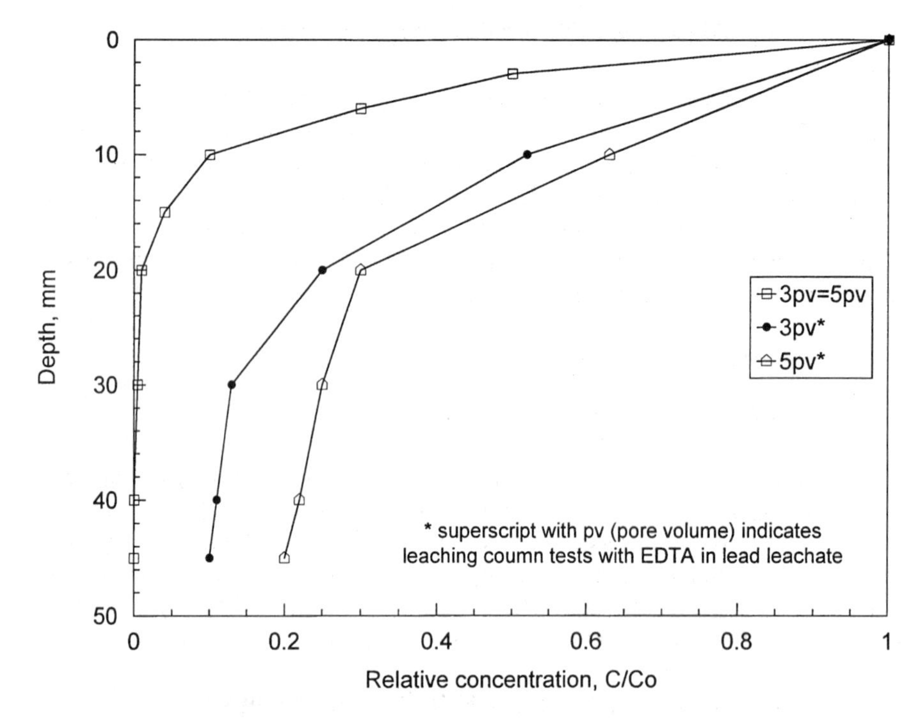

Figure 7.2 Difference in porewater Pb concentration profiles due to high affinity sorption by EDTA in leachate. Note that * superscript with pore volume (pv) signifies samples leached with EDTA in Pb leachate (basic data from Darban, 1997).

It is not only effective in non-reductive and reductive dissolution of certain oxides, but also serves as a complexing agent for heavy metals. The high affinity of EDTA for heavy metals is shown in the sorption-partitioning results obtained in a simple set of leaching column experiments (Figure 7.2). The soil used in the leaching experiments is the same kaolinite and silica gel mixture used for the results shown previously in Figure 5.23. Comparison of Pb concentration in the porewater given in Figure 7.2 is between a Pb leachate with and without EDTA. From data given in Darban (1997), a decrease in the k_d values of the order of two to four magnitudes has been calculated between the Pb-without and Pb-with (EDTA) tests. Confirmation of the EDTA high affinity for Pb leachate, demonstrated via competitive sorption of Pb is seen not only in the dramatic decrease in the k_d values, but also in the higher concentration of Pb in the porewater.

Soil pH and the pE status of the microenvironment are major factors in the control of the fate of HMs. To very large extent, they also control the *availability* of the HMs. We define the *availability of metals* as the potential for release of metals sorbed by the soil fractions, under circumstances which are within the influence of the immediate microenvironment. These influences are generally manifested not only in the form of pH and/or pE changes, but also through fluxes originating from external sources. The solubility of the metals, for example, and their ability to form chelates is by and large directly affected by the soil pH. In addition, resultant release of metals can also be obtained by oxidation of sulphide minerals and soil organic matter. This can occur through a change in the microenvironment from anoxic to oxic, or through bacterial activity.

Forces of attraction between uncharged molecules that are neutral but in which the centres of positive and negative charge are separated are called *dipoles*. Certain orientations of adjacent dipoles are statistically preferred, resulting in a net attraction. Even molecules that are not polar can be considered as instantaneous dipoles because of changes in the instantaneous positions of the electrons in the atomic shells. This instantaneous dipole induces an in-phase dipole in an adjacent molecule, with a resultant net attraction. The maximum attraction results when the dipoles are oriented in an end-to-end configuration.

The energy by which contaminants (pollutants and non-pollutants) are held within the soil matrix can be theoretically calculated using the modified DDL model or the DLVO method as discussed in Chapter 3. Because the assumptions used require some idealization of particle arrangements and contaminant species, the calculated energies of interaction developed between contaminants and soil particles can at best be considered as approximate. Nevertheless, the calculations are useful since they can provide some insight into the level of effort required to obtain pollutant release from the soil fractions. In particular, the use of electrokinetics to extract HMs (Section 7.4) is a direct use of a knowledge of energies of interactions.

7.3 DETERMINATION OF POLLUTANT RELEASE

Determination of treatments required to obtain pollutant release from contaminated soils is a prudent requirement in the development of remediation-treatment methods for removal of pollutants from contaminated ground. In essence, this type of determination seeks to establish the potential and extent of pollutant release from the soil fractions, and can be identified as a *pollutant release potential* (PRP) determination (see Figure 7.3). This should not be confused with *treatability studies* which are basically screening tests used to determine (a) the rate and extent of biodegradation of specific biological treatments, and (b) the effectiveness (rate and extent of pollutant removal) of the applied chemical, electrical, etc., remediation treatments. The most common techniques used in PRP determination include: (a) batch equilibrium studies which include desorption-isotherm type and SSE-type experiments; (b) leaching column desorption-type experiments; and (c) bench-top studies.

7.3.1 Batch Equilibrium Studies

Batch equilibrium studies are generally conducted with soil suspensions. At least two types of soil suspension studies can be conducted: (a) determination of desorption characteristics of the candidate polluted soil, and (b) soil washing-type studies designed to weaken bonding relationships, as discussed in the previous section. While desorption-type soil suspension studies have not been as widely conducted as adsorption-type studies, it is nevertheless instructive and useful to obtain desorption-based information. At the very least, this permits one to determine how strongly the pollutants of interest are held by the soil fractions.

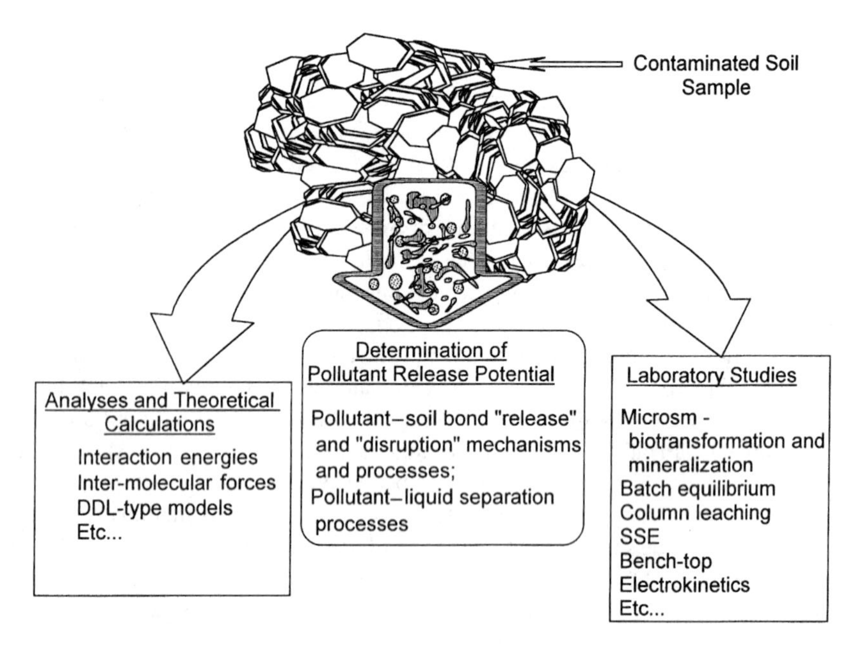

Figure 7.3 Procedures for determination of *pollutant release potential* (PRP).

The simplest procedure in determining whether metal release can be obtained from metal-polluted soils is to conduct extraction tests on the polluted soil — using soil suspensions. Desorption-extraction tests using extractants at various pH levels would provide information on the extractability of the metals in the metal-polluted soil. The desorption test results shown in Figure 7.4 for a Pb-polluted illite soil (Li, 1997) show that the proportion of metal released from the polluted soil is dependent on the initial state of the soil, i.e., concentration of sorbed Pb in the polluted soil. In addition, the number of washings needed to achieve metal release from the soil is also dependent on the initial state of the soil.

Determination of desorption characteristics and properties of polluted soils is generally conducted in one of two procedures. In the first instance, the results of successive multiple washings of polluted soil samples obtained at the various stages of a regular adsorption isotherm determination are used to construct the resultant desorption curve. The sketch shown in Figure 7.5 illustrates this procedure. The sample points on the adsorption isotherm are used as the source of the multiple desorption tests. The aqueous-washing solution used in the desorption determination can range from water to a chemical reagent.

In the second instance, each sample used for determination of the adsorption isotherm is in turn used as the source sample for single desorption tests. In this manner, an equal number of desorption points are obtained. As in the first instance, the choice of solution used for the desorption test is guided by the objective of the desorption experiment. The temptation to identify these curves as desorption isotherms should be carefully avoided. Strictly speaking, these desorption curves are

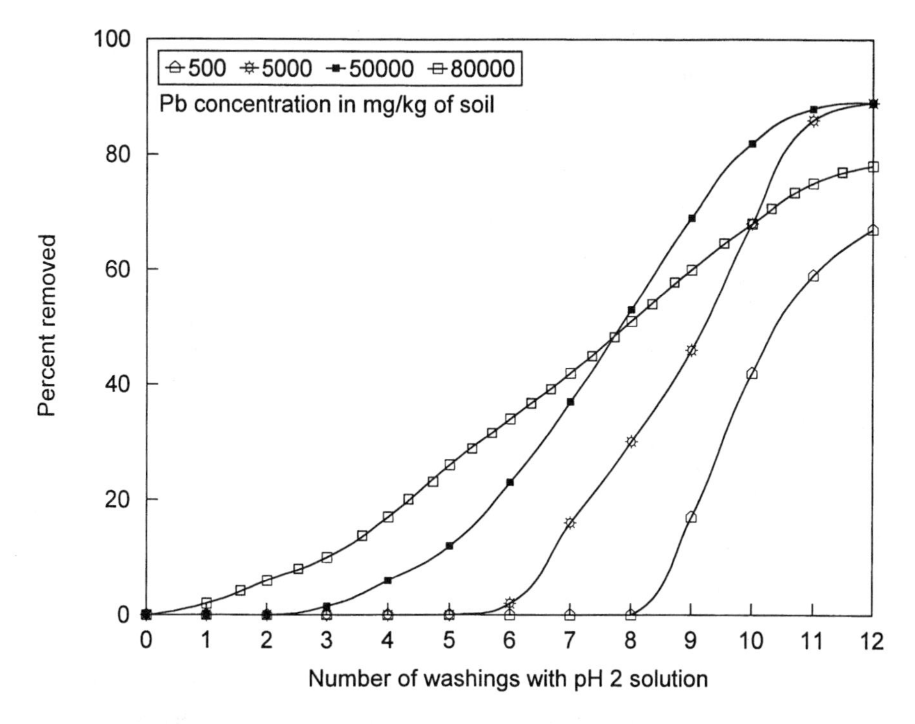

Figure 7.4 Desorption-extraction of Pb from Pb-polluted illite soil. Dilute HNO_3 at pH 2 used in extraction process (data from Li, 1997).

projected desorption curves which are determined from desorption test results conducted on polluted soil samples. It is clear that the desorption characteristics are highly operationally defined.

Taking a cue from selective sequential extraction (SSE) studies (Chapter 5), neutral salts are generally used as the reagent solution of interest in the evaluation of desorption of heavy metals. The metals held by the soil fractions by mechanisms associated with the CEC constitute the easily extractable metals. These are expected to be the most mobile. Unlike SSE experiments, the use of reagents in the desorption characterization studies are principally designed to seek extraction of the pollutants attached to the soil fractions with less regard to the source of pollutant release. Regular SSE studies are necessary if the metal pollutant release source is to be determined (Mulligan et al., 2001).

Similar-type desorption experiments can be conducted on soils polluted with organic chemicals. The desorption characteristics of the 2-methyl naphthalene using the multiple extraction (washing) is shown in Figure 7.6. This is the same PAH previously used in the determination of adsorption isotherms of typical PAHs (Figure 6.6). As in the case of desorption characterization of heavy metal-polluted soils, the desorption curves obtained for the organic chemicals are highly dependent on the extraction medium used and the nature of the source samples. Also in common with the metal polluted soils is the fact that the soil suspension-type studies provide

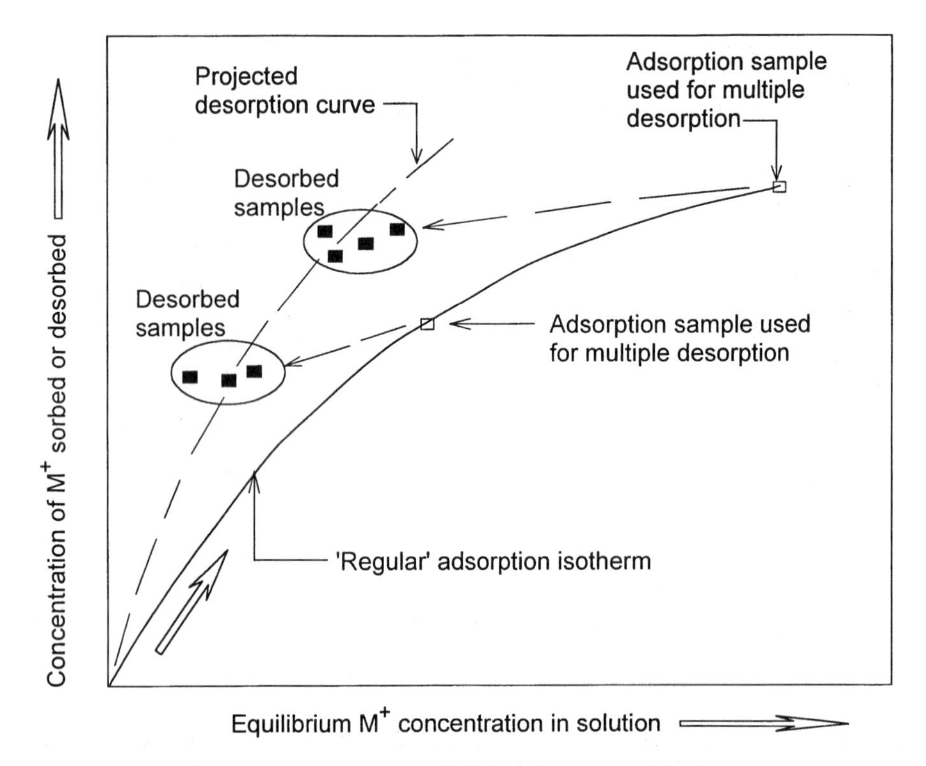

Figure 7.5 Desorption characterization procedure. Note that the desorption curve obtained is a *projected* desorption curve and that M^+ denotes metal pollutant.

the greatest opportunity for reagents to interact with the pollutants because of the nature of the technique used. All results obtained must be considered to be operationally defined.

7.3.2 Column Tests

As in column leaching tests designed to determine sorption performance of soils, desorption column tests are conducted with similar equipment and techniques as discussed in Chapter 5. Desorption column tests are useful if information on the capability of various target extractants is needed as a screening procedure. These types of column tests are also very useful for permeation with rainwater or groundwater as a means for determination of the environmental mobility of the pollutants in the soil.

For desorption tests, polluted soil samples are placed in the soil columns (Figure 5.14) and the target extractant is used as the permeant. All other test procedures, as in the sorption tests, remain the same. An example of the kind of results obtained from such a procedure is shown in Figure 7.7, using basic data reported by Darban (1997). The Pb in the porewater, released from the soil after leaching with 3 pore volumes of EDTA at a concentration of 0.01 mmol/L at a pH of 4.5, is

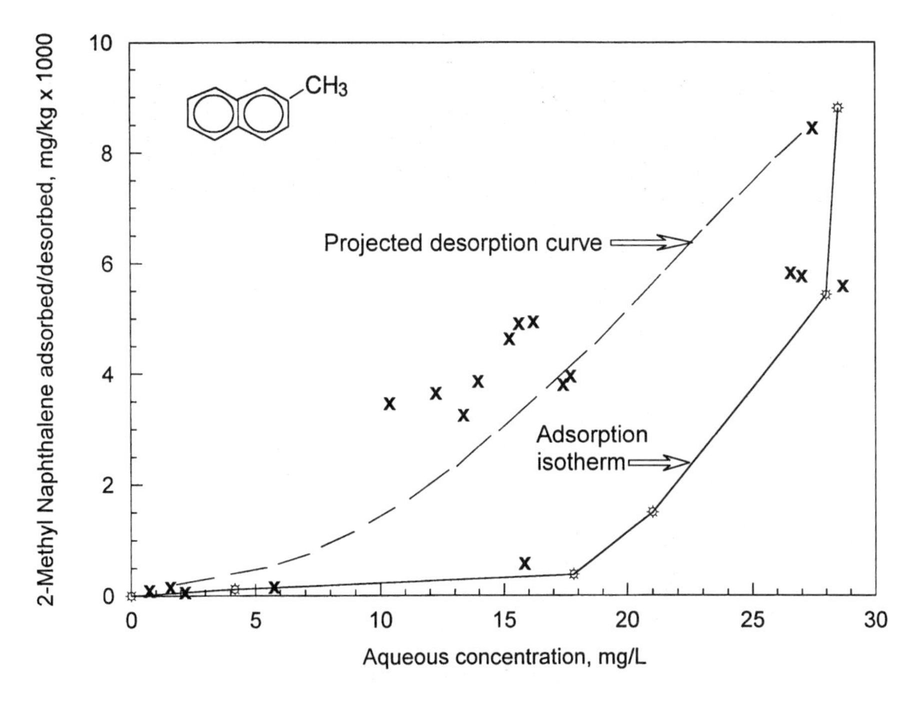

Figure 7.6 Desorption characterization of 2-methyl naphthalene. (Data from Hibbeln, 1996.)

also shown in the figure. Desorption of Pb from the upper portion of the sample due to leaching will accumulate in the lower portion of the sample. Continued leaching with EDTA will continue to detach more Pb at a diminishing quantity and rate. Whilst the Pb-EDTA complexes in the porewater can be managed simply in laboratory experiments, i.e., they can be removed from the leached columns as effluents, field application and technology must be structured to remove these complexes in the porewater in the field. Some of these will be addressed in the next chapter.

Determination of the environmental mobility of a DNAPL requires accountability for losses due to processes associated with degradation and volatilization. The results of DNAPL-polluted leaching column tests using water as the permeant are shown in Figure 7.8 (data from Mohammed, 1994), using mass balance calculations to determine the losses due to volatilization and other degradative effects. Determination of the chemical compounds in the porewater together with the DNAPL product remaining sorbed onto the soil fractions will be necessary to complete the requirements for inventory information, using procedures similar to those prescribed for column leaching tests in HM migration studies (e.g., Figure 5.14).

7.3.3 Selective Sequential Analyses

Selective sequential extraction (SSE) studies are very useful for evaluation of the ease of removal of HM pollutants in contaminated (polluted) soil samples. The

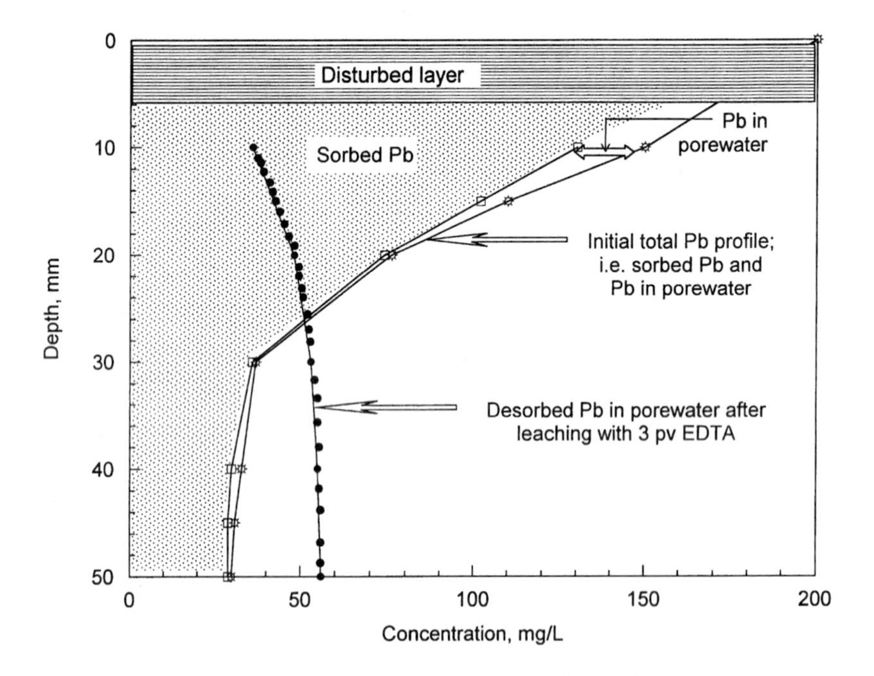

Figure 7.7 Pb in porewater obtained from desorption of Pb-polluted soil column with EDTA. Initial Pb profile in soil column is from permeation of soil column with Pb-leachate. (Basic data from Darban, 1997.)

sets of typical reagents used as extractants to remove heavy metal pollutants in SSE studies, shown in Figure 7.9, can provide some guidance into actual field application. Further elaboration of the reagents used can be found in Table 5.3. The information presented in the figure can be viewed in terms of the ease of removal of sorbed heavy metal pollutants — or, conversely, how strongly the contaminants are held by the soil fractions. Mulligan et al. (2001) used the procedure to evaluate the effectiveness of various kinds of biosurfactants in removal of sorbed heavy metals in a soil sediment contaminated with both organic chemicals and heavy metals.

The amount of Pb held by each soil fraction (minerals, natural organic matter, oxides/hyroxides, carbonates) in the illitic soil, in relation to the pH of the system, is determined by mechanisms which range from ion-exchange to precipitation and/or co-precipitation. This provides a simple appreciation of the bonding picture, i.e., how strongly the metals can be held in relation to the type of soil fraction. The dominant mechanisms responsible for accumulation of heavy metal pollutants are sensitive to pH of the immediate environment because of the solubility of the hydroxide species of the heavy metals, as can be seen in the figure. When the pH in the porewater increases to a certain level (generally seen to be near the precipitation pH of the metal contaminant), Pb begins to form hydroxy species, resulting

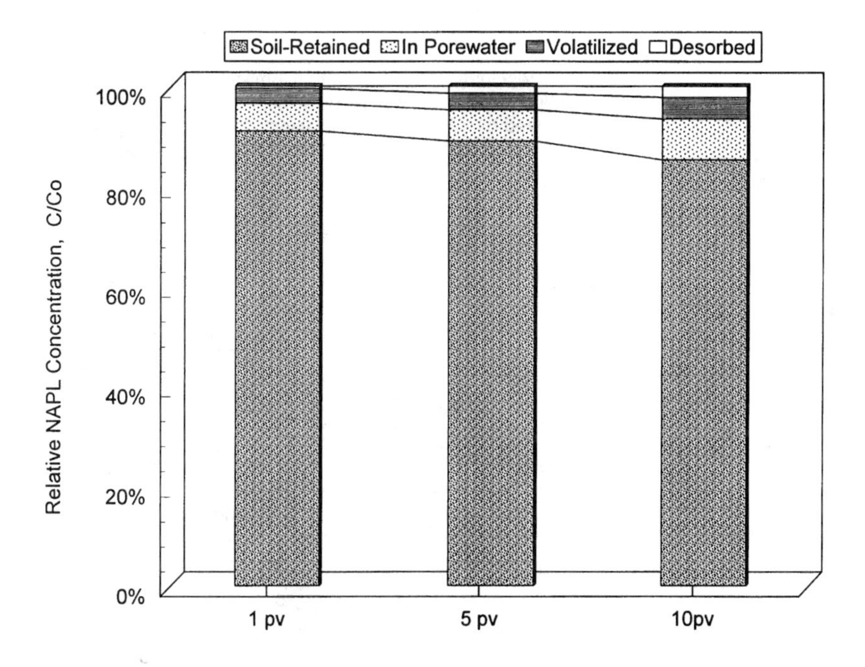

Figure 7.8 Distribution of DNAPL in a DNAPL-polluted soil column due to leaching with water after 1, 5, and 10 pv leaching. (Data from Mohammed, 1994.)

in the onset of Pb retention by the hydroxide fractions. The Pb precipitated or co-precipitated as natural carbonates can be released if the immediate environment is acidified.

Because bonding between Pb and amorphous or poorly crystallized Fe, Al, and Mn oxides is relatively strong, it is more difficult to extract the Pb from the soil. The degree of crystallization of minerals and other soil fractions plays a significant role in the bonding of contaminants and pollutants. For poorly crystallized fractions such as oxides, the kinds of bonding mechanisms with heavy metal pollutants include exchangeable forms via surface complexation with functional groups (e.g., hydroxyls, carbonyls, carboxyls, amines, etc.) and interface solutes (electrolytes), moderately fixed via precipitation and co-precipitation (amorphous), and relatively strongly bound.

Summing up what we know from the HM pollutants in soil-water systems, and the knowledge of what the interactions are between the pollutants and the soil fractions, we see that:

- pH sensitivity can be used to our advantage in removing the metal solutes from solution (porewater). This means not only that the precipitation pH of the various metal pollutants in the contaminated site needs to be determined, but also the pH

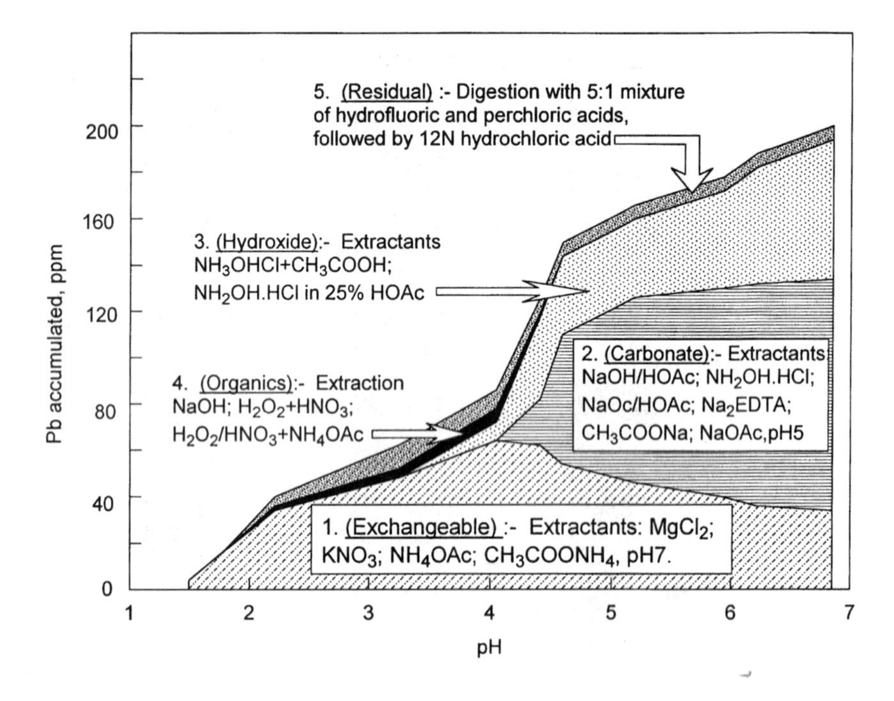

Figure 7.9 Typical extractants used as reagents to extract HM pollutants sorbed by the various soil fractions. The example shown is for Pb sorbed by an illitic soil.

relationship for sorption by the soil fractions. The results shown in Figure 7.10 illustrate not only pH sensitivity, but also the sorption distribution amongst the soil fractions and the role of preferential sorption of the heavy metals (previously discussed in Chapter 5).

- So long as low pH values can be maintained, exchange methods can be used to remove the heavy metals held as exchangeable ions. Neutral salts such as $MgCl_2$, $CaCl_2$, and $NaNO_3$, are commonly used as ion-displacing extractants to promote the release of ions physically bound by electrostatic attraction to the negatively charged sites on the soil particle surfaces. Because of the affinity of group II and II cations (valence of 2 or 3) for most surface sites, the cations in the extractant solution must be present in larger concentrations than the metal being subjected to extraction. In practice, concentrations higher than 1 *M* are widely used, although lower concentrations are sometimes favoured if natural leaching conditions are to be exploited.
- The presence of poorly crystallized oxides and organics in the soil will make it more difficult to remove the HM pollutants because these will be strongly bonded to these soil fractions. Those heavy metals that are attached to amorphous or poorly crystallized Fe, Al, and Mn oxides can be removed, e.g., amorphous ferromanganese oxyhydrates can be dissolved under the effect of redox gradients. It is necessary to choose an extraction method capable of differentiating between amorphous and crystalline oxides.

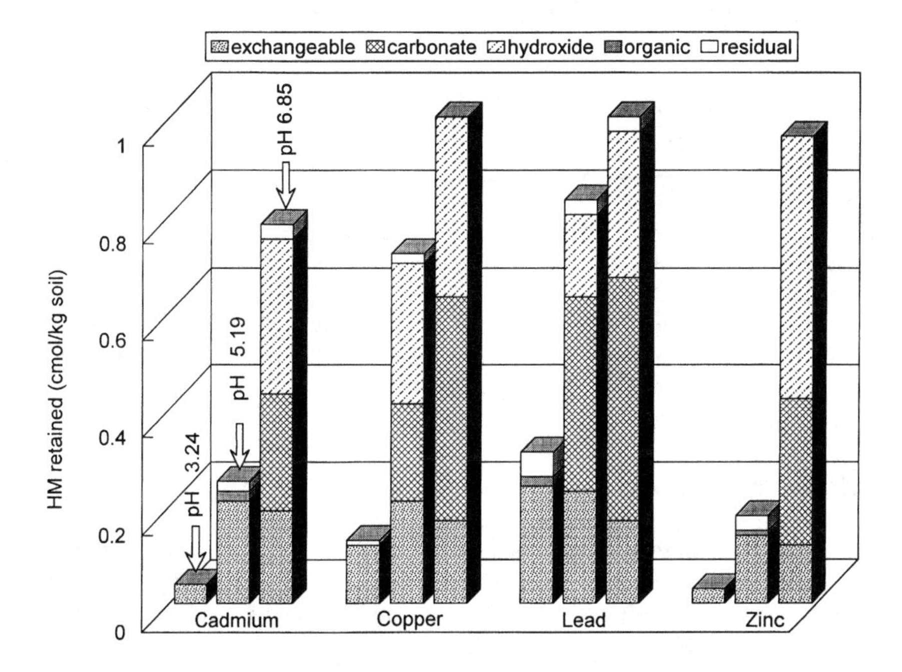

Figure 7.10 Influence of pH on distribution of HMs by illitic soil fractions. Note differences in HM species retained by the various soil fractions in relation to pH.

7.3.4 Bench-top Tests

We define *bench-top tests* as those that involve the use of compact soil masses and treatment methods designed to obtain pollutant release from the soil fractions. These are generally grouped into categories which separate along the lines of how well the generic treatment process will remove the pollutants. A good example of this is bioremediation treatments which address organic chemical pollutants, and chemical treatments which alter the bond relationships between inorganic/organic pollutants and soil fractions. Bench-scale reactor studies classify under the former category, and chemical reagent injections into bench-top samples classify under the latter category. Of the many kinds of bench-top tests, the electrokinetic types of tests have received considerable attention, to a very large extent because of the sets of driving forces developed in electrokinetic processes. The fundamental principles underlying the use of the electrokinetic remediation technique are discussed in the next section.

Bench-top studies are useful validation exercises for recipes and treatments structured from screening tests. The use of scale-model treatment walls to study interaction between reactive agents embedded in the walls and pollutants in the permeating leachate plume is a very good example of the type of bench-top studies that provide a screening procedure and the necessary input for scale-up modelling. Information obtained permits determination of efficiency of treatment and development of scale-up technology and protocols.

7.4 ELECTRODICS AND ELECTROKINETICS

The electrified interfaces in soil-water systems, established between the surfaces of soil solids and counterions which have been discussed in Chapter 3, together with the treatment of the DDL models in the same chapter, provide us with the basis for examining the use of electrodics in treatment and removal of pollutants in the soil-water system. It is not uncommon to consider only electrokinetic phenomena in relation to the properties of the interfaces and the double layer — as witness the development of technologies exploiting the electro-osmotic behaviour of the system. However, the introduction of electrodes into the soil-water system to create an electric field within which electro-osmotic flow (electrophysics) can occur will also produce chemical transformations (electrochemistry) that can significantly change the interaction properties of the pollutants. It may not be prudent to ignore the electrochemistry component.

7.4.1 Electrodics and Charge Transfer

The general principles of electrochemistry which apply to the situation of electrodes in soil-water systems is embodied in the study of *electrodics*. The introduction of electrodes in a soil-water system to create an electric field not only produces electronic conduction, but also provides for charge transfer (electron transfer) between the electrodes and solids in the soil-water system. These solids may be soil fractions, water molecules, simple ions, or pollutants.

The transfer mechanisms are expressed as *electronation* and *de-electronation*. Charge transfer, i.e., electron transfer, implies the existence of an electric current across the interface separating the electrode and the solids. Figure 7.11 shows the basic elements of the transfers involving electrodes, electron acceptors (oxidants), and electron donors (reductants). In the single electrode system shown in the figure, charge (electron) transfer or movement from the electron source into the soil-water system occurs if the electrode (shown in the diagram) is a cathode. The transfer of electrons to electron acceptors (oxidants) in the soil-water system is defined as the *electronation* process, and is seen to be a reductive process.

If, on the other hand, the single electrode shown in Figure 7.11 is an anode, this renders it an electron sink. The transfer of electrons from electron donors (reductants) in the soil-water system to the electron sink (anode) is called a *de-electronation* process, and is an oxidative process. The transfer of electrons to or from the electrode bears directly on the direction of (electric) current flow. The direction of current flow at the electrode interface is controlled by whether electronation or de-electronation is greater. If the sum of the electronation and de-electronation processes is zero, no net current flow from charge transfer occurs. These processes have direct relevance in control of the fate of pollutants since these are either electron acceptors or electron donors. By accepting an electron from the cathode or donating an electron to the anode, the valence or oxidation state of the pollutant will be changed. To a certain extent therefore, charge transfer produced by electrodes in a soil-water system can result in chemical transformation of those elements that accept or donate electrons.

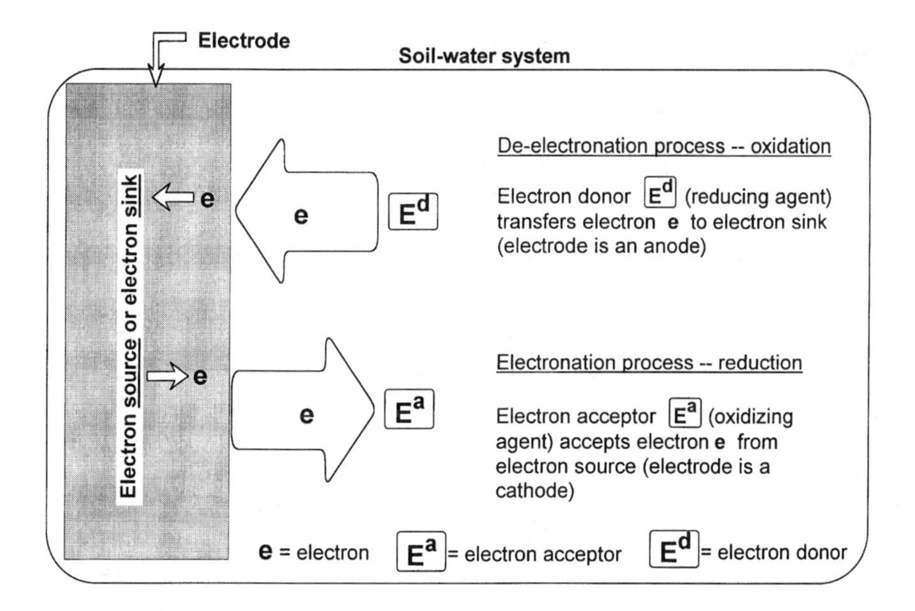

Figure 7.11 Charge transfer between electrode and electron acceptor (electronation) and between electron donor and electrode (de-electronation).

By providing a two-electrode system with a source of direct current, i.e., a constant unidirectional flow of electrons, as shown in Figure 7.12, the current flow can be directed. Electrons exiting from the cathode provide the means for electronation, i.e., the electrons transferred to those solids (water molecules, compounds, ions, etc.) that are electron acceptors will result in reduction of the solids. Using an ion as an example, electronation will result in a reduction of the positive charge of the ion. The flow of charges in the system is defined by the nature of the charge carrier. In the electrode, the charge carrier is the electrons. However, in the soil-water system, the carrier is the ions (positive and negative). To maintain current flow, the charge carrier changes from electrons in the cathode to ions (positive and negative) in the soil-water system — and vice versa when the ions change to electrons at the anode. At the anode, the de-electronation that occurs when the electron donors (reductants) donate their electrons to the anode will result in the oxidation of the electron donors. In the portions of the soil-water system away from the electrodes, movement of the electron donors or electron acceptors to the respective electrodes will ensure that the transfers occur as a continuous process. The amount and rate of movement of the current across the soil-water system is thus seen to be composed of the intensity of the input from the cathode (DC power supply) and the nature of the constituents that define the soil-water system.

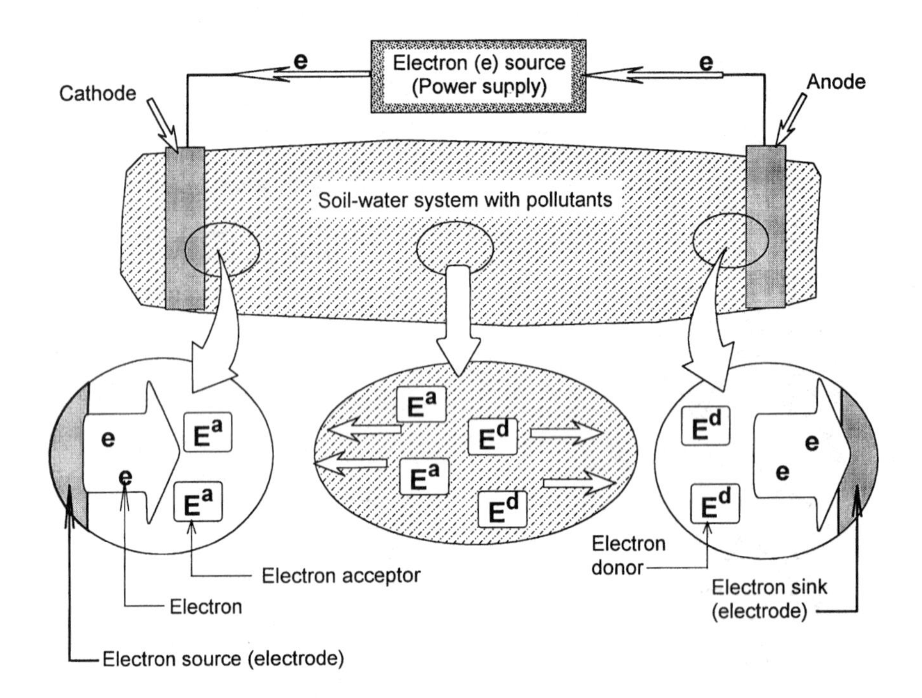

Figure 7.12 Cathode-anode electrodic system. Current flow rate will be defined by DC power supply and nature of constituents in the soil-water system.

The dissociation reactions occurring in water in conjunction with charge transfer will produce resultant effects which can seriously impact the efficiency of pollutant removal for situations where electrokinetic phenomena are used. In particular, the dissociation reactions at the cathodes and anodes in respect to water molecules need to be considered. At the *cathode*, we will obtain:

$$2H_2O + 2e^- \rightleftharpoons H_2(g) + 2OH^-\,(aq)$$

and at the *anode*, the following will be obtained:

$$H_2O \rightleftharpoons ½O_2(g) + 2H^+(aq) + 2e^-$$

A change in the acidity of the soil-water system at the anode will result from the production of the H^+ ions (i.e., dissociated H^+ ions). These could interact with the sorbed cations and would move toward the cathode under the influence of concentration gradients. At the other electrode (cathode), the production of OH^- ions will also cause a change in the acidity in the immediate vicinity.

7.4.2 Electrokinetics and Pollutant Removal

We classify electrokinetic removal of pollutants in soils as a special case of bench-top tests because it requires special equipment constructed to impart the electrodic potentials. It is also a special case because the application of electrokinetics is a direct recognition of the electrokinetic phenomena which is well demonstrated in the DDL models discussed previously in Chapter 3. The various types of electrokinetic phenomena associated with charged surfaces of soil fractions in a soil-water system include:

- **Electro-osmosis** — Electric potential-driven movement of liquid relative to a stationary charged body;
- **Electrophoresis** — Electric potential-driven movement of charged particles relative to a stationary liquid (reverse of electro-osmosis);
- **Streaming potential** — Development of electric potential due to liquid flow across stationary charged bodies; and
- **Sedimentation potential** — Development of electric potential due to movement of charged bodies relative to a stationary liquid.

The electro-osmotic and electrophoretic types of electrokinetic phenomena are the principal types involved in HM pollutant removal considerations. This assumes that advective flow is of little consequence, thus obviating the need for consideration of streaming potentials, and also assumes that compact soils are the subject of interest. In the case of sediments, however, the sedimentation potential can be of some consequence.

The basic elements of electrokinetic phenomena involved in soil-water systems have long been exploited in general geotechnical engineering practice, as technological application of the electro-osmotic process to groundwater dewatering and consolidation of soft ground. Application of the technology for removal of pollutants will be addressed in the next chapter. For the moment, we want to examine the basic conditions established when electrokinetics are utilized as the basis for pollutant removal, as shown in the top diagram in Figure 7.13 in respect to the electro-osmotic flow of electrolytes in a capillary tube. We make use of the fact that the positive ions in an electrolytic solution in an electric field will move to a cathode — with negative ions moving to the anode. The imposition of an electric field in the horizontal capillary tube shown in the top diagram will produce a resultant flow of electrolytes, i.e., the potential difference across the capillary tube causes the electro-osmotic flow. The electrolytes and their associated water molecules will move as bulk water. This movement of the bulk phase (liquid) relative to the capillary tube is defined as an *electrokinetic phenomenon*. The viscous force resisting flow will be a function of the viscosity of the bulk phase and the velocity.

The bottom diagram shown in Figure 7.13 illustrates the diffuse ion-layer bulk flow of the M^{nx} ions with their hydration shells. The thickness of this layer is given by the inverse of the Debye-Hückel parameter χ, i.e., thickness = $1/\chi$. The viscous resistance is a direct function of the viscosity of the fluid phase that constitutes the

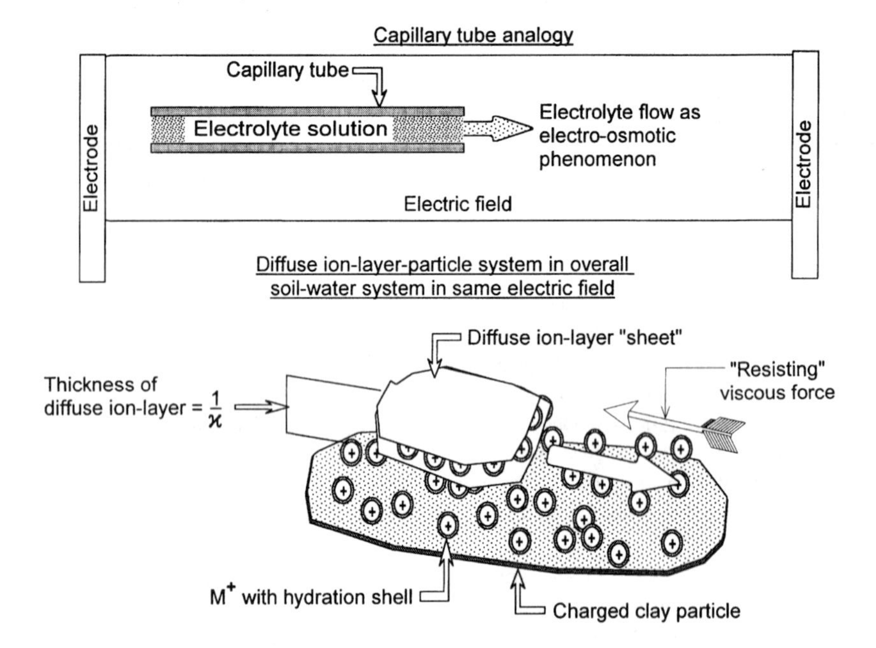

Figure 7.13 Electro-osmotic flow induced by an electric field. Top diagram shows the capillary tube analogy using an electrolytic solution, and the bottom diagram shows the "planar" situation represented by the diffuse-ion layer and the associated Stern layer-particle system.

porewater and the velocity of flow. Resistance to flow is also offered by the ψ_{ohp}. If we consider the confining boundaries of the bulk flow as the *ohp* (outer Helmholtz plane), the velocity is nearly zero at the boundaries and increases in a parabolic manner to a maximum at the midpoint between the boundaries.

The several processes that occur in the application of an electric field to generate electrokinetic flow include:

- Dissociation and ionization;
- Changes in the redox potential and acidity at the electrodic regions;
- Interactions between the H^+ ions and the M^{nx} sorbed onto the soil particle surfaces;
- Movement of the H^+ and OH^- ions;
- Solubilization and precipitation of the metal pollutants;
- Oxidation-reduction reactions in the vicinity of the electrodes, resulting in changes in the oxidation state of the reacting solids (ions, molecules, etc.); and
- Electrophoretic flow.

For compact soil-water systems such as those which constitute the substrate material in a contaminated site, i.e., in situ soil, eletrophoretic flow is generally considered to be negligible. The results of the various reaction-transformation processes occurring in a pollutant-soil-water system in an electric field have been obtained in bench-top tests. Published results show that not all of the processes contributing to the movement of the target pollutants have been considered in the investigations. In particular, the consequences of electronation and de-electronation processes and oxidation-reduction reactions of the soil fractions and pollutants have not been addressed. Consideration of all these independent and interdependent processes in the application of electrokinetic phenomena as a pollutant removal tool would be required. For this reason, bench-type studies can prove to be very useful.

7.5 BIOCHEMICAL REACTIONS AND POLLUTANTS

As in Chapter 6, the study of transformations of organic chemical compounds by processes associated with the activities of microorganisms has been comprehensively addressed in the many textbooks dealing with soil microbiology, environmental organic chemistry, biogeochemistry, etc. The interested reader should consult these for detailed treatments of the various processes leading to transformation and mineralization of organic chemical compounds. For the present, we are interested in highlighting some of the issues of pollutant-soil interaction in regard to the problem of elimination or reduction of organic chemical pollutant concentration (mitigation).

Attachments formed between organic chemicals and soil solids are very dependent on the nature of the chemical compounds and the soil constituents. These attachments and the nature of the chemical will control the type of decontamination procedure used in site remediation. Because organic chemical pollutants are particularly susceptible to biotic/abiotic processes, reactions, and transformations, the bonds formed between these chemicals and soil fractions can change as a result of these processes and changes. Determination or assessment of the fate of these chemicals must account for the changes in the properties, characteristics of the chemicals, and the changes in the bonding relationships. Organic chemical pollutants include organic acids and bases, and non-aqueous phase liquids (NAPLs). Some of the basic mechanisms of interaction and bonding include: (a) ion exchange involving organic cations (anions) and reactive surfaces of soil fractions; (b) hydrogen bonding; (c) π bonding; (d) covalent bonding; (e) ion-dipole and coordination; and (f) van der Waals forces.

The use of microorganisms in mitigation of pollution by inorganic and organic pollutants has been recognized as one of the important tools that can contribute significantly to the remediation of contaminated soils. Pollution mitigation (partial or complete removal of the pollutants) by biochemical means can take the form of reduction in concentration of the pollutants, transformation of the inorganic and/or organic chemical compounds, and mineralization of the organic chemical compounds, i.e., converted to CO_2 and H_2O. Changes in toxicity of the pollutants which can also result from application of specific biochemical treatments is not considered within the scope of this discussion.

The soil-water system in its natural state has shown itself to be capable of supporting a considerable variety of microhabitats. The distribution of the various soil fractions including soil organic matter, and the availability of macro- and micronutrients provide for the variety in the microhabitats. The distribution and nature of the microbial activities are affected by the nature of the interfaces presented by the soil fractions and SOM, and the matric potential ψ_m of the soil-water system. The optimal soil bacteria activity is generally found to exist when ψ_m is greater than –10 Mpa. Soil bacteria participate not only in the oxidation-reduction reactions of nitrogen, sulphur, phosphorous, and many other elements, but also in the degradation and transformation of organic chemicals.

7.5.1 Nitrogen and Sulphur Cycles

Nitrogen and sulphur cycles in the biosphere have been extensively studied and reviewed. Some significant portions of the various cycles are directly related to the activities of microorganisms in the soil-water system. These impact on the pollutant-soil relationships which are of direct concern in pollution-mitigation and pollutant-removal considerations. Both aerobic (micro-aerophilic) and anaerobic microorganisms contribute to the biochemical reactions and the fate of the pollutants. While *eukaryotes* (cells with true nucleus; generally <2 μm) and *prokaryotes* (cells without true nucleus; generally >2 μm) utilize terminal electron acceptors, prokaryotes constitute the larger proportion of microorganisms that utilize a greater range of terminal electron acceptors — e.g., NO_3^-, SO_4^{2-}, Fe^{3+}.

Although nitrogen N can exist in valence forms that range from a valence of +5 (represented by NO_3^-) to a valence of –3 (represented by NH_4^+), it is the reduced form that is most relevant to the problem at hand. Ammonification of biomass, which relates directly to the microbial decomposition of the biomass, provides the most reduced form of nitrogen, NH_3 or NH_4^+. Oxidation of NH_4^+ to NO_3^- by chemolithotrophic prokaryotes, known otherwise as nitrifying bacteria, occurs under oxic conditions. This process, which is generally known as *nitrification*, is sensitive to redox and temperature. Redox values lower than 200 mV are considered to be inhibitory to nitrification. Oxidized nitrogen compounds result from dissimilatory nitrate reduction. These can serve as terminal electron acceptors by other microorganisms under anoxic conditions.

As with nitrogen, sulphur S can exist also in valence forms that range from a valence of +6 (represented by SO_4^{2-}) to a valence of –2 (represented by S^{2-}). Amino acids such as sulphydril thiol groups (R–SH), which are sulphur analogues of alcohol, can be *desulphydrated* by both eukaryotes and prokaryotes, releasing S^{2-}. Abiotic and biotic oxidation of the S^{2-} will result in the production of SO_4^{2-}. In turn, reduction of the sulphate provides for the formation of iron sulphides and pyrite (FeS_2). The sulphides have a tendency to form on metal surfaces, and have detrimental effects on electrodes used in processes designed to elicit metal extraction via application of electrokinetic phenomena. In contrast to the phenomena and problems introduced by sulphate reducing bacteria, the documented problems introduced by S-oxidizing bacteria in the form of *acid mine drainage* (AMD) is perhaps the most publicized. Acid production and transport due to the oxidation of FeS_2 to H_2SO_4 constitutes one of the major problems of metal mining industries:

$$FeS_2 + H_2O + 3\tfrac{1}{2}O_2 \rightarrow FeSO_4 + 2H_2SO_4$$

7.5.2 Pollutant–Soil Bond Disruption

Application of chemical procedures for removal of organic chemical pollutants can benefit from a knowledge of the interactions between the functional groups of the organic chemicals and the soil fractions. As we have seen in Section 6.2, chemical properties of the functional groups of the soil fractions will influence the acidity of the soil particles — a significant property of the soil, since surface acidity is very important in the adsorption of ionizable organic molecules of clays. Surface acidity is a major factor in clay adsorption of amines, *s*-triazines, amides and substituted ureas where protonation takes place on the carbonyl group. Many organic molecules (amine, alcohol, and carbonyl groups) are positively charged by protonation and are adsorbed on clays, the extent of which depends on the CEC of the clay minerals, the amount of reactive surfaces, and the molecular weight of the organic cations. Large organic cations are adsorbed more strongly than inorganic cations because they are longer and have higher molecular weights.

Polymeric hydroxyl cations are adsorbed in preference to monomeric species not only because of the lower hydration energies, but also because of the higher positive charges and stronger interactive electrostatic forces. The hydroxyl groups in organic chemical compounds comprise two broad classes of compounds, alcohols (ethyl, methyl isopropyl, etc.), and phenols (monohydric and polyhydric), and the two types of compound functional groups are those having a *C–O* bond (carboxyl, carbonyl, methoxyl, etc.) and the nitrogen-bonding group (amine and nitrile).

Carbonyl compounds possess dipole moments as a result of the unsymmetrically shared electrons in the double bond, and are adsorbed on clay minerals by hydrogen bonding between the *OH* group of the adsorbent and the carbonyl group of the ketone or through a water bridge. On the other hand, the carbonyl group of organic acids such as benzoic and acetic acids interacts directly with the interlayer of cation or by forming a hydrogen bond with the water molecules (water bridging) coordinated to the exchangeable cation of the clay complex.

The NH_2 functional group of amines can protonate in soil, thereby replacing inorganic cations from the clay complex by ion exchange. The phenolic function group, which consists of a hydroxyl attached directly to a carbon atom of an aromatic, can combine with other components such as pesticides, alcohol, and the hydrocarbons to form new compounds. The sulfoxide group, which is a polar organic functional group, forms complexes through either the sulphur or oxygen atom. Complexes are formed with transition metals and with exchangeable cations.

The chemical treatment processes for removal of organic contaminants that appear to be feasible include:

- **Base-catalyzed hydrolysis** — Using water with lime or *NaOH*. For application to contaminants such as esters, amides, and carbamates;
- **Polymerization** — Using catalyst activation, i.e., conversion of the compound to a larger chemical multiple of itself. For application to aliphatic, aromatic, and oxygenated monomers;
- **Oxidation** — For application to benzene, phenols PAHs, etc.; and
- **Reduction** — For application to nitro and chlorinated aromatics.

Removal of sorbed pollutants (onto soil fractions' surfaces) can be facilitated through disruption of the bonds established between the pollutants and soil particles. For example, the release of organic pollutants sorbed onto the surfaces of soil fractions has been generally obtained through chemical washing. The objective of such a technique is to obtain bond disruption or breakage through dissolution of the functional groups or surfaces at the bond-soil interface. Solvents are the chemical agents that have shown good success as soil-washing agents. However, the problems arising therefrom are not dissimilar to the initial pollution problem, i.e., the solvents themselves become pollutants and will obviously require application of treatment and/or recovery technology. These are addressed in the next chapter.

We generally associate the use of surfactants with organic chemical pollutants in soil. However, surfactants for reducing or disrupting organic pollutant-soil bonding relationships can prove to be very useful in releasing sorbed organic pollutants, and particularly in reducing the surface tension of the aqueous phase. Thus whilst they will increase the apparent solubility of the pollutant in water, they will also reduce the interfacial tension between the pollutants and the reactive surfaces of the soil fractions. Other advantages include enhancement of in situ biodegradation of hydrophobic organic chemical pollutants.

In general, surfactants are chemical compounds containing a hydrophobic and hydrophilic moiety. They are pictured as a compound with a hydrophilic head and a hydrophobic tail (Figure 7.14). The hydrophilic head can be negatively charged (cationic), positively charged (anionic), neutral (nonionic), or zwitterionic (plus and minus charge). The compositional features render them favourable for alignment at interfaces. This facilitates removal of inorganic and organic chemical pollutants — through micellular solubilization and emulsion formation in combination with lowered surface tension (Mulligan, 1998). These surfactants can consist of adhesives, wetting agents, de-emulsifiers, foaming agents, and biosurfactants which are produced from yeast or bacteria (Mulligan and Gibbs, 1993; Mulligan et al., 1999b). Selection of the type of surfactant depends on the nature of the soil fractions and pollutants, site specific conditions, and recoverability of the surfactants. The principal objective in the use of surfactants is the reduction of the interfacial energy at the interface between the pollutant and the soil fraction surface.

The results and discussion regarding the use of biosurfactants in enhancing pollutant removal through disruption of pollutant-soil bonding draws heavily on the work of Mulligan (1998), and Mulligan et al. (1999a, 1999b). Disruption of pollutant-soil bonding is indirectly obtained by the lowering of the surface tension of the medium within which the surfactants are found. For example, Mulligan (1998) reports that reduction in surface tension of water from 72 to 35 mN/m has been obtained, and that in the case of water against *n*-hexadecane, a reduction in interfacial tension from 40 to 1 mN/m has been recorded. The mechanism by which these changes in interfacial energies are obtained relies on the concentration of the surfactants at the interfaces between immiscible bodies, e.g., liquid-solid, liquid-liquid, and vapour-liquid. A correlated relationship exists between the concentration of surfactants and surface tension. As the concentration increases, surface tension is reduced until a minimum value is obtained. The concentration of surfactants at that point is defined as the critical micelle concentration (CMC).

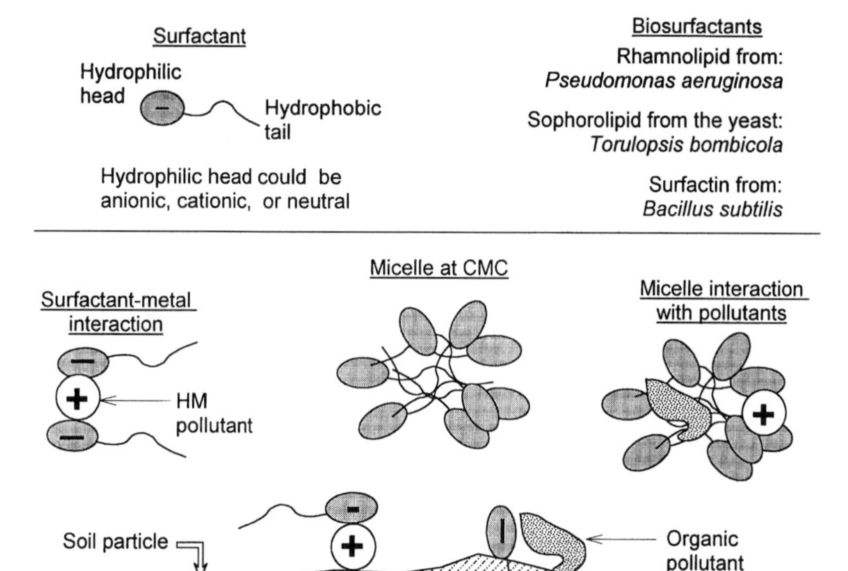

Figure 7.14 Some biosurfactants and mechanism of interaction with pollutants in the soil-water system. Note CMC = critical micelle concentration. (Adapted from Mulligan, 1998.)

Biosurfactants are synthesized as metabolic byproducts of yeasts and/or bacteria. These can be categorized into groups known as glycolipids, phospholipids, fatty acids and neutral acids, and are predominantly anionic or neutral (Biermann et al., 1987). As in the case of synthetic surfactants, biosurfactants are amphiphilic, with hydrophilic heads and hydrophobic tails. The diagram in Figure 7.14 shows the three biosurfactants obtained by Mulligan (1998) together with the likely mechanisms of entrapment of pollutants in the soil-water system. The efficiency of one of the biosurfactants (i.e., surfactin produced through fermentation of *Bacillus subtilis,* Figure 7.14) in removal of HMs in a soil heavily contaminated with both HMs and hydrocarbons is shown in Figure 7.15. Increased efficiency is obtained when surfactin is used in combination with NaOH.

7.5.3 Biotic Redox and Microcosm Studies

Biotic redox conditions in the environment defined by the soil-water system will produce activities from different microorganisms, depending on the kinds of electron acceptors available. Those electron acceptors that can offer the greatest positive reduction will provide the controlling microorganism. Depletion of the particular "greatest positive reduction" electron acceptors will lead to the utilization of other

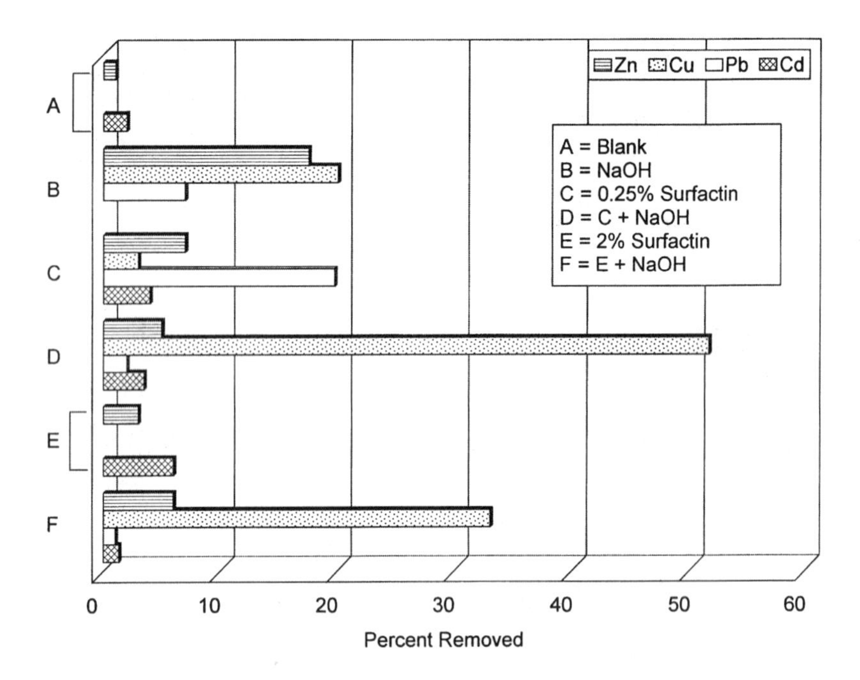

Figure 7.15 Biosurfactant treatment for removal of soil contaminated with HMs and organic chemical pollutants. (Data from Mulligan and Yong, 1999.)

alternative electron acceptors by other microorganisms. One moves from aerobic respiration where molecular oxygen serves as the terminal electron acceptor, to anaerobic processes that utilize oxidizing agents such as iron (III) and manganese (III/IV). The redox sequence that begins from the aerobic condition has been categorized as *oxic* (oxygen-available) at the aerobic end, to *anoxic* (oxygen-deprived) at the other end (anaerobic condition). In between, the term *suboxic* has been used to include such processes as denitrification and sulphate reduction (Section 7.5.1). We should note that microorganisms are always in competition with each other, and that the greater or lesser presence of the various species will affect or control the conditions in their environment.

As previously discussed in Section 6.4, transformation of organic chemical pollutants results from processes associated with: (a) oxidation — using electrophiles; (b) reduction — using nucleophiles; and (c) hydrolysis — which occurs through enzymatically mediated nucleophilic attack (Schwarzenbach et al., 1993). The resultant transformations provide compounds that would be more metabolically suitable for the available microorganisms. Mineralization of organic chemical pollutants can only be achieved through biotic transformation processes. Microcosm studies are useful since they can provide for a controlled environment for evaluation of the degradation of the organic chemical pollutant under study. For example, the mineralization of a PAH in soil could be examined as a process that includes the presence

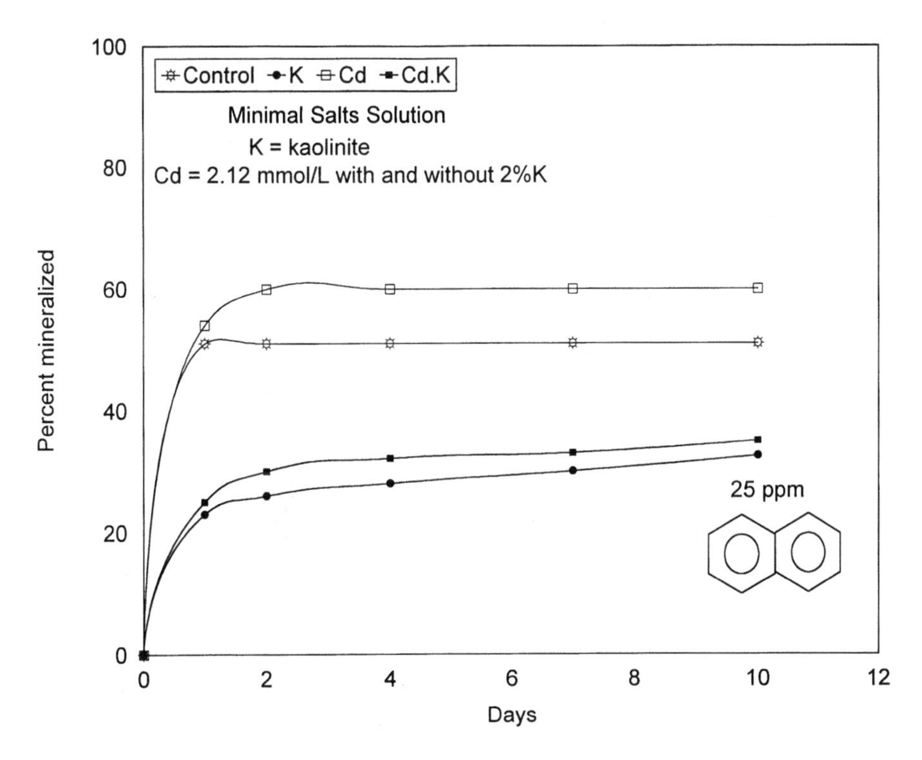

Figure 7.16 Mineralization of naphthalene by *Pseudomonas putida*. Note influence of Cd and kaolinite soil on mineralization (data from Hibbeln, 1996).

of a heavy metal as a possible inhibitory factor in microbial metabolism. The mineralization results shown in Figures 7.16 and 7.17 for naphthalene and 2 methyl naphthalene are obtained from the microcosm studies reported by Hibbeln (1996). These show the effects of presence of Cd as an inhibitory factor to mineralization. The results show degree of mineralization by *Pseudomonas putida* of the 2 PAHs with and without the presence of Cd and/or kaolinite soil.

The mineral salts medium (MSM) used by Hibbeln (1996) provided for a buffered system that could compensate for the H^+ ion concentration released during degradation of the organic chemical. It would appear that the heavy metal Cd, although not ordinarily identified as an essential element for growth of microorganisms, did function as such — as witness the increase in mineralization rate. This is curious since Cd is expected to affect the respiration of microorganisms, and has been reported to decrease the rate of mineralization. The normal species of trace elements commonly recognised as essential elements for growth and function include Mn, Fe, Cu, Co, Zn, and Mo. On the other hand, the results show that the presence of the reactive surfaces of the kaolinite clay soil provided for sorption of some of the salts in the MSM, contributing thereby to the reduction in the rate of mineralization of the organic chemicals. The salts Na, K, and Mg are commonly associated with the bulk elements required for growth of microorganisms.

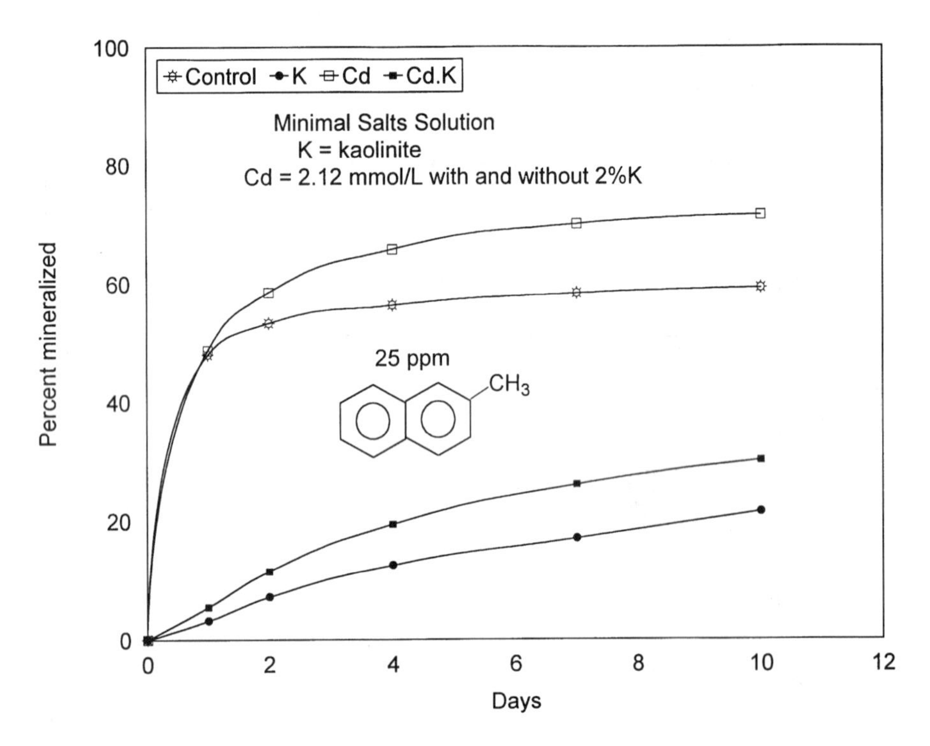

Figure 7.17 Mineralization of 2-methyl naphthalene by *Pseudomonas putida*. Note influence of Cd and kaolinite soil on mineralization (data from Hibbeln, 1996).

7.6 ASSESSMENT, SCREENING, AND TREATABILITY

The term *treatability* has most often been used in conjunction with remedial treatment (biodegradability) of organic chemical pollutants. The term *remedial treatment* is used deliberately to emphasize that the treatment procedures used are designed to reduce and/or eliminate the pollutants (contaminants) from the contaminated soil. Strictly speaking, *treatability* refers to how well or how effective remedial treatment of contaminated soil can be achieved — independent of the type of pollutant. In the broader definition which includes treatability of heavy metals, for example, we need to be concerned with the removal of the HM from the porewater as well as removal (detachment) of the HM from the soil fractions. Assessment of treatability of a contaminated soil requires knowledge of whether the pollutant can be extracted or degraded, as the case may be. Assessment of treatability of pollutants in the porewater is performed as a water treatment procedure — separate from a soil-water system. This means to say that: (a) pollutants are assessed in respect to whether they can be detached from the soil solids, and (b) pollutants in the porewater are subject to procedures of treatment assessment in common with water treatment protocols. A simple scheme for assessment would include the elements shown in Figure 7.18. The objective of screening tests is to arrive at the best combination of parameters and factors for remedial treatment. When applied to determination of

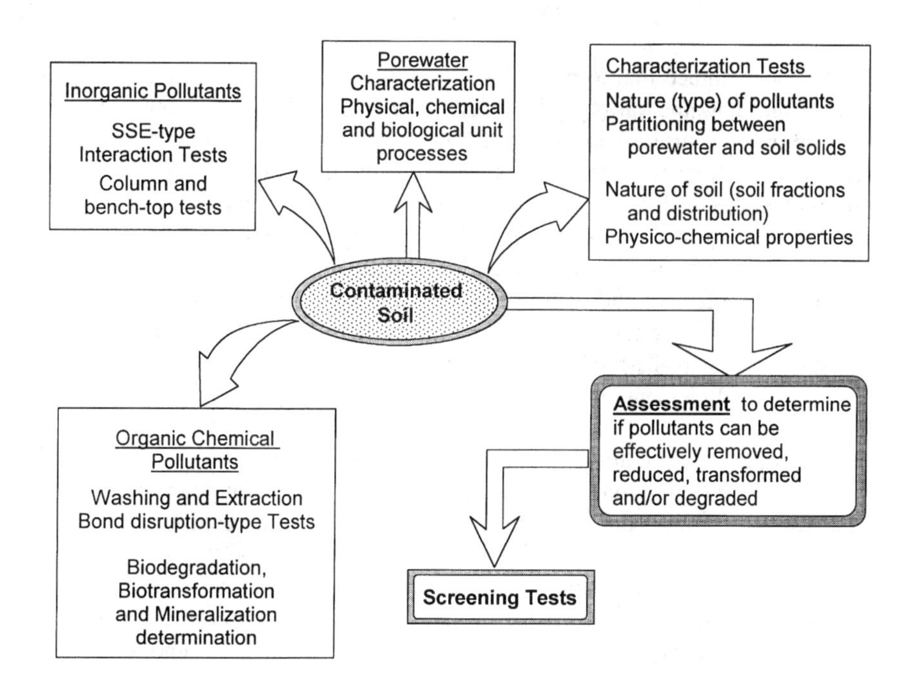

Figure 7.18 Tests and analyses required to support assessment of potential for remedial treatment of a contaminated soil.

best bioremediation technique for ground contamination by organic chemical pollutants, tests include determination of microbial population that can be metabolically active, oxygen uptake, pH, plate counts, respiration, nutrient conditions etc. In the case of inorganic pollutants such as heavy metals, best or most efficient HM extraction screening will test for the most efficient sets of reagents at various concentrations and pH levels. It is these sets of parameters and factors (from screening tests) that will be used to structure implementation of remedial treatment technology for application to site conditions.

Treatability studies are the prototype experiments and studies that can either be conducted at bench level or in the field. Since scale-up quantities, rates, etc. are required as final output from treatability studies, it is necessary to structure the study that will provide information on unit quantities, reaction rates, environmental influences, site-specific conditions, and physical (pollutant) removal capabilities. Mass balance reckoning is important.

Studies on biodegradability, for example, show that petroleum hydrocarbons (PHCs) are amenable to degradation, and that the lighter fractions are more rapidly degraded in contrast to the heavier fractions. Solubility is a key factor, as is the structure of the compound. MAHs degrade much easier than PAHs. Residual levels obtained in degradation are not uniform, again depending on the complexity of the structure and the solubility of the compound.

7.7 CONCLUDING REMARKS

It has been argued that to provide for effective site remediation, it is important to understand how the various pollutants are retained in the soil-water (substrate) system. Why? Because a knowledge of how the pollutants are retained by the soil fractions will tell us what techniques should be used to fully remove them from these soil factions. This means obtaining the proper information relating to the kinds and distribution of pollutants, and the manner by which they are held within the substrate. Until such is achieved, it is difficult to devise a clean-up strategy that would be both effective and compatible, i.e., consistent with the manner in which the contaminants are retained within the soil system. What we need to avoid is application of the "black-box" technique for site remediation.

The essential elements of ground contamination, and the problems of seeking and implementing rehabilitation and treatment technology compatible with the type and extent of ground pollution, require a proper appreciation of what constitutes the basis for pollutant retention in the soil. Until the processes that determine the fate of the pollutants in the contaminated site are known, it would be difficult to arrive at the best method required to remove the contaminants. The "shopping list" of generic techniques for detachment of pollutants from soil solids and porewater, given below, utilizes a working knowledge of the interactions and processes involved in retention of pollutants in the soil. These basic techniques include, amongst others:

> Solvent and surfactant extraction, air stripping, steam stripping, acid extraction treatment, electrodialysis, ion exchange and neutralization, electrokinetics, abiotic oxidation and reduction reactions, biotic oxidation and reduction, reverse osmosis, membrane processes, infrared irradiation, chemical dechlorination, bioscrubbing, and adsorptive filtration.

Application of any of the techniques in the "shopping list" for remedial treatment of contaminated soil requires development of appropriate and compatible technology which must satisfy environmental, land use, and economic requirements. This is a challenging task.

CHAPTER 8

Remediation and Pollution Mitigation

8.1 INTRODUCTION

Implementation of effective techniques and procedures for treatment of contaminated sites to remove or minimize the concentration of pollutants constitutes the fundamental aim of remediation and pollution mitigation programs. The previous chapter has addressed the need for development of effective and compatible techniques for site decontamination (pollutant removal) based upon a proper understanding of the nature of the problem, and the processes involved in pollutant fate determination.

In this chapter, some of the generic procedures for pollutant removal will be examined insofar as they relate to the pollutant-removal and pollution-mitigation issues. In addition to the present standard procedures available for treatment of contaminated sites, innovative procedures and technologies are continuously being developed. It is recognized that it is not always necessary to completely remove all pollutants from a contaminated site. It is not unusual to find that complete pollutant removal could be prohibitively expensive, and may not be necessary since residual pollutant concentrations (i.e., pollutants remaining after clean-up) would be considerably below regulatory limits and limits defined by health-protection standards. Reduction of pollutant concentration below critical limits (i.e., pollution mitigation) is therefore a serious alternative.

There are many ways of approaching site remediation implementation. It is useful to follow a protocol of procedures that would eliminate inefficient procedures, as shown, for example by the requirements and procedures developed in Figures 7.1 and 7.18. These are captured in Figure 8.1. The sets of general information and protocols needed for assessment of site contamination, and treatment required to provide for effective clean-up are shown in the diagram.

8.2 POLLUTANTS AND SITE CONTAMINATION

Experience shows that very few contaminated sites are contaminated by one species (type) of pollutants. Generally, one finds various kinds of organic or inorganic contaminants (pollutants) or mixtures of these in contaminated soil, thus making it

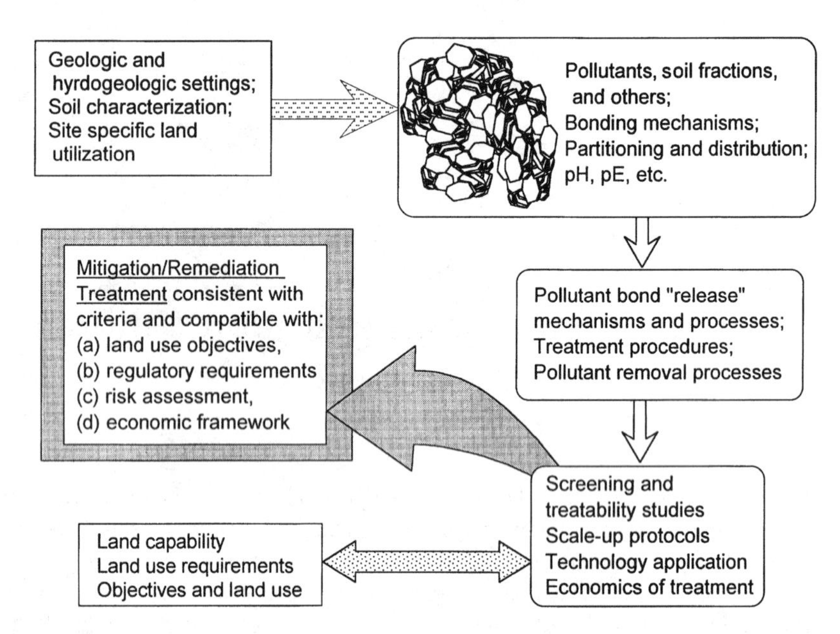

Figure 8.1 Requirements and procedures in assessment of remediation-treatment of a contaminated site.

difficult to structure a one-step remedial treatment technique that can effectively remove the spectrum of pollutants in the contaminated soil. For remedial treatments to be effective, it is essential to match the treatment technique with the nature of the pollutants in the site and their bonding with the soil fractions. The use of treatment procedures as "black-box procedures" is not prudent since it is likely that this:

- Would limit improvement of decontamination capability;
- Would limit introduction of innovative techniques;
- Could lead to application of inappropriate and incompatible technology; and
- Could develop unexpected and perhaps adverse reactions or treatment products.

8.2.1 Pollution Mitigation, Elimination, and Management

The first and foremost requirement in remedial treatment of a contaminated site is to eliminate the health and environmental threats posed by the presence of pollutants in the contaminated site. This requires management of the pollutants in the contaminated site, and can be achieved by:

- **Total removal of all the pollutants** — This meets the requirement of a pristine site. Both aggressive remedial treatment and the traditional "dig and dump" (to be replaced by clean fill) are likely candidate procedures. Removal of all sorbed pollutants and also all pollutants transferred to (and originally in) porewater will

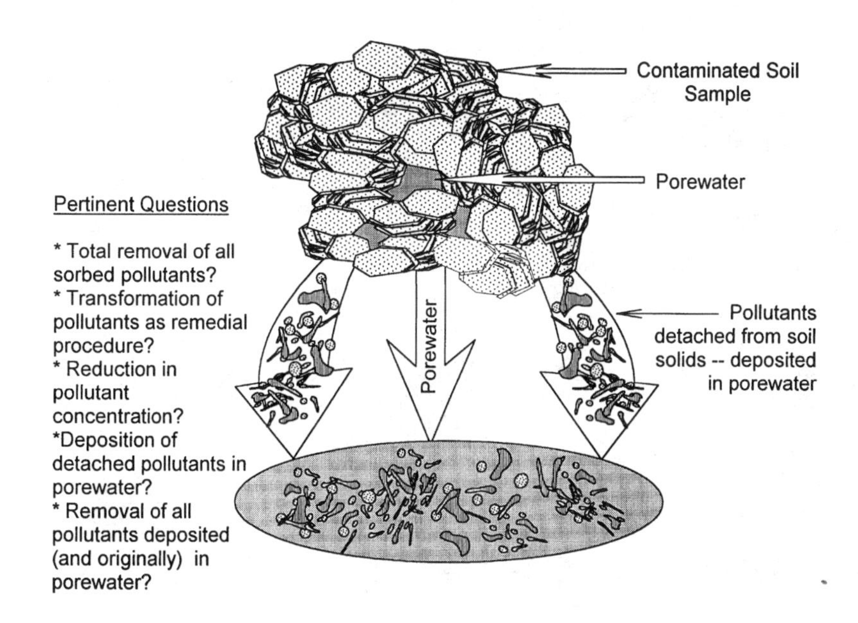

Figure 8.2 Principal elements in consideration of in situ and ex situ remedial treatment.

be required. Measurements of likelihood of presence of residual pollutants is required (Figure 8.2);

- **Reduction of concentration of the pollutants to levels below critical (allowable) levels** — This requires remedial treatments and measurements of "residual" concentrations of pollutants and assurance that they would not become environmentally mobile;
- **Immobilizing the pollutants to ensure no movement of the pollutants from their fixed (immobilized) positions** — Solidification and stabilization procedures are the most likely candidate procedures. Monitoring is a key requirement; and
- **Containment of the pollutants in situ** — By constructing impermeable cells or barriers to contain the pollutants. Management of pollutant transport through the cell walls or barriers is a prime requirement, together with monitoring.

It is clear that "return to pristine conditions" is an objective that will never be easily met. This is due to either one or both of the following: (a) technical requirements and available technology, and (b) economics of required treatment. The basic elements shown in Figure 8.2 demonstrate that in the initial stages, detached pollutants (from soil solids) will be transferred to the porewater. Removal of all pollutants from the porewater will be required as an integral element of the total remedial treatment process. It should be fairly clear that the remedial treatment process will

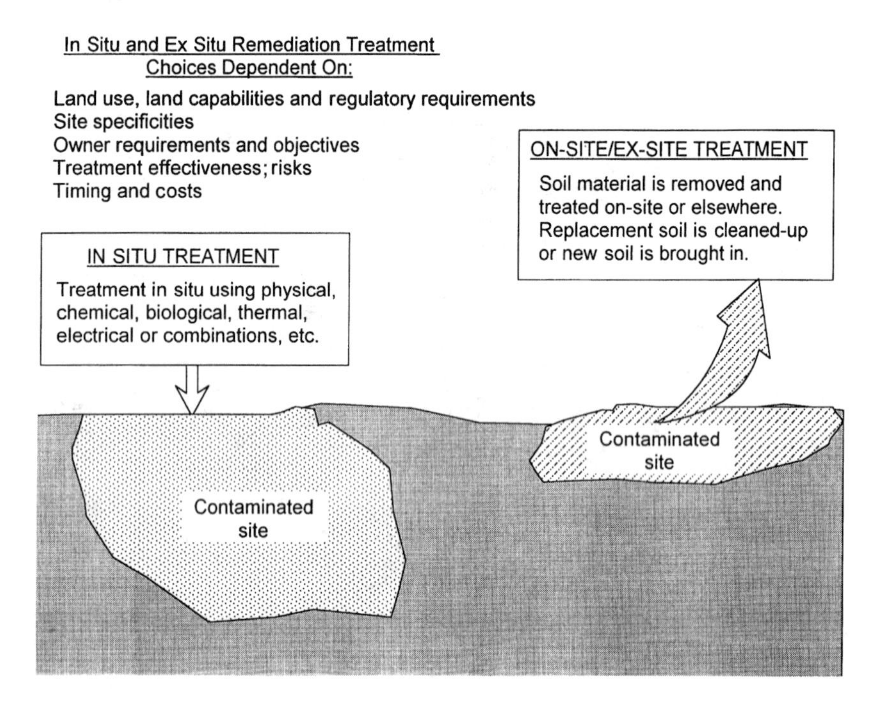

Figure 8.3 Principal elements in in situ and ex situ remedial treatment of a contaminated site.

not be a one-step process. "Return to pristine conditions" and even the "pollutant concentration reduction" objective are treatment objectives that require integration of multi-step processes. For these reasons, and for reasons associated with requirements for long-term performance predictions, risk assessment and risk management are necessary tools in pollution management.

8.2.2 In situ and Ex situ Remedial Treatment

The choice of in situ and/or ex situ remedial treatment options is most often dictated by such considerations as: (a) requirements and objectives set forth by land use policies; (b) regulatory requirements; (c) site specificities; (d) land capability; (e) ownership objectives, requirements and expectations; (f) timing; and (g) economics — as illustrated in Figure 8.3. As will be seen, there are basically three options: (a) total remediation in situ; (b) removal of the contaminated soil substrate material for treatment elsewhere (off-site); and (c) removal of the contaminated soil material for treatment above ground but remaining on-site. There are other ramifications to the basic three options. These will be evident when the generic techniques are addressed.

The basic factors considered in determining whether on-site ex situ, off-site ex situ, or in situ remediation technology and procedures for remedial treatment of contaminated sites should be used include:

- **Contaminants/Pollutants** — Type, concentration, and distribution in the ground;
- **Site** — Site specificities, i.e., location, site constraints, substrate soil material, lithography, stratigraphy, geology, hydrogeology, fluid transmission properties, etc.;
- **Rehabilitation** — Intended land use, land suitability/capability, local zoning regulations, and requirements for clean-up remediation;
- **Economics and Timing** — Economics and compatible technology, efficiency, time and penalties;
- **Regulatory Requirements** — Regulations, constraints, etc.; and
- **Risks** — Risk management.

The first three factors are required in the evaluation of the technical feasibility for site decontamination, and determination of the best available technology for site decontamination and rehabilitation. The final choice is generally made in accord with other governing considerations, e.g., risk, treatment effectiveness, benefits, and permanency of treatment. Regulations and requirements become very important considerations. In summary, we note that the choice of remediation/decontamination technique requires one not only to consider the many scientific and technological aspects of the problem, but also hazard identification, toxicity and exposure, and risk characterization or evaluation.

8.3 BASIC SOIL DECONTAMINATION CONSIDERATIONS

The simplest basic requirement in in situ clean-up of contaminated sites pays attention to remedial treatment procedures that will: (a) remove the offending contaminants (pollutants) in the substrate, and/or (b) immobilise the pollutants in the substrate — to prevent them from moving in the substrate. In the first case, removal of the pollutants can be achieved either by treatment processes which will remove (detach) them from the soil solids and subsequently from the porewater, or by physically removing the substrate material. At the very least, ex situ treatment requirements pay attention to the first case (removal of pollutants).

Immobilization of contaminants is generally achieved by processes that fix the pollutants in the substrate (i.e., stabilization and solidification), or by virtual thermal destruction. If the end-point objectives specified in regulatory requirements for remediation and rehabilitation of the contaminated sites are known, the required treatment technology can be developed in conjunction with geotechnical engineering input to produce the desired sets of actions. The general techniques that support the end-point objectives can be broadly grouped as follows:

- (Group 1) **Physico-Chemical** — e.g., techniques relying on physical and/or chemical procedures for removal of the pollutants, such as precipitation, desorption, soil washing ion exchange, flotation, air stripping, vapour/vacuum extraction, demulsification, solidification stabilization, electrochemical oxidation, reverse osmosis, etc.;
- (Group 2) **Biological** — i.e., generally bacterial degradation of organic chemical compounds, biological detoxification; bioventing, aeration, fermentation, in situ biorestoration;

- (Group 3) **Thermal** — e.g., vitrification, closed-loop detoxification, thermal fixation, pyrolysis, super critical water oxidation, circulating fluidized-bed combustion;
- (Group 4) **Electrical-Acoustic-Magnetic** — e.g., techniques involving electrical, acoustic, and/or magnetic, procedures for decontamination such as electrokinetics, electrocoagulation, ultrasonic, electroacoustics, etc; and
- (Group 5) **Combination** of any or all of the preceding four groups, e.g., laser-induced photochemical, photolytic/biological, multi-treatment processes, treatment trains, reactive walls, etc.

8.4 PHYSICO-CHEMICAL TECHNIQUES

8.4.1 Contaminated Soil Removal and Treatment

The simplest physical procedure for decontamination of a contaminated site is an ex situ procedure which involves removal of the contaminated soil in the affected region, and replacement with clean soil — i.e., the "dig, dump, and replace" procedure. For contaminated sites that are limited in spatial size and depth, this procedure is very popular because of the obvious simplicity in site rehabilitation. The removed contaminated soil is relocated in a prepared waste containment (landfill) site, or is treated by any of the means covered under Groups 1 through 4 listed in the preceding section. The simplest general treatment procedure for dislocated contaminated soil is a soil washing procedure as shown in Figure 8.4. This is best suited for contaminated soils that do not have significant clay contents. Granular soils with little clay contents, which are contaminated with inorganic pollutants, will present the best candidates for washing procedures.

Using heavy metal (HM) pollutants as an example, we note that HM sorption mechanisms associated with the reactive surfaces of clay fractions, such as those listed in Table 5.1, render the washing-extraction procedure more difficult — in the sense that chemical treatments will need to be introduced to detach the sorbed pollutants from the surfaces of the clay soil solids. The retention mechanisms listed in Table 5.1 make it very difficult to remove the HM pollutants without resorting to aggressive chemical treatments in the wash process. In addition to the preceding set of problems, dispersants will need to be introduced in the *grinding and wet slurry preparation* stage of the process shown in Figure 8.4 to disperse the soil solids for chemical washing to achieve effective HM pollutant removal.

For soils contaminated with organics, incineration of the soil is most often recommended — for destruction of the contaminants. However, if removal of the organic chemicals is warranted, as, for example, in instances where the organic chemical contents are high, extraction of the chemicals using the process shown in Figure 8.5 may be necessary. For soils containing soil fractions with little reactive surfaces, the product leaving the extractor should contain little extractant residue. For soils where the reactive surfaces of the soil solids are a significant factor, the choice of extractant(s) used becomes very critical. Two particular actions can be considered: (a) use of solvents, surfactants, biosurfactants, etc. as extractants, and (b) use of a secondary washing process that would remove the residual extractants.

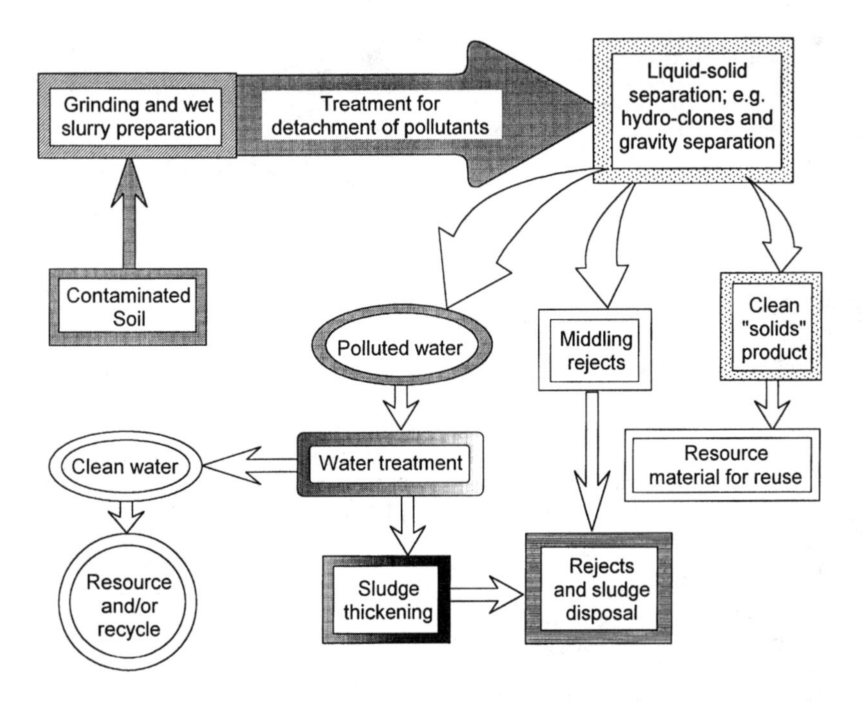

Figure 8.4 Basic elements in ex situ soil washing treatment of granular soils contaminated by inorganic pollutants.

Option (a) is the more useful course of action. The merits of choosing an effective biosurfactant have been shown in Chapter 7.

8.4.2 Vacuum Extraction — Water and Vapour

Vacuum extraction, which is commonly used to obtain contaminated groundwater for cleaning, is generally classed as a physical technique, in the same manner of reasoning as physical removal of contaminated soil. For obvious reasons, application of this extraction technique is limited in respect to subsurface depth. The treatment of the extracted groundwater, which is required before discharge, can be achieved by several means, not the least of which are the standard wastewater chemical and biological treatment techniques and air stripping. Standard wastewater treatment will not be discussed herein.

Application of the vacuum technique for soil vapour extraction is sometimes identified as *air sparging* when it includes extraction of volatilized groundwater pollutants, i.e., volatilized VOCs in the groundwater. This technique is best suited for treatment of soils contaminated by volatile and semi-volatile organic compounds. Biosparging, which is sometimes included with air sparging, relies on enhanced biodegradation as a contribution to the total vapour product being removed. The biodegradation of the less volatile and higher molecular weight of the VOCs and

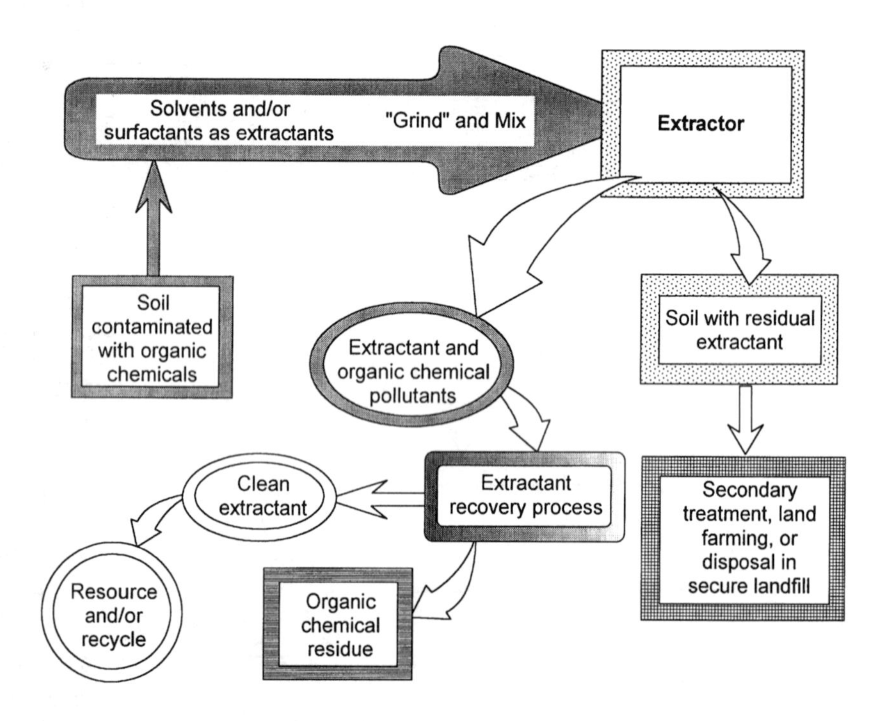

Figure 8.5 Multi-step process for removal of a soil heavily contaminated with organic chemicals.

the removal of the vapour phase allows for a degree of remedial treatment of the VOC-contaminated soil. Soil venting and bioventing are considered to be essentially similar to air sparging and biosparging in respect to the removal or mass transfer of the volatile compounds from the VOCs. The basic elements for soil water and vapour extraction of VOCs (volatile organic compounds) is shown in Figure 8.6. The extraction probe is located in the vadose zone. The tendency of the VOCs to volatilize from water into air is an important factor in the structuring of the remediation technique. If oxygen is used in place of nitrogen as the injecting medium, it not only promotes volatilization, but also contributes to the aerobic biodegradation processes. The first part of the technique is considered to be a physical technique (i.e., soil water and soil vapour extraction), and the second part of the technique where cleaning of the soil water and soil vapour occurs is not necessarily "physical" since one generally uses water treatment procedures (for water) and a packed tower containing activated carbon or synthetic resins to facilitate interphase mass transfer.

Soil-structural features that impede flow of fluid and vapour can be significant. Not only must the delivery of the injected nitrogen or oxygen be effective, but the exiting conditions for the products must also be minimally impeded. Once again, granular soils permit better transmissivity, and soils with high clay and SOM content will present difficulties in transmission of both fluid and vapour. High density soils and high water contents in the unsaturated zone do not provide for good transmission properties. In particular, soils with SOM will show good VOC retention capability.

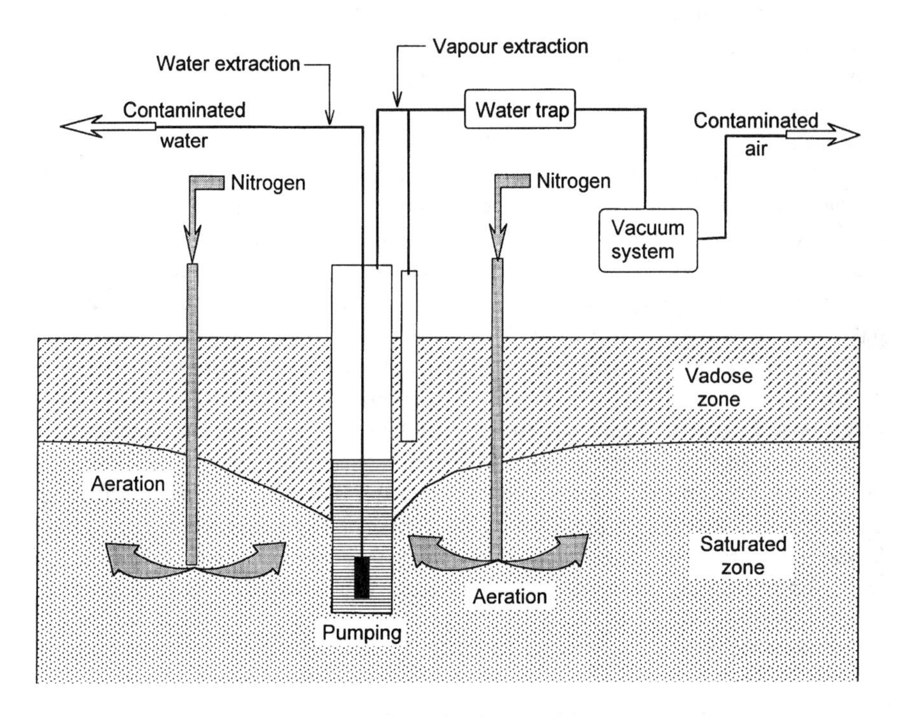

Figure 8.6 Elements of vacuum extraction of water and vapour in a VOC contaminated site. Treatments of contaminated water and air are not shown in the diagram.

In other words, complexes formed between the organic chemicals and soil fractions (particularly SOM) will inhibit volatilization.

Properties of the VOCs are also important considerations. Solubility, sorption and partitioning coefficients, vapour pressure and Henry's law constant, and concentration of the VOCs are important factors which will affect withdrawal of the vapours. Preconditioning of the contaminated soil to obtain better transmission of water and vapour, and also to obtain release of the VOC will provide for a better treatment process.

8.4.3 Electrokinetic Application

The use of electrokinetics for containment or treatment of sites with inorganic contaminants has attracted considerable attention, partly because of previous experiences with electro-osmotic procedures in soil dewatering, and partly because of the relatively "simplicity" of the field application method. This is generally considered a physico-chemical technique because of the field application methods, i.e., the use of electrodes and current energy. For the more granular types of soils (silts), the procedure can be effective. However, in the case of clay soils, diffuse double-layer mechanisms developed in the soils can pose several problems, not the least of which are the energy requirements needed to maintain ionic movement.

The basic principles involved in the use of electrokinetics in pollutant-removal processes have been discussed in Section 7.4 and will not be repeated here. In application of electrokinetic technology, one introduces similar procedures used in electro-osmotic dewatering, i.e., anodes and cathodes are inserted into the soil to produce movement of cations and anions to their respective receiving electrodes. In soils that have significant surface activity, i.e., where interpenetration of diffuse double layers are prominent, one needs to move the pollutants from the region dominated by diffuse double layers. The amount of energy required will need to be greater than the interaction energies established between the contaminant ions and the soil particles. Development of dissociation reactions (see Section 7.4) can seriously impair the useful life of the electrodes.

Capitalizing on the electro-osmosis and ion migration effects when the direct current is established between electrode pairs, and benefitting from pre-conditioning of the soil to permit easier release of pollutants, in-field electrokinetics can be successfully applied. However, treatability studies are necessary for determination of the necessary pre-treatment procedures and the reagents to be used at the electrodes to facilitate removal of the pollutants. These can take the form of conditioning fluids that will improve the electrochemistry (of interactions) at the electrodes, as discussed in Section 7.4. "Fouling" of the electrodes is a serious consideration.

8.4.4 Solidification and Stabilization

Techniques for "fixing" pollutants in their sorbed environment, i.e., pollutants sorbed to the soil solids and pollutants in the porewater, require an end product that ensures the pollutants are totally immobilized. Present application of stabilization-solidification (SS) techniques are either single-step or two-step processes. In the two-step process, the first step is the stabilization process where the polluted soil is rendered insoluble. This is followed by the second procedure which is a solidification process — to render the insoluble soil-pollutant mass solid. The single-step process uses a "binder-fix" that is designed to produce the same effect as the two-step process. The economics of the remedial treatment is best justified for toxic pollutants.

In situ SS process application is limited by the permeability of the soil substrate being treated. Since application of the binder mixture is generally made with the aid of injectors which work similarly to a hollow-stem auger, penetration (propagation) of the binder mixture into the surrounding soil will be controlled by the transmission characteristics of the penetrated soil mass, the viscosity of the binder, and the "set" time of the binder. High densities, clay soils, presence of soil organic matter and amorphous oxides all render application of in situ SS application highly problematic.

Ex situ application of SS processes are more effective if the contaminated soil is in a dispersed state. As in soil washing processes, the excavated material is broken up by grinders, pulverizers, etc. prior to application of the binder mixture. The greater the cohesive nature of the soil, the greater will be the effort needed to grind the material to the kinds of sizes needed for best application of the binder mixture. Disposal of the resultant SS material will still be needed. Since the solidified or stabilized material still contains the toxic pollutants, the SS material will need to be contained in a secure landfill.

The question of whether one only needs to produce a stabilized product — as opposed to the solidified product — is a question that is resolved by regulatory requirements. In general, the requirements of pollutant fixation in a soil mass are such that the treated material, i.e., the solidified product, must undergo and pass aggressive leaching tests together with other types of tests such as wet/dry, freezing/thawing, abrasion, strength, etc. as specified by the regulatory agencies. Typical types of inorganic binders used include: cement, lime, kiln dust, flyash, clays, zeolites, and pozzolonic materials. Typical types of organic binders include: bitumen products, epoxy, polyethylene, resins. The organic-type binders are favoured for binding soils contaminated by organic chemicals. There is no assurance that stabilization, or even solidification after stabilization would produce remediated (solidified) products that would successfully pass all the test requirements and standards.

8.5 CHEMICAL TECHNIQUES

8.5.1 Inorganic Pollutants (HM Pollutants)

Innovative chemical decontamination technologies are continuously being developed. To apply the appropriate chemical technique, it is necessary to first determine the type of bonding established between contaminants (pollutants) and soil constituents — to prescribe the proper sets of processes to detach or release the sorbed pollutants. The efficiency of chemical reagents used to detach sorbed heavy metal pollutants has been discussed in Sections 5.4.1 and 7.3.3. It has been stressed that it is important to recognize that the results obtained from the use of SSE for evaluation of partitioning and distribution of sorbed HM pollutants (Section 5.4.1) are only valid qualitatively. This is because: (a) it is not possible to ascertain or to ascribe all recorded detached HMs as originating from a particular target source; (b) the amount of HM pollutants extracted can be influenced by the type and concentration of extractant used; and (c) degradation of soil solids from reactions with the extractants will obviously affect the release of sorbed HM pollutants, and will also release structural Fe, Mn, Al, etc. For these reasons, the quantitative use of these results could lead to serious errors in specification of the exact distribution of partitioned HM pollutants.

However, in the case of evaluation of the procedures for detachment of HM pollutants from soils, the value of SSE analyses lies in the portrayal of the relative proportions of heavy metals sorbed by the various soil fractions. In addition, treatments used to detach the HMs can also be evaluated through SSE-type studies (Mulligan et al., 2001). The degree of aggressive chemical treatment required to detach the sorbed HM pollutants from the hydrous oxides and SOM can be well appreciated (see Figure 7.9). In general, the types of extractant reagents that need to be used include concentrated inert electrolytes, weak acids, reducing agents, complexing agents, oxidizing agents, and strong acids. Application of any of these, singly or in combination, will be a function of the concentrations of the HM pollutants, and the nature of the soil affected by the HM pollutants.

In situ application of HM extractants (reagents) through injectors or similar probes will detach the HM pollutants and deposit them in the porewater. Treatment of the porewater which contains the reagents and HM pollutants requires either: (a) extraction of the porewater for treatment on surface before discharge (pump and treat), or (b) passing the porewater through a permeable reactive wall. Water treatment of extracted contaminated groundwater (pump and treat) will seek to recover the chemical reagents and the HMs. As for the intercepting permeable reactive walls, the materials in the walls will capture the HM through exchange, complexation, and precipitation mechanisms. These can be achieved relatively easily by providing the appropriate soil material and pH environments in the reactive walls such that precipitation of the HMs would occur. Simple calculations concerning transmission time through the wall (controlled by the hydraulic conductivity of the material in the wall) and precipitation reaction time should inform one about the thickness of the various kinds of walls required to allow for complete reactions and precipitation of the HMs carried by the porewater. The specification of materials to be used in the permeable reactive walls is conditioned by the types of HMs in the contaminated site — recognizing that ion exchange, complexation, and the precipitation pH of the various metals acting singly and in conjunction with others will be variable.

If the HM-removed porewater is still considered to be contaminated with the chemical reagents, this can be extracted by a secondary row of extraction wells located behind the reactive wall. In that manner, it might be possible to seek recovery of the chemical reagents. A simplified scheme showing the essential elements is seen in Figure 8.7. The secondary row of extraction wells after the permeable reactive wall is not shown in the diagram.

8.5.2 Treatment Walls

The successful use of treatment walls as part of an overall remedial treatment procedure in a contaminated site, such as the permeable reactive wall shown in Figure 8.7, relies upon the movement of the contaminated groundwater into and through the wall. Left by itself, the treatment wall does not play an active role in the remedial treatment of the contaminated soil as a whole, i.e., it is essentially a passive component in the remediation exercise. The treatment wall only becomes an active remedial agent when it is contacted by a contaminant or pollutant. In other words, the treatment wall needs to be strategically located such that it intercepts the contaminant plume, and/or the contaminant plume must be channeled to flow through the treatment wall. Figure 8.8 shows the basic elements that illustrate its function.

There are many ways in which the contaminant plumes can be channeled to flow through reactive walls. A basic knowledge of the hydrogeological setting is needed to determine how effective channelization can be performed. The *funnel-gate* technique is one of the more common techniques. In this technique, the contaminant plume is essentially guided to the intercepting reactive wall by a funnel. This funnel, which is constructed or placed in the contaminated ground, is composed basically

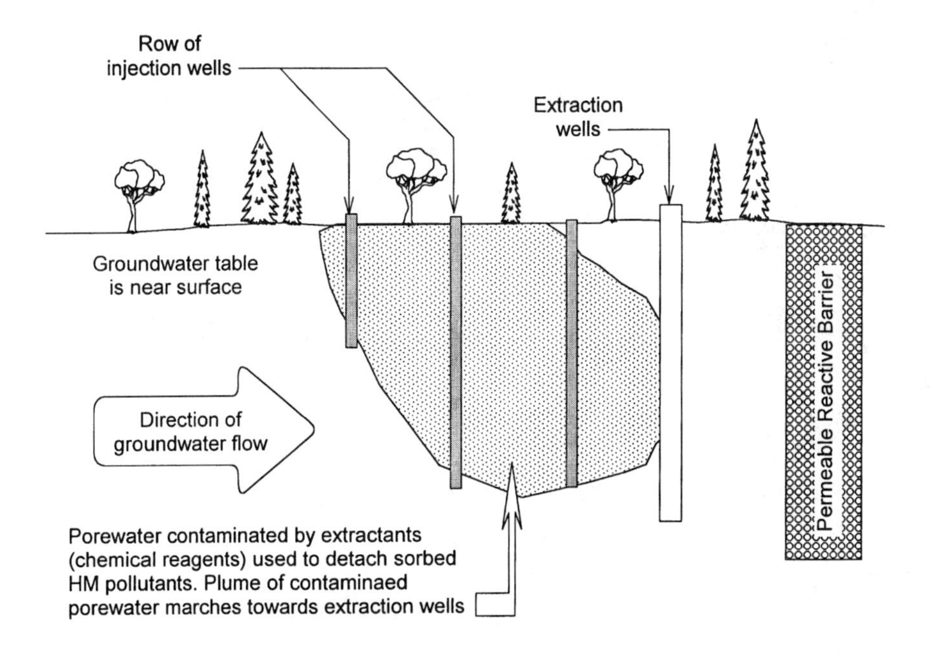

Figure 8.7 Porewater contaminated by HMs and chemical reagents used to extract the sorbed HMs are extracted through extraction wells for treatment. Alternatively, the contaminated porewater plume is intercepted by the permeable reactive wall which captures the pollutants.

of confining boundaries of impermeable material (e.g., sheet pile walls) which narrow toward the funnel mouth where the reactive wall is located. Other variations of the funnel-gate technique exist, obviously in accord with site geometry and site specificities.

The basic principles governing the efficacy of treatment walls are precisely those that have been addressed in our considerations of pollutant-soil interaction. With the proper sets of reactive materials in the walls, and the proper sets of circumstances provided for the reaction kinetics to function efficiently (i.e., achieve equilibrium or close-to-equilibrium conditions), treatment walls can function as agents for various processes which will remove the pollutants from the contaminant plume, or trap the pollutants in the wall. The removed and trapped pollutants retained in the treatment wall can subsequently be removed by renewing the materials in the wall. Some of the major pollutant-removal and immobilization processes in the treatment wall include the following:

- **Inorganic pollutants** — Sorption, precipitation, substitution, transformation, complexation, oxidation and reduction; and
- **Organic pollutants** — Sorption, abiotic transformation, biotransformation, abiotic degradation, and biodegradation.

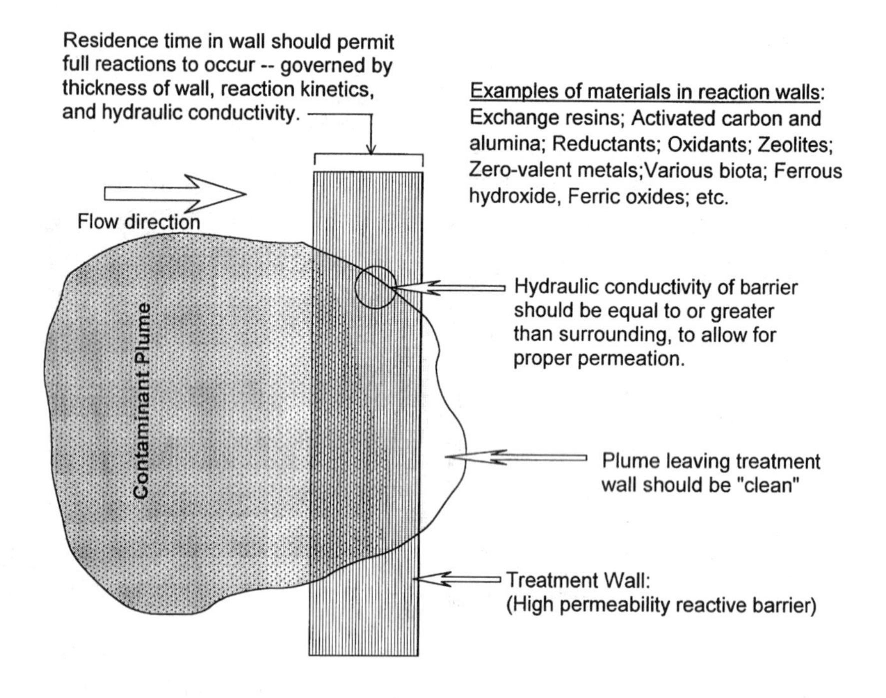

Figure 8.8 Basic elements for treatment walls.

The types of reagents, compounds, and microenvironment in the treatment walls include a range of oxidants and reductants, chelating agents, catalysts, microorganisms, zero-valent metals, zeolite, reactive clays, ferrous hydroxides, carbonates and sulphates, ferric oxides and oxyhydroxides, activated carbon and alumina, nutrients, phosphates, and soil organic materials. The choice of reagents and compounds, and the manipulation of the pH-pE microenvironment in the treatment walls will need to be made on the basis of site-specific knowledge of the interaction processes between pollutants and soil fractions. Nevertheless, the use of treatment walls is a very direct confirmation of the need and usefulness for a greater appreciation of the various processes involved in determination of the fate of pollutants. Whilst the remedial treatment process is directed toward pollutants in the contaminant plume, it is nevertheless a very important component in the total remediation of contaminated sites.

8.5.3 Organic Chemical Pollutants

Remedial treatment of organic chemical pollutants in soils are most often dealt with by: (a) removal of the contaminated soil for treatment off-site (or on-site); (b) application of bioremediation procedures in situ; or (c) through in situ chemical treatments. Case (a) has been addressed previously, and bioremediation of sites contaminated by organic chemical compounds will be briefly considered in the next section. Abiotic (chemical) techniques for in situ remedial treatment of organic chemical pollutants most often rely on extraction or detachment of the organic chemical compounds through the use of solvents and surfactants. The application technology ranges from the "pump and treat" to solvent or surfactant flushing in combination with treatment walls or pump-out sequences. The intent of the use of cosolvents and surfactants is to increase the solubility of the pollutants and to reduce the interfacial tension between the organic chemical pollutants and the reactive surfaces of the soil fractions. The discussion on the processes involved is summarized in Section 7.5.2.

The use of procedures that rely on transformations and degradation resulting from acid-base and oxidation-reduction reactions appears to be minimal at best, perhaps because of the time frame for treatment and the need for extraction of the contaminated porewater. Reaction kinetics in relation to such processes, and those initiated by the catalytic action of soils resulting in abiotic transformation, are considered to be relatively slow. Practical considerations appear to suggest that if accommodation is to be made for the time-frame required in abiotic transformations, it would be expedient to consider biodegradative means to achieve the "transformation and degradation" route.

8.6 BIOLOGICAL TECHNIQUES

Bioremediation of soil contaminated by organic chemical pollutants benefits considerably from the use of soil microorganisms to metabolize the organic chemical compounds. Table 4.1 in Section 4.6 shows the closely similar types of natural and synthetic organic chemical compounds. Thus for example, the aromatic natural soil organics such as vanillin, lignin, and tannin are closely similar to the synthetic aromatic organic compounds represented by benzene, toluene, PAHs, etc. It is natural therefore to expect that there would be a naturally occurring consortia of microorganisms — ranging from bacteria and fungi to viruses — available to successfully address the synthetic organic chemicals since they would be expected to be well adapted to the specific habitat. The available energy sources and all the other microenvironmental factors such as pH, temperature, water content, etc. will produce the suites of biomass that have adapted to the microenvironment.

In the event that the naturally occurring microorganisms do not contain all the enzymes necessary for degradation of the synthetic organic compounds introduced into their habitat, genetically engineered microorganisms would be required. In such cases, these should contain the necessary suite of enzymes for degradation of the organic chemical pollutants in the habitat. Because these are not naturally occurring, we would expect competition in the habitat, and we should ensure that these are not pathogenic to plants nor should they produce undesirable effects, e.g., toxins.

Considerable study and reporting of the application of a whole range of bioremediation techniques for remedial treatment of contaminated soils can be found in the textbooks and specialized symposia dedicated to this subject. These concern, for example, the various microbial preparations that can address different types of organic chemical compounds and will reduce acclimation times. It is evident from a knowledge of soil catalysis that the control of biotic redox reactions cannot be studied without attention to both the pollutants and the nature of the soil fractions — in addition to the usual factors that govern the metabolic processes of the microorganisms. Application or selection of a bioremediation technique or procedure for remedial treatment of a contaminated soil requires consideration of the biological and chemical factors of the problem at hand. Many of these have been examined in Chapter 7 and in the preceding sections. As with other remedial treatment procedures, the "state of the art" is fast evolving and the interested reader is advised to consult these dedicated publications. A listing of some of these is given in the Reference section.

Many different kinds of technologies fall within the broad umbrella of bioremediation. However, all of these serve to satisfy one simple goal, i.e., the use of microorganisms to biodegrade the organic chemical pollutants through their metabolic processes. As in the previous chapter, a "shopping list" of various techniques for bioremediation of contaminated soils can be offered. If such is done, the "shopping list" would include (amongst others):

> Biosparging, bioventing, biostripping, biofiltration, biostimulation, biotransformation, biotraps, biodegradation, biorestoration, land farming, and composting.

All of the above utilize in one form or another the various processes that include microbial degradation, hydrolysis, substitution, aerobic and anaerobic transformations and degradations, biotic redox reactions, mineralization, and volatilization. Manipulation of the microenvironment — including macro- and micronutrients — as part of the enhancement procedures is a requirement that is examined in conjunction with screening and treatability studies. It is fairly clear that most of the applied techniques require a co-treatment procedure for removal of the biotreated product. Thus for example, bioventing or biosparging requires the removal of the volatilized products via vacuum or pump techniques. The use of co-treatment processes is not unusual since, as we have pointed out before, the detachment of pollutants from their sorbed status from the soil solids will invariably lead to deposition of these detached pollutants in the porewater.

There are some problems which attend the use of bioremediation techniques. These are not necessarily technological, but more so in relation to risks or threats to human health and the environment. One of these (risks) has been addressed previously in Chapter 6 under the topic of persistence and fate of organic chemical compounds. We refer to the intermediate products or intermediary metabolites that result from incomplete biodegradation of the parent organic chemical compound, demonstrated in Figure 6.13. The toxicity, persistence, and mobility of the intermediary metabolites (which can accumulate) are concerns that need to be fully addressed.

The other risks are more difficult to quantify or fully establish. These arise when unknown results are obtained from interactions between the genetically engineered microorganisms and the various chemicals in the contaminated ground. The use of microorganisms grown in uncharacterized consortia, which include bacteria, fungi, and viruses can produce toxic metabolites (Strauss, 1991). In addition, the interaction of chemicals with microorganisms may result in mutations in the microorganisms themselves, and/or microbial adaptions.

8.7 MULTIPLE TREATMENTS AND TREATMENT TRAINS

The use of multiple treatments applied in sequence or as co-treatment procedures is common in in situ remedial treatment of contaminated sites. To a very large extent, this is because very few contaminated sites (soils) contain only one type of pollutant. In addition, as has been discussed many times previously, removal of pollutant from soil solids does not mean removal from the site itself. Detached (desorbed) pollutants will be transferred to the porewater which will need to be treated. Thus, we will at the very least have a two-step process for site remediation, assuming that the pollutant detachment process is a one-step process. Movement of the removed pollutants to the ground surface most often requires a different set of procedures. Figure 8.9 shows a summary view of some of the main multiple treatment techniques.

While the general category of *multiple treatments* has been shown in the diagram, some popular classification schemes can be found in the literature, e.g., *layered treatments* and *treatment trains*. The question of which component treatment (of a multiple treatment scheme) comes first is the issue that needs to be addressed when

MULTIPLE TREATMENTS

Two-Treatment-Sequence

[1] First... "conditioning" treatment, Second... "removal" treatment.

[2] Two-part process where each treatment addresses its own part , e.g., detachment of sorbed pollutants followed by removal of pollutants from porewater

Combination Treatment (layered treatment)

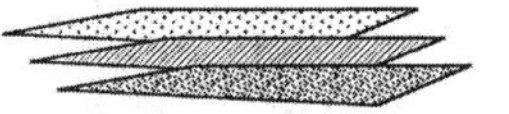

Combination of electro-osmosis, hydrofracturing, and in situ treatment for remediation of VOCs in saturated clayey silty soils

Combination of co-metabolic bioventing for VOCs and phytoremediation for metals in VOCs-metal-contaminated soil

TREATMENT TRAINS

Sequential application of different treatment processes. e.g., SVE followed by in situ flushing or soil washing

Treatment trains using more than 2 processes are possible. Question of timing and operational costs

Figure 8.9 Multiple treatments and treatment trains.

structuring a multiple treatment process. Multiple treatments can be implemented as: (a) *combined treatments* where the two or more remedial treatment schemes are implemented together in a combined scheme, or (b) *sequence treatments* where individual treatments are applied to detach and finally remove the pollutants from the contaminated soil.

Some very good examples of multiple treatment techniques include preconditioning as the primary treatment process. This is part of the sequence treatment scheme where another treatment is needed to detach the pollutants from the soil solids. Application of preconditioning techniques could mean using solvents to solubilize the organic chemical pollutants, or surfactants to reduce interfacial tensions between pollutants and soil solids, and also to reduce the viscosity of the medium. Changing the redox or pH environment as a means of facilitating abiotic and/or biotic redox reactions will also fall under the category of conditioning. Provision of macronutrients in addition to changes in the pH environment will also be considered as preconditioning.

The secondary treatment techniques that follow from the preconditioning phase will involve procedures that seek to detach the pollutants from the surfaces of the soil solids. The use of electrokinetics is a good example of such a procedure. While there may be a question as to whether enhanced biodegradation can be strictly labeled as the secondary phase of a multiple treatment program, it is nevertheless a process which benefits from control of the microenvironment as a preconditioning exercise. A possible compromise in terminology is offered through classification of enhanced biodegradation as a combined treatment process. We can consider biostimulation as an example of this combined treatment process since this requires the addition of nutrients and/or electron acceptors to the contaminated region. Thus, anaerobic degradation can proceed with the availability of nitrates, Fe(III) oxides, Mn(IV) oxides, sulphates, and CO_2.

Removal of the detached and/or transformed pollutants in the porewater is a necessary requirement. Following from the preconditioning and primary sequence treatments, this "removal phase" is the third treatment procedure in the sequence treatment process. This could involve, for example, pumping (out) of contaminated water as a "treatment" process. As such, it will constitute the tertiary treatment technique for the multiple (sequence) treatment procedure. All of these various combinations and sequences of treatments which are necessary for removal of pollutants from the contaminated substrate can be lumped under the general category of *treatment trains*.

8.8 CONCLUDING REMARKS

The choice of treatment technology involves a process that begins with site and contaminant specificities — as shown in Figure 8.1. It is not always a simple matter of "black box" technology since the applied technique must accommodate the type of contaminants involved in interaction with the soil material, and the end-point objectives. Thus for example, we know that incineration has been successfully applied on-site and off-site for destruction of organic contaminated materials. Thermal processes rely on high temperature breakdown of pollutants through combustion or pyrolysis. Application is best performed as ex situ treatment and is best applied for destruction of organic chemical pollutants.

Experience shows that a combination treatment technique is generally more beneficial in site rehabilitation. This is primarily because most contaminated sites consist of a whole variety of organic and inorganic contaminants. Using techniques that address only inorganic or only organic contaminants will not be satisfactory. We have a variety of physical and chemical options that can be used on-site, off-site and in situ, which can be developed into application techniques. These have been detailed in Section 7.7.

Laboratory treatability studies are mandatory, and pilot testing should always be implemented if circumstances permit. Scaling from laboratory and pilot tests will always remain as the most challenging task, particularly if new technology is to be developed. The scale-up procedures suffer not only from scale effects, but also from lack of control of soil and contaminant compositions and uniformity, and local physical/chemical control. Biological, chemical, and physical reactions do not appear to scale linearly, and interactive relationships are likewise affected.

Treatment in situ can become complicated when complex mixtures of contaminants are encountered. Aeration or air stripping, steam stripping, soil vapour extraction, and thermal adsorption are techniques that are suited for removal of volatile organics. Chemical precipitation and soil washing can be used for removal of many of the heavy metals in the soil-water in in situ treatment procedures. However, complete removal will be difficult because of high affinity and specific adsorption of the contaminant ions. A good working knowledge of contaminant-soil bonding would provide for better structuring of appropriate options and compatible technology for soil decontamination and site remediation.

It has not been the intent of this chapter to enter into the argument that asks "How clean is clean," nor is it within the scope of this chapter (or book) to provide the final sets of technology for complete pollutant removal from a contaminated site. The former (argument) leads to endless debates and the latter (provision of final sets of technology) is too presumptuous. The state of the art in remedial treatment of contaminated ground is fast evolving, and there will undoubtedly be great strides made in various ways in which contaminated ground can be properly and effectively remediated. A good example of this is the emerging *phytoremediation* treatment technique.

Many plants have the ability to extract and concentrate certain kinds of elements in the soil. Their root systems absorb and accumulate the necessary nutrients (and water) to sustain their growth. While metal-tolerant plants have some tolerance for toxic metal ions uptake, by and large, their tolerance level for such metals is very low. However, hyperaccumulating plants have higher levels of tolerance for toxic metal ions and can take HM ions up to several percent of their dry weight (Chaney, 1995; Bradley, 1997). Schnoor et al. (1995) report that some plants can uptake organic pollutants and accumulate nonphytotoxic metabolites. Much research remains to be conducted. The source of hyperaccumulating plants has yet to be made more available.

The important points that need to be communicated at this juncture (in this book) are those relating to the mechanisms and processes by which pollutants are retained in the soil substrate system. When these processes are well appreciated, procedures and associated technology can be developed to provide for pollutant removal (or reduction) to eliminate the health and environmental risks posed by these pollutants in the ground.

References and Suggested Reading

Aiken, G.R., McKnight, R.L., and MacCarthy, P., (1985), *Humic Substances in Soil, Sediments, and Water,* John Wiley & Sons, New York.

Alammawi, A.M., (1988), "Some aspects of hydration and interaction energies of montmorillonite clay", Ph.D. Thesis, McGill University.

American Society for Testing and Materials, (1998), "Special procedure for testing soil and rock for engineering purposes", Philadelphia.

Anderson, D., Brown, K.W., and Green, J.W., (1982), "Effect of organic fluids on the permeability of clay soil liners", in *Land Disposal of Hazardous Waste., Proc. of the 8th Annu. Res. Symp.,* EPA-600/9-82-002, pp. 179–190.

Anderson, D.C., (1981), "Organic leachate effects on the permeability of clay soils", M. Sc. Thesis, Soil and Crop Sciences Department, Texas A & M University, College Station.

Aral, M.M., (1989), *Ground Water Modelling in Multilayer Aquifers,* Lewis Publishers, Ann Arbor, MI, 143p.

Arnold, P.W., (1978), "Surface-Electrolyte Interactions", in *The Chemistry of Soil Constituents,* D.J. Greenland, and M.H.B. Hayes, (eds.), John Wiley & Sons, New York, pp. 355–401.

Baes, C.F., and Messmer, R.E., (1976), "The Hydrolysis of cations", John Wiley & Sons, New York.

Bailey, G.W., and White, J.L., (1964), "Soil-pesticide relationships: review of adsorption and desorption of organic pesticides by soil colloids with implications concerning pesticide bioactivity", *J. Agric. Food Chem.* 12:324–332.

Baker, D.H., and Bhappu, R.B., (1974), "Specific environmental problems associated with the processing of minerals", in *Extraction of Minerals and Energy: Today's Dilemmas,* R.A. Deju, (ed.), Ann Arbor Science, Ann Arbor, MI, 301p.

Baker, E.G., (1962), "Distribution of hydrocarbons in petroleum", *Bull. Am. Assoc. Pet. Geologists,* 46:76–84.

Barone, F.S., Mucklow, J.P., Quigley, R.M., and Rowe, R.K., (1991), "Contaminant transport by diffusion below an industrial landfill site", CGS First Can. Conf. on Env. Geotech. pp. 81–90.

Barone, F.S., Yanful, E.K., Quigley, R.M., and Rowe, R.K., (1989), "Effect of multiple contaminant migration on diffusion and adsorption of some domestic waste contaminants in a natural clayey soil", *Can. Geotech. J.,* 26:189–198.

Bear, J., (1972), *Dynamics of Fluids in Porous Media,* Elsevier Scientific, Amsterdam, 764p.

Belzile, B., Lecomte, P., and Tessier, A., (1989), "Testing readsorption of trace elements during partial chemical extractions of bottom sediments", *Environ. Sci. Technol.* 23:1015–1020.

Benjamin, M.M., and Leckie, J.O., (1981), "Multiple-site adsorption of Cd, Zn, and Pb on amorphous iron oxyhydroxide", *J. Colloid Interface Sci.* 79:209–221.

Benjamin, M.M., and Leckie, J.O., (1982), "Effects of complexation by Cl, SO_4, S_2O_4, on adsorption behaviour of Cd on oxide surfaces", *Environ. Sci. Technol.* 16:152–170.

Bergna, H.E., (1994), "The colloidal chemistry of silica", Am. Chem. Soc., *Advances in Chemistry,* Series 234, 695p.

Bermond, A., and Malenfant, C., (1990), "Estimation des cations métalliques liés à la matière organique à l'aide de réactifs chimiques: approche cinétique", *Science du sol,* 28:43–51.

Bernal, J.D., and Fowler, R.H., (1933), "A theory of water and ionic solution with particular reference to hydrogen and hydroxyl ions", *J. Chem. Phys.* 1:515–548.

Bhatt, H.G., (1985), *Management of Toxic and Hazardous Waste,* Lewis Publishers, Ann Arbor, MI, 418p.

Biddappa, C.C., Chino, M., and Kumazawa, K., (1981), "Adsorption, desorption, potential and selective distribution of heavy metals in selected soils of Japan", *J. Environ. Sci. Health, Part B,* 156:511–528.

Biermann, M., Lange, F., Piorr, R., Ploog, U., Rutzen, H., Schindler, J., and Schmidt, R., (1987), in *Surfactants in Consumer Products, Theory, Technology and Application,* J. Falbe (ed), Springer Verlag, Heidelberg.

Bockris, J. O'M., and Reddy, A.K.N., (1970), *Modern Electrochemistry,* Plenum Press, New York, Vols. 1 and 2.

Bohn, H.L., (1979), *Soil Chemistry,* John Wiley & Sons, New York, 329p.

Bolt, G.H., (1955), "Analysis of the validity of the Gouy-Chapman theory of the electric double layer", *J. Colloid Sci.,* 10:206–218.

Bolt, G.H., and Bruggenwert, M.G.M., (1978) *Soil Chemistry, Part A: Basic Elements,* Elsevier Scientific, 281p.

Bolt, G.H., (1979), *Soil Chemistry, Part B: Physico-Chemical Models,* Elsevier Scientific, 479p.

Bosma, T.N.P., van der Meer, J.R., Schraa, G., Tros, M.E., and Zehnder, A.J.B., (1988), "Reductive dechlorination of all trichloro- and dichlorobenzene isomers", *FEMS Microbiol. Ecol.* 53:223–229.

Bowden, J.W., Posner, A.M., and Quirk, J.R., (1980), "Adsorption and charging phenomena in variable charge soils", in *Soils with Variable Charge,* B.K. Theng (ed.), New Zealand Society of Soil Science, 147p.

Boyd, S.A., Lee, J.F., and Mortland, M.M., (1988), "Attenuating organic contaminant mobility by soil modification", *Nature,* 333:345–347.

Boyd, S.A., Mortland, M.M., and Chiou, C.T., (1988), "Sorption characteristics of organic compounds on hexadecyl trimethyl-ammomium-smectite", *Soil Sci. Soc. Amer. J.,* 52:652–657.

Brady, N.C., (1984), *The Nature and Properties of Soils,* 9th Ed., Macmillan, New York.

Bradley, T., (1997), "The phytoremediation of heavy metals", Unpublished technical report, Environmental Technology Program, Algonquin College, Ottawa, 48p.

Brady, P.V., Cygan, R.T., and Nagy, K.L., (1998), "Surface charge and metal sorption to kaolinite", Chap. 17, in *Adsorption of Metals by Geomedia,* E.A. Jeanne (ed.), Academic Press, San Diego, CA.

Briggs, G.G., (1974), "A simple relationship between soil adsorption of organic chemicals and their octanol/water partition coefficient", *Proc. 7th Br. Insect. Fung. Conf.,* Brighton, pp. 83–86.

Buckingham, D.A., (1977), "Metal-OH and its ability to hydrolyze (or hydrate) substrates of biological interest", in *Biological Aspects of Inorganic Chemistry,* W.S. Addison, W.R. Cullen, D. Dolphin, and B.R. James (eds.), Wiley Interscience, New York.

Buckingham, E., (1907), "Studies on the movement of soil moisture" U.S. Dep. Agr., Bur. Soils, Bull., 38:61 p.

Buckman, H.O., and Brady, N.C., (1969), *The Nature and Properties of Soils,* 7th Ed. Macmillan, London, 653p.

Buffle, J., (1988), *Complexation Reactions in Aquatic Systems: An Analytical Approach,* Ellis Harwood, Chichester.

Buol, S.W., Hole, F.D., and McCracken, R.J., (1980), *Soil Genesis and Classification,* 2nd Ed., Iowa State University Press.

Butler, J.N., (1964), *Ionic Equilibrium: A Mathematical Approach,* Addison-Wesley, New York.

Button, K.K., (1976), "The influence of clay and bacteria on the concentration of dissolved hydrocarbon in saline solution", *Geochim. Cosmochim. Acta,* 40:435–440.

Cabral, A.R., (1992), "A study of compatibility to heavy metal transport in permeability testing", Ph.D. Thesis, McGill University.

Cadena, F., (1989), "Use of tailored bentonite for selective removal of organic pollutants", *ASCE J. Environ. Engr.,* 115:756–767.

Callahan, M., Slimak, M., Gabel, N., May, I., Fowler, C., Freed, R., Jennings, P., Durfee, R., Whitmore, F., Maestri, B., Mabey, W., Holt, B., and Gould, C., (1979), "Water-related environmental fate of 129 priority pollutants", Vol. II, Office of Water Planning and Standards, Office of Water and Waste Management, Washington, D.C., U.S.EPA (EPA 440/4-79-029b).

Carter, D.L., Mortland, M.M., and Kemper, W.D., (1986), "Specific surface", in *Methods of Soil Analysis,* A. Klute (ed.), American Society of Agronomy, pp. 413–423.

Casagrande, A., (1947), "Classification and identification of soils", Proc. ASCE, pp. 783–810.

Chan, J., (1993), "A comparative study of three computerized geochemical models", M. Eng. Thesis, McGill University.

Chaney, R., (1995), "Potential use of metal accumulators", *Min. Env. Manage.,* 3:9–11.

Chang, A.C., Page, A.L., Warneke, J.E., and Grgurevic, E., (1984), "Sequential extraction of soils heavy metals following a sludge applications", *J. Environ. Qual.,* 13:33–38.

Chao, T.T., (1972), "Selective dissolution of manganese oxides from soils and sediments with acidic hydroxylamine hydrochloride" *Soil Sci. Soc. Amer. Proc.,* 36:764–768.

Cherry, J.A., [guest editor], (1983), "Migration of contaminants in groundwater at a landfill: a case study", *J. Hydrology, Special Issue,* Elsevier, 197p.

Cherry, J.A., Gillham, R.W., and Barker, J.F., (1984), "Contaminants in groundwater: chemical processes", in *Groundwater Contamination: Studies in Geophysics,* National Academy Press, pp. 46–64.

Chester, R., and Hughes, R.M., (1967), "A chemical technique for the separation of ferro-manganese minerals, carbonate minerals and adsorbed trace elements from Pelagic sediments", *Chem. Geol.,* 2:249–262.

Chhabra, R., Pleysier, J., and Cremers, A., (1975), "The measurement of the cation exchange capacity and exchangeable cations in soil: a new method", *Proc. Int. Clay Conf.,* Applied Publishing, IL, pp. 439–448.

Chiou, G.T., Freed, V.H., Schmedding, D.W., and Kohnert, R.L., (1977), "Partition coefficient and bioaccumulation of selected organic chemicals", *Environ. Sci. Technol.,* 11:5.

Chiou, G.T., Schmedding, D.W., and Manes, M., (1982), "Partition of organic compounds on octanol-water system", *Environ. Sci. Technol.,* 16:4–10.

Church, B.W., (1997), "Dose assessment considerations for remedial action on plutonium-contaminated soils", *J. Soil Contamination,* 6:257–170.

Clevenger, T.E., (1990), "Use of sequential extraction to evaluate the heavy metals in mining waste" *Water, Air, Soil Pollut. J.,* 50:241–254.

Cloos, P., Leonard, A.J., Moreau, Herbillon, A., and Fripiat, J.J., (1969), "Structural organization in amorphous silico-aluminas", *Clays and Clay Miner.,* 17:279–285.

Coles, C.A., (1998), "Sorption of lead and cadmium by kaolinite, humic acid and mackinawite", Ph.D. Thesis, McGill University.

Coles, C.A., Rao, S.R., and Yong, R.N., (2000), "Lead and cadmium interactions with Mackinawite: retention mechanisms and role of pH", *Environ. Sci. Technol.,* 34:996–1000.

Crank, J., (1975), *The Mathematics of Diffusion,* 2nd Ed., Oxford University Press, 414p.

Crooks, V.E., and Quigley, R.M., (1984), "Saline leachate migration through clay: a comparative laboratory and field investigaton", *Can. Geotech. J.,* 21:349–362.

Crosby, D.G., (1972), "Photodegradation of pesticides in water", *Adv. Chem. Ser.,* 111:173–188.

Darban, A.K., (1997), "Multi-component transport of heavy metals in clay barriers", Ph.D. Thesis, McGill University.

Darcy, H., (1856), "Les fontaines publiques de la ville de Dijon", Dalmont, Paris, 674p.

Davidson, J.M., Rao, P.S.C., and Nkedi-Kizza, P., (1983), "Physical processes influencing water and solute transport in soils", in *Chemical Mobility and Reactivity in Soil Systems,* SSSA Spec. Publ. No. 11, pp. 35–47.

Davis, A.P., Mantange, D., and Shokouhiank, M., (1998), "Washing of cadmium (II) from a contaminated soil column", *J. Soil Contamination,* 7:371–394.

De Vries, W., and Breeuwsma, A., (1987), "The relation between soil acidification and element recycling", *Water, Air, Soil Pollut. J.,* 35:293–310.

Debye, P., (1929), "Polar Molecules", Reinhold, New York, 172p.

Derjaguin, B., and Landau, L.D., (1941), "Acta Physicochim", U.R.S.S. 14:635; *J. Exp. Theor. Phys.* (U.S.S.R.), 11:802.

Desjardins, S., (1996), "Investigation on montmorillonite-phenol interactions", Ph.D. Thesis, McGill University.

Doner, H.E., (1978), "Chloride as a factor in mobilities of Ni (II), Cu(II), and Cd(II) in soil", *Soil Sci. Soc. Amer. J.,* 42:882–885.

Dowdy, R.H., and Volk, V.V., (1983), "Movement of heavy metals in soils", *Proc. Amer. Soc. Agronomy and Soil Science Soc. Amer.,* Atlanta, pp. 229–239.

Dragun, J., (1988), "The soil chemistry of hazardous materials", The Hazardous Materials Control Research Institute, Silver Springs, MD, 458p.

Egozy, Y., (1980), "Adsorption of cadmium and cobalt on montmorillonite as a function of solution composition", *Clays and Clay Miner.,* 28:311–318.

Ehlers, W., Letey, J., Spencer, W.F., and Farmer, W.J., (1969), "Lindane diffusion in soils: I. Theoretical considerations and mechanism of movement", *Soil Sci. Soc. Amer. J.,* 33:504–510.

Einstein, A., (1905), "Uber die von der Molekularkinetischen theorie der Warme Geforderte Bewegung von in Ruhenden Flussigkeiten Suspendierten Teilchen", *Annalen der Physick,* 4:549–660.

Elliott, H.A., Liberati, M.R., and Huang, C.P., (1986), "Competitive adsorption of heavy metals by soils", *J. Environ. Qual.,* 15:214–219.

Eltantawy, I.N., and Arnold, P.W., (1972), "Adsorption of n-alkanes on Wyoming montmorillonite", *Nature Phys. Sci.,* pp. 225–237.

Eltantawy, I.N., and Arnold, P.W., (1973), "Reappraisal of ethylene glycol mono-ethyl ether (EGME) method for surface area estimations of clays", *Soil Sci.,* 24:232–238.

Elzahabi, M., and Yong, R.N., (1997), "Vadose zone transport of heavy metals", in *Geoenvironmental Engineering — Contaminated Ground: Fate of Pollutants and Remediation,* R.N. Yong, and H.R. Thomas, (eds.), Thomas Telford, London, pp. 173–180.

Elzahabi, M., (2000), "The effect of soil pH on heavy metal transport in the vadose zone", Ph.D. Thesis, McGill University.

Emmerich, W.E., Lund, L.J., Page, A.L., and Chang, A.C., (1982), "Solid phase forms of heavy metals in sewage sludge-treated soils", *J. Environ. Qual.,* 11:178–181.

Engler, R.N., Brannon, J.M., Rose, J., and Bigham, G., (1977), "A practical selective extraction procedure for sediment characterization", in *Chemistry of Marine Sediments*, T.F. Yen (ed.), Ann Arbor Science, MI.

Farmer, V.C., (1978), "Water on particle surfaces" Chapter 6, *The chemistry of soil constituents*, D.J. Greenland, and M.H.B. Hayes (eds.), John Wiley & Sons, New York, pp.405-448.

Farrah, H., and Pickering, W.F., (1976), "The sorption of copper species by clays. I. Kaolinite", *Aust. J. Chem.,* 29:1167–1176.

Farrah, H., and Pickering, W.F., (1976), "The sorption of copper species by clays. II. Illite and montmorillonite", *Aust. J. Chem.,* 29:1177–1184.

Farrah, H., and Pickering, W.F., (1976), "The sorption of zinc species by clay minerals", *Aust. J. Chem.,* 29:1649–1656.

Farrah, H., and Pickering, W.F., (1977a), "Influence of clay-solute interactions on aqueous heavy metal ion levels", *Water, Air, Soil Pollut. J.,* 8:189–197.

Farrah, H., and Pickering, W.F., (1977b), "The sorption of lead and cadmium species by clay minerals", *Aust. J. Chem.,* 30:1417–1422.

Farrah, H., and Pickering, W.F., (1978), "Extraction of heavy metal ions sorbed on clays", *Water, Air, Soil Pollut. J.,* 9:491–498.

Farrah, H., and Pickering, W.F., (1979), "pH effects in the adsorption of heavy metal ions by clays", *Chem. Geol.,* 25:317–326.

Fernandez, F., and Quigley, R.M., (1984), "Hydraulic conductivity of natural clays permeated with simple liquid hydrocarbons", *Can. Geotech. J.,* 22:205–214.

Fernandez, F., and Quigley, R.M., (1988), "Viscosity and dielectric constant controls on the hydraulic conductivity of clayey soils permeated with water soluble organics", *Can. Geotech. J.,* 25:582–589.

Ferris, A.P., and Jepson, W.B., (1975), "The exchange capacities of kaolinite and the preparation of homoionic clays", *J. Colloid Interface Sci.,* 52:245–259.

Fetter, C.W., (1993), *Contaminant Hydrogeology,* Macmillan, New York, 458p.

Figura, P. and McDuffie, B., (1980), "Determination of labilities of soluble trace metals species in aqueous environmental samples by anodic stripping voltammetery and Chelex column and batch methods", *Anal. Chem.,* 52:1433–1439.

Flegmann, A.W., Goodwin J.W., and Ottewill, R.H., (1969), "Rheological studies on kaolinite suspensions", *Proc. Brit. Ceramic Soc.*, pp. 31–44.

Forbes, E.A., Posner, A.M., and Quirk, J.P., (1974), "The specific adsorption of inorganic Hg (II) species and Co (II) complex ions on geothite", *J. Colloid Interface Sci.,* 49:403–409.

Forstner, U., and Wittmann, G.T.W., (1984), *Metal Pollution in Aquatic Environment,* Springer-Verlag, New York.

Freeze, R.A., and Cherry, J.A., (1979), *Groundwater,* Prentice-Hall, London, 604p.

Frenkel, M., (1974), "Surface acidity of montmorillonites", *Clays and Clay Miner.,* 22:435–441.

Frost, R.R., and Griffin, R.A., (1977), "Effect of pH on adsorption of copper, zinc, and cadmium from landfill leachate by clay minerals", *J. Environ. Sci. Health,* Part A, 12 (4&5):139–156.

Fuller, W.H., and Warrick, A.W., (1985), *Soils in Waste Treatment and Utilization,* Volumes 1 & 2, CRC Press, Boca Raton, FL.

Garcia-Miragaya, J., and Page, A.L., (1976), "Influence of ionic strength and inorganic complex formation on the sorption of trace amounts of Cd by montmorillonite", *Soil Sci. Soc. Amer. J.,* 40:658–663.

Gibson, M.J., and Farmer, J.G., (1986), "Multi-step sequential chemical extraction of heavy metals from urban soils", *Environ. Pollut. Bull.,* 11:117–135.

Gieseking, J.E., (1975), *Soil Components Vol. 1, Organic Components,* Springer-Verlag, Heidelberg.

Gillham, R.W., and Cherry, J.A., (1982), "Contaminant migration in saturated unconsolidated geologic deposits", in *Recent Trends in Hydrogeology,* T.N. Narasimhan (ed.), Geological Society of America, Special Publication 1989, Boulder, CO, pp.31–62.

Gillham, R.W., Robin, M.L.J., Dytynyshyn, D.J., and Johnston, H.M., (1984), "Diffusion of nonreactive and reactive solutes through fine-grained barrier materials", *Can. Geotech. J.,* 21:541–550.

Gillham, R.W., (1987), "Processes of contaminant migration in groundwater", Proc. CSCE Centennial Conf. on *Mangement of Waste Contamination of Groundwater,* R.N. Yong (ed.), pp. 239–269.

Goodall, D.C., and Quigley, R.M., (1977), "Pollution migration from two sanitary landfills near Sarnia, Ontario", *Can. Geotech. J.,* 14:223–236.

Gouy, G., (1910), "Sur la constitution de la charge electrique à la surface d'un electrolyte", *Ann. Phys.,* (Paris), Série, 4(9):457–468.

Gouy, G., (1917), "Sur la function electrocapillaire", *Ann. Phys.* (Paris), Série, 9(7):129–184.

Grahame, D.C., (1947), "The electrical double layer and the theory of electocapillarity", *Chem. Rev.,* 41:441–501.

Green-Kelly, R., (1955), "Sorption of aromatic organic compounds by montmorillonite, Part I, Orientation studies", *Trans. Faraday Soc.,* 51:412–424.

Greenland, D.J., (1963), "Adsorption of polyvinyl alcohols by montmorillonite", *J. Colloid Sci.,* 18:647–664.

Greenland, D.J., and Hayes, M.H.B., (eds.), (1981), *The Chemistry of Soil Processes,* John Wiley & Sons, Chichester, 714p.

Greenland, D.J., and Hayes, M.H.B., (eds.), (1985), *The Chemistry of Soil Constituents,* John Wiley & Sons, Chichester, 469p.

Greenland, D.J., and Mott, C.J.B., (1985), "Surfaces of soil particles", in *The Chemistry of Soil Constituents,* D.J. Greenland, and M.H.B. Hayes (eds.), John Wiley & Sons, Chichester, pp. 321–354.

Griffin, R.A., and Jurinak, J.J., (1974), "Kinetics of phosphate interaction with calcite", *Soil Sci. Soc. Amer. J.,* 38:75–79.

Griffin, R.A., Shimp, N.F., Steele, J.D., Ruch, R.R., White W.A., and Hughes, G.M., (1976), "Attenuation of pollutants in municipal leachate by passage through clay", *Environ. Sci. Technol.,* 10:1262–1268.

Griffith, S.M., and Schnitzer, M., (1975), "Analytical characteristics of humic and fulvic acids extracted form tropical soils", *Soil Sci. Soc. Amer. J.,* 39:861–869.

Grim, R.E., Bray, R.H., and Bradley, W.F., (1937), "The mica in argillaceous sediments", *Am. Mineral,* 32:813–829.

Grim, R.E., (1953), *Clay Mineralogy,* McGraw-Hill Inc. New York, 296p.

Guy, R.D., Chakrabarti, C.L., and McBain, D.C., (1978), "An evaluation of extraction of copper and lead in model sediments", *Water Resources,* 12:21–24.

Haines, W.B., (1930), "The studies in the physical properties of soils", *V. J. Agr. Sci.,* 20:97–116.

Hamaker, H.C., (1937), "The London-Van der Waals attraction between spherical particles" *Physica,* 4:1048–1072.

Hamaker, J.W., and Thompson, J.M., (1972), " Adsorption", Chapter 3, in *Organic Chemicals in the Soil Environment,* C.A.I. Goring, and J.W. Hamaker (eds.), Marcel Dekker, New York, 1:49–144.

Hamaker, J.W., (1975), "Interpretation of soil leaching experiments", in *Environmental Dynamics of Pesticides,* R. Haque, and V.H. Freed (eds.), Plenum Press, New York.

Harter, R.D., (1979), "Adsorption of copper and lead by Ap and B2 horizons of several northeastern United States soils", *Soil Sci. Soc. Amer. J.,* 43:679–683.

Harter, R.D., (1983), "Effect of soil pH and adsorption of lead, copper, zinc and nickel", *Soil Sci. Soc. Amer. J.,* 47:47–51.

Hartton, D., and Pickering, W.F., (1980), "The effect of pH on the retention of Cu, Pb, Zn and Cd by clay-humic acid mixtures", *Water, Air, Soil Pollut. J.,* 14:13–21.

Hatcher, P.G., Schnitzer, M., Dennis, L.W., and Maciel, G.E., (1981), "Aromaticity of humic substances in soils", *Soil Sci. Soc. Amer. J.,* 45:1089–1094.

Hesse, P.R., (1971), *A Textbook of Soil Chemical Analysis,* William Clowes and Sons, London, 519p.

Hibbeln, K.S., (1996), "Effect of kaolinite and cadmium on the biodegradation of naphthalene and substituted naphthalenes", M. Sc. Thesis, McGill University.

Hitchon, B., and Wallick, E.I., (eds.), (1984), "Practical applications of groundwater geochemistry" *Proc. First Canadian/American Conference on Hydrogeology,* 323p.

Hitchon, B., and Trudell, M., (eds.), (1985), "Hazardous wastes in groundwater, A soluble dilemma", *Proc. Second Canadian/American Conference on Hydrogeology,* 255p.

Hoffman, R.F., and Brindley, G.W., (1960), "Adsorption of nonionic aliphatic molecules from aqueous solutions on montmorillonite. Clay organic studies II", *Geochim. Cosmochim. Acta,* 20:15–29.

Hogg, R., Healy, T.W., and Fuerstenay, D.W., (1966), "Mutual coagulation of colloidal dispersions", *Trans. Faraday Soc.,* 62:1638–1651.

Hopper, D.R., (1989), "Remediation's goal: protect human health and the environment", *Chemical Engineering,* August, pp. 94–110.

Jackson, M.L., (1956), "Soil chemical analysis — advance course", published by the author, University of Wisconsin, Madison.

Jackson, M.L., (1958), *Soil Chemical Analysis,* Prentice-Hall Inc., Englewood Cliffs, N.J.

Johnson, O., (1955), "Acidity and polymerization activity of solid acid catalysis", *J. Phys. Chem.,* 59:827–830.

Johnson, R.L., Cherry J.A., and Pankow, J.F., (1989), "Diffusive contaminant transport in natural clay: a field example and implication for clay-lined water disposal sites", *Environ. Sci. Technol.,* 23:340–349.

Jones, G., and Dole, M., (1929), "The viscosity of aqueous solutions of strong electrolytes with special reference to barium chloride", *J. Am. Chem. Soc.,* 52:29–50.

Jones, L.H.P., and Jarvis, S.C., (1981), "The fate of heavy metals," in *The Chemistry of Soil Processes,* D.J. Greenland, and M.H.B. Hayes (eds.), John Wiley & Sons, Chichester, pp. 593–620.

Jost, W., (1960), *Diffusion in Solids, Liquids, Gases,* Academic Press, New York.

Karickhoff, S.W., Brown, D.S., and Scott, T.A., (1979), "Sorption of hydrophobic pollutants on natural sediments", *Water Res.,* 13:241–248.

Karickhoff, S.W., (1984), "Organic pollutants sorption in aquatic system", *J. Hydraulic Engineer.,* 110:707–735.

Karickhoff, S.W., and Morris, K.R., (1985), "Sorption dynamics of hydrophobic pollutants in sediments suspension", *Environ. Toxicol. Chem.,* 4:469–479.

Keller, W.D., (1968), *Principles of Chemical Weathering,* Lucas Brothers, Los Angeles, 111p.

Kemper, W.D., and Van Schaik, J.C., (1966), "Diffusion of salts in clay-water systems", *Soil Sci. Soc. Amer. J.,* 30:534–540.

Kenaga, E.E., and Goring, C.A.I., (1980), "Relationship between water solubility, soil sorption, octanol-water partitioning and concentration of chemicals in biota", *ASTM-STP* 707, pp. 78–115.

Kinniburgh, D.G., Jackson, M.L., and Syers, J.K., (1976), "Adsorption of alkaline earth, transition and heavy metal cations by hydrous oxide gels of iron and aluminum", *Soil Sci. Soc. Amer. J.,* 40:796–799.

Kinzelbach, W., (1986), *Groundwater Modelling: An Introduction with Sample Programs in BASIC,* Elsevier, Amsterdam, 333p.

Kruyt, H.R., (1952), *Colloid Science, Vol. I,* Elsevier, Amsterdam, 389p.

Laflamme, R.E., and Hites, R.A., (1978), "The global distribution of polycyclic aromatic hydrocarbons in recent sediments", *Geochim. Cosmochim. Acta,* 42:289–303.

Langmuir, I., (1916a), "The evaporation, condensation and reflection of molecules and the mechanism of adsorption", *Phys. Rev.* 8 (2nd Series), pp. 149–176.

Langmuir, I., (1916b), "The constitution and fundamental properties of solids and liquids. Part I. Solids", *J. Am. Chem. Soc.,* 38:2221–2295.

Langmuir, I., (1918), "The adsorption of gases on plane surface of glass, mica and platinum", *J. Am. Chem. Soc.,* 40:1361–1403.

Langmuir, I., (1938), "The role of attractive and repulsive forces in the formation of tactoids, thixotropic gels, protein crystals, and coacervates", *J. Chem. Phys.,* 6:873–896.

Larson, R.A., and Weber, E.J., (1994), *Reaction Mechanisms in Environmental Organic Chemistry,* Lewis Publishers, Boca Raton, FL, 433p.

Lee, J.F., Mortland, M.M., Chiou, C.T., Kile, D.E., and Boyd, S.A., (1990), "Adsorption of benzene, toluene and xylene by two tetraethylammonium smectites having different charge densities", *Clays and Clay Miner.,* 38:113–120.

Lee, M., and Fountain, J.C., (1999), "The effectiveness of surfactants for remediation of organic pollutants in the unsaturated zone", *J. Soil Contamination,* 8:39–62.

Leo, A., Hansch, C., and Elkins, D., (1971), "Partition coefficients and their uses", *Chemical Rev.,* 71:525–616.

Lerman, A., (1979), *Geochemical Processes: Water and Sediment Environments,* John Wiley & Sons, New York, 481p.

Lewis, G.N., (1923), "Valences and the structure of atoms and molecules", The Chemical Catalogue, New York.

Li, R.S., (1997), "A study of the efficiency and enhancement of electro-kinetic extraction of a heavy metal from contaminated soils", Ph.D. Thesis, McGill University.

Li, R.S., Yong, R.N., and Li, L.Y., (1999), "Use of an interaction energy model to predict Pb removal from illite", in *Geoenvironmental Engineering — Ground Contamination: Pollutant Management and Remediation,* R.N. Yong, and H.R. Thomas (eds.), Thomas Telford, London, pp.230–237.

Li, Y.H., and Gregory, S., (1974), "Diffusion of ions in sea water and in deep-sea sediments", *Geochim. Cosmochim. Acta,* 38:603–714.

Ludwig, R.D., (1987), "A study of post-dehydration bonding and ion adsorption in a bauxite waste", Ph.D. Thesis, McGill University.

MacDonald, E., (1994), "Aspects of competitive adsorption and precipitation of heavy metals by a clay soil", M. Eng. Thesis, McGill University.

MacDonald, E., (2000), "Pb and Cu retention by an illite clay soil: a soil multicomponent study", Ph.D. Thesis, McGill University.

MacDonald, E.M., and Yong, R.N., (1997), "On the retention of lead by illitic soil fractions", in *Geoenvironmental Engineering — Contaminated Ground: Fate of Pollutants and Remediation,* R.N. Yong, and H.R. Thomas (eds.), Thomas Telford, London, pp. 117–127.

Mackay, D., Shiu, W.Y., and Ma, K.C., (1992), *Illustrated Handbook of Physical-Chemical Properties and Environmental Fate for Organic Chemicals,* Vols. 1, 2, and 3, Lewis Publishers, Boca Raton, FL.

Magdoff, F.R., and Barlett, R.J., (1985), "Soil pH buffering revisited", *Soil Sci. Soc. Amer. J.,* 49:145–148.

Maguire, M., Slavek, J., Vimpany, I., Higginson, F.R., and Pickering, W.F., (1981), "Influence of pH on copper and zinc uptake by soil clays", *Aust. J. Soil Res.,* 19:217–29.

Manahan, S.E., (1990), *Environmental Chemistry,* 4th Ed., Lewis Publishers, Ann Arbor, MI, 612p.

Manassero, M., Van Impe, W.F., and Bouazza, A., (1996), "Waste disposal and containment", *Proc. 2nd. Int. Congr. Env. Geotech.,* Japan, 3:1425–1474.

Manicotti, K., and Cohere, M., (1969), "Photochemical degradation products of pentachlorophenol", *Residue Rev.,* 25:13–23.

Mathioudakis, M., (1988), "Numerical simulation of contaminant transport", M. Eng. Thesis, McGill University.

McBride, M.B., (1982), "Cu2+ adsorption characteristics of aluminium hydroxides and oxyhydroxides", *Clays and Clay Miner.,* 30:21–28.

McBride, M.B., (1989), "Reactions controlling heavy metal solubility in soils", *Adv. Soil Sci.,* 10:1–56.

McCarty, P.L., Reinhard, M., and Rittman, B.E., (1981), "Trace organics in groundwater", *Environ. Sci. Technol.,* 15; 1:40–51.

McCarty, P.L., and Semprini, L., (1994), "Ground-water treatment for chlorinated solvents", in *Bioremediation of Ground Water and Geologic Material: A Review of In-Site Technologies*, Government Institutes, Inc., MD, Section 5.

Means, J.C., Wood, S.G., Hassett, J.J., and Banwart, W.L., (1982), "Sorption of amino and carboxyl-substituted polynuclear aromatic hydrocarbons by sediments and soils", *Environ. Sci. Technol.,* 15(2):93.

Metcalf, R.L., Booth, G.M., Schuth, C.K., Hansen, D.J., and Lu, P.Y., (1973), "Uptake and fate of di-ethylhexyphthalate in aquatic organisms and in a model ecosystem", *Environmental Health Perspectives.*

Meyers, P.A., and Oas, T.G., (1978), "Comparison of associations of different hydrocarbons with clay particles in simulated seawater", *Environ. Sci. Technol.,* 132:934–937.

Meyers, P.A., and Quinn, J.G., (1973), "Association of hydrocarbons and mineral particles in saline solution", *Nature,* 244:23–24.

Mohammed, L.F., (1994), "Assessment of soil stabilization of oil residue and its environmental implications", Ph.D. Thesis, McGill University.

Montgomery, J.H., and Welkom, L.M., (1991), *Groundwater Chemicals Desk Reference,* Lewis Publishers, Ann Arbor, MI, 640p.

Mooney, R.W., Keenan, A.C., and Wood, L.A., (1952), "Adsorption of water vapour by montmorillonite, II: Effect of exchangeable ions and lattice swelling as measured by x-ray diffraction", *J. Am. Chem. Soc.,* 74:1371–1374.

Mortland, M.M., (1970), "Clay-organics complexes and interactions", *Adv. Agron.,* 22:75–117.

Mortland, M.M., and Raman, K.V., (1968), "Surface acidity of smectites in relation to hydration, exchangeable cation, and structure", *Clays and Clay Miner.,* 16:393–398.

Mourato, D., (1990), "The influence of polysaccharides on sub-surface soil properties and interactions", Ph.D. Thesis, McGill University.

Mulligan, C., (1998), "On the capability of biosurfactants for the removal of heavy metals from soil and sediments", Ph.D. Thesis, McGill University.

Mulligan, C.N., and Gibbs, B.F., (1993), "Factors influencing the economics of biosurfactants", in *Biosurfactants, Production, Properties, Applications*, N. Kosaric (ed.), Marcel Dekker, New York, pp. 329–371.

Mulligan, C.N., and Yong, R.N., (1997), "The use of biosurfactants in the removal of metals from oil-contaminated soil", in *Geoenvironmental Engineering — Contaminated Ground: Fate of Pollutants and Remediation*, R.N. Yong, and H.R. Thomas (eds.), Thomas Telford, London, pp. 461–466.

Mulligan, C.N., and Yong, R.N., (1999), "On the use of biosurfactants for the removal of heavy metals from oil-contaminated soil", *J. Environ. Prog.,* 18:50–54.

Mulligan, C.N., Yong, R.N., and Gibbs, B.F., (1999a), "Removal of heavy metals from contaminated soil and sediments using the biosurfactant surfactin", *J. Soil Contamination,* 8:231–254.

Mulligan, C.N., Yong, R.N., and Gibbs, B.F., (1999b), "A review of surfactant-enhanced remediation of contaminated soil", in *Geoenvironmental Engineering — Ground Contamination: Pollutant Management and Remediation,* R.N. Yong, and H.R. Thomas (eds.), Thomas Telford , London, pp. 441–449.

Mulligan, C.N., Yong, R.N., and Gibbs, B.F., (2001), "The use of selective extraction procedures for soil remediation", *Proc. Int. Symp. on Suction, Swelling, Permeability and Structure of Clays,* Balkema, Rotterdam.

Munakata, K., and Kuwahara, M., (1969), "Photochemical degradation products of pentachlorophenol", *Residue Rev.,* 25:13–23.

Murray, J.W., (1975), "The interaction of metal ions at the manganese dioxide solution interface", *Geochim. Cosmochim. Acta,* 39:505–519.

Nernst, W., (1888), "Zur knetik der in losung befinlichen korper", *Zeitschrift fur Physikalishe Chemie,* 2:613–637.

Norrish, K., (1954), "The swelling of montmorillonite" *Discussion, Faraday Soc.,* 18:120–134.

Nyffeler, U.P., Li, Y.H., and Santschi, P.H., (1984), "A kinetic approach to describe trace-element distribution between particles and solution in natural aquatic systems", *Geochim. Cosmochim. Acta,* 48:1513–1522.

Ogata, A., (1970), "Theory of dispersion in a granular medium", U.S. Geological Survey, Professional Paper 411–1.

Ogata, A., and Banks, R.B., (1961), "A solution of the differential equation of longitudinal dispersion in porous media", U.S. Geological Survey Paper 411-A.

Ohtsubo, M., Yoshimuara, A., Wada, S., and Yong, R.N., (1991), "Particle interaction and rheology of illite-iron oxide complexes", *Clays and Clay Miner.,* 39:347–354.

Oliver, B.G., (1984), "Distribution and pathways of some chlorinated benzenes in the Niagara River and Lake Ontario", *Water Pollut. Res. J. Can.,* 19:47–59.

Oliver, B.G., and Nicol, K.D., (1982), "Chlorobenzenes in sediments, water, and selected fish from Lakes Superior, Huron, Erie, and Ontario", *Environ. Sci. Technol.,* 16(8):532–536.

Oliver, B.G., and Nicol, K.D., (1984), "Chlorinated contaminants in the Niagara river, 1981-1983", *Sci. Total Environ.,* 39:57–70.

Oliver, B.G., and Pugsley, C.W., (1986), "Chlorinated contaminants in St. Clair river sediments", *Water Pollut. Res. J. Can.,* 21:368–379.

Oliver, B.G., Charlton, M.N., and Durham, R.W., (1989), "Distribution, redistribution, and geochronology of polychlorinated biphenyl congeners and other chlorinated hydrocarbons in Lake Ontario sediments", *Environ. Sci. Technol.,* 23:200–208.

Olsen, R.I., and Davis, A., (1990), "Predicting the fate and transport of organic compounds in groundwater, Part 1", *Hazardous Mater. Control,* 3:38–64.

Onsager, L., (1931), "Reciprocal relation in irreversible processes, II", *Physics Rev.,* 38:2265–2279.

Ontario Ministry of the Environment, Policy No. 15-08, (1986), "The incorporation of the reasonable use concept into the environment".

Ouhadi, V.R., (1997), "The role of marl components and ettringite on the stability of stabilized marl", Ph.D. Thesis, McGill University.

Parks, G.A., (1965), "The isoelectric points of solid oxides, solid hydroxides, and aqueous hydroxy complex systems", *Chem. Revs.,* 65:177–198.

Pask, J.A., and Davis, B., (1954), "Thermal analysis of clays and acid extraction of alumina from clays", Differential Analysis, U.S. Bureau of Mines, Denver, CO, 6p.

Pavilonsky, V.M., (1985), "Varying permeability of clayey soil linings", *Proc. Int. Conf. Soil Mech. Found. Eng.,* San Francisco, 2:1213–1216.

Pearson, R.G., (1963), "Hard and soft acids and bases", *J. Am. Chem. Soc.,* 85:3533–3539.

Pearson, R.G., (1968), "Hard and soft acid and bases, Part I", *J. Chem. Educ.,* 45:581–587.

Pearson, R.G., (1968), "Hard and soft acid and bases, Part II", *J. Chem. Educ.,* 45:643–648.

Perkins, T.K., and Johnston, O.C., (1963), "A review of diffusion and dispersion in porous media", *J. Soc. of Petroleum Engr.,* 17:70–84.

Phadungchewit, Y., (1990), "The role of pH and soil buffer capacity in heavy metal retention in clay soils", Ph.D. Thesis, McGill University.

Pickering, W.F., (1986), "Metal-ion speciation — soils and sediments (a review)", *Ore Geol. Rev.,* 1:83–146.

Pierce, R.H., Gower, S.A., and Victor, D.M., (1980), "Pentachlorophenol and degradation products in lake sediment", in *Contaminants and Sediments*, R.A. Baker (ed.), 2:43–56.

Plastourgou, M., and Hoffmann, M.R., (1984), "Transformation and fate of organic esters in layered-flow systems: the role of trace metal catalysis", *Environ. Sci. Technol.,* 18:756–764.

Potter, H.A.B., (1999), "A study of the retention of heavy metals by amorphous iron-aluminum oxides and kaolinite", Ph.D. Thesis, McGill University.

Potter, H.A.B., and Yong, R.N., (1999), "Influence of iron/aluminum ratio on the retention of lead and copper by amorphous iron-aluminum oxides", *Appl. Clay Sci.,* 14:1–26.

Puls, R.W., and Bohn, H.L., (1988), "Sorption of cadmium, nickel, and zinc by kaolinite and montmorillonite suspensions", *Soil Sci. Soc. Amer. J.,* 52:1289–1292.

Pusch, R., and Güven, N., (1990), "Electron microscopic examination of hydrothermally treated bentonite clay", *J. Engr. Geol.,* 28:303–324.

Pusch, R., and Karnland, O., (1996), "Physico-chemical stability of smectite clays", *J. Engr. Geol.,* 41:73–86.

Quigley, R.M., (1980), "Geology, mineralogy, and geochemistry of Canadian soft clays: a geotechnical perspective", *Can. Geotech. J.,* 17:261–285.

Quigley, R.M., Fernandez, F., and Lowe, R.K., (1988), "Clayey barrier assessment for impoundment of domestic waste leachate (Southern Ontario) including clay-leachate compatibility by hydraulic conductivity testing", *Can. Geotech. J.,* 25:574–581.

Quigley, R.M., Fernandez, F., Yanful, E.K., Helgason, T., and Margaritis, A., (1987), "Hydraulic conductivity of contaminated natural clay directly below a domestic landfill", *Can. Geotech. J.,* 24:377–383.

Quigley, R.M., Sethi, A.J., Boonsinsuk, P., Sheeran, D.E., and Yong, R.N., (1985), "Geologic control on soil composition and properties, Lake Ojibway clay plain, Matagami, Quebec", *Can. Geotech. J.,* 22:491–500.

Quirk, J.P., (1968), "Particle interaction and soil swelling", *Isr. J. Chem.,* 6:213–234.

Rand, B., and Melton, I.E., (1977), "Particle interaction in aqueous kaolinite suspensions, I. Effect of pH and electrolyte upon the mode of particle interaction in homoionic sodium kaolinite suspensions", *J. Colloid Interface Science,* 60:308–320.

Rao, P.S.C., and Davidson, J.M., (1980), "Estimation of pesticide retention and transformation parameters required in nonpoint source pollution models", in *Environmental Impact of Nonpoint Source Pollution*, M.R. Overcash, and J.M. Davidson (eds.), Ann Arbor Science, MI, pp.23–27.

Richards, L.A., (1949), "Methods of measuring soil moisture tension", *Soil Sci.,* 68:95–112.

Robinson, R.A., and Stokes, R.H., (1959), *Electrolyte Solutions,* 2nd Ed., Butterworths, London.

Samani, H.M.V., (1987), "Mathematical modeling of contaminant transport through clay soils using irreversible thermodynamics", Ph.D. Thesis, McGill University.

Sawyer, C.N., McCarty, P.L., and Parkin, G.F., (1994), *Chemistry for Environmental Engineering,* 4th Ed., McGraw-Hill, New York, 658p.

Shackelford, C.D., (1996), "Modelling and analysis in environmental geotechnics: an overview of practical applications", *Proc. 2nd Int. Congr. Environ. Geotech.,* 3:1375–1404.

Shackelford, C.D., and Redmond, P.L., (1995), "Solute breakthrough curves for processed kaolin at low flow rates", *ASCE J. Geotech. Engr.,* 121(1).

Schnitzer, M., (1969), "Reactions between fulvic acid, a soil humic compound, and inorganic soil constituents", *Soil Sci. Soc. Amer. Proc.,* 33:75–81.

Schnitzer, M., and Khan, S.U., (1978), *Soil Organic Matter,* Elsevier Scientific, Amsterdam.

Schnitzer, M., and Khan, S.U., (1972), *Humic Substances in the Environment,* Marcel Dekker, New York.

Schnitzer, M., Ortiz de Serra, M.I., and Ivarson, K., (1973), "The chemistry of fungal acid-like polymers and of soil humic acids", *Soil Sci. Amer. Proc.,* 7:229–326.

Schnitzer, M., and Skinner, S.I.M., (1967), "Organo-metallic interaction in soils: 7. Stability constants of Pb^{++}-, Ni^{++}-, CO^{++}- and Mg^{++}-fulvic acid complexes", *Soil Sci.,* 103:247–252.

Schnoor, J.J., Licht, L.A., McCutcheon, S.C., Wolfe, N.L., and Carreira, L.H., (1995), "Phytoremediation of organic and nutrient contaminants", *Environ. Sci. Technol.,* 29(7):318–323.

Schofield, R.K., (1946), "Ionic forces in thick films of liquid between charged surfaces", *Trans. Faraday Soc.,* 42B:219–228.

Schwarzenbach, R.P., Gschwend, P.M., and Imboden, D.M., (1993), "Environmental organic chemistry", John Wiley & Sons, New York, 681p.

Schwarzenbach, R.P., and Westall, J., (1981), "Transport of non-polar organic compounds from surface water to groundwater: laboratory sorption studies", *Environ. Sci. Technol.,* 15(11):1360–1367.

Segalen, P., (1968), "Note sur une methode de determination des produits mineraux amorphes dans certains sols a hydroxides tropicaux", *Cah. Orstom ser. Pedol.,* 6:105–126.

Singh, U., and Uehara, G., (1986), "Electrochemistry of the double layer principles and applications to soils", Chapter 1 in *Soil Physical Chemistry,* D.L. Sparks (ed.), CRC Press, Boca Raton, FL, pp.1–38.

Skempton, A.W., (1953), "The colloidal activity of clays", *Proc. 3rd Int. Conf. On Soil Mechanics and Foundation Engineering,* Switzerland, 1:57–61.

Soil Science Society of America, (1983), "Chemical mobility and reactivity in soil systems", SSSA Spec. Publ. No. 11, 262p.

Solomon, D.H., and Murray, H.H., (1972), "Acid-base interactions and properties of kaolinite in non-aqueous media", *Clays and Clay Miner.,* 20:135–141.

Sparks, D.L., (1995), "Kinetics of metal sorption reactions", in *Metal Speciation and Contamination of Soil,* H.E. Allen, C.P. Huang, G.W. Bailey, and A.R. Bowers (eds.), Lewis Publishers, Boca Raton, FL, pp. 35–58.

Spooner, A.J., and Giusti, L., "Geochemical interactions between landfill leachate and sodium bentonite", in *Chemical Containment of Waste in the Geosphere,* R. Metcalfe, and C.A. Rochelle (eds.), Geological Society, London, Special Publ. 157:131–142.

Sposito, G., (1981), *The Thermodynamics of Soil Solutions,* Oxford University Press, New York, 223p.

Sposito, G., (1984), *The Surface Chemistry of Soils,* Oxford University Press, New York, 234p.

Sposito, G., (1989), *The Chemistry of Soils,* Oxford University Press, New York, 277p.

Starkey, H.C., Blackmon, P.D., and Hauff, P.L., (1984), "The routine mineralogical analysis of clay-bearing samples", U.S. Geological Survey Bulletin 1563, Washington, D.C., pp. 2–18.

Stern, O., (1924), "Zur Theorie der elektrolytischen Doppelschicht", *Z. Elektrochem.*, 30:508–516.

Stevenson, F.J., (1977), "Stability constants of Cu^{2+}, Pb^{2+}, and Cd^{2+} complexes with humic acids", *Soil Sci. Soc. Amer. J.*, 40:665–672.

Stone, A.T., (1989), "Enhanced rates of monophenyl terephthalate hydrolysis in aluminum oxide suspension", *J. Colloid Interface Sci.*, 127(2):429–441.

Strauss, H., (1991), "Final report: an overview of potential health concerns of bioremediation", Env. Health Directorate, Health Canada, Ottawa, 54 p.

Stumm, W., and Morgan, J.J., (1981), *Aquatic Chemistry: An Introduction Emphasizing Chemical Equilibria in Natural Waters,* 2nd Ed., John Wiley & Sons, New York, 780p.

Sullivan, P.J., (1977), "The principle of hard and soft acids and bases as applied to exchangeable cation selectivity in soils", *Soil Sci.*, 124:117–121.

Suquet, H., de la Calle, C., and Pezerat, H., (1975), "Swelling and structural organization of saponite", *Clays and Clay Miner.*, 23:1–9.

Tecsult Inc. et Roche Ltée., (1993), "Environmental evaluation of the Lachine canal decontamination project", Parks Canada, Vol 1: 112 p., Vol. 2: 178 p.

Tessier, A., Campbell, P.G.C., and Bisson, M., (1979), "Sequential extraction procedure for the speciation of particulate trace metals", *Anal. Chem.*, 51:844–851.

Tessier, A., and Campbell, P.G.C., (1987), "Partitioning of trace metals in sediments: relationships with bioavailability", *Hydrobiologia,* 49:43–52.

Tessier, A., and Campbell, P.G.C., (1988), "Partitioning of trace metals in sediments", in *Metal Speciation: Theory, Analysis and Applications,* J. Kramer, and H.E. Allen (eds.), Lewis Publishers, Ann Arbor, MI, 183–200.

Theng, B.K.G., (1974), *The Chemistry of Clay-Organic Reactions,* John Wiley & Sons, New York.

Theng, B.K.G., (1979), *Formation and Properties of Clay-Polymer Complexes,* Elsevier Scientific, Amsterdam, 362p.

Theng, B.K.G., (1982), "Clay-polymer interactions: summary and perspectives", *Clays and Clay Miner.*, 30:1–10.

Thornton, S.F., Bright, M.I., Lerner, D.N., and Tellam, J.H., (1999), "The geochemical engineering of landfill liners for active containment", in *Chemical Containment of Waste in the Geosphere,* R. Metcalfe and C.A. Rochelle (eds.), Geological Society, London, Special Publ. 157:143–157.

Tinsley, I., (1979), *Chemical Concepts in Pollutant Behaviour,* John Wiley & Sons, New York, 265p.

Tousignant, L.P., (1991), "Disappearance of polycyclic aromatic hydrocarbons from a soil using biostimulation techniques", M. Eng. Thesis, McGill University.

U.S. EPA (Environmental Protection Agency), (1992), "Metal equilibrium speciation model: MINTEQA2", Instruction material for Workshop, Athens, GA.

U.S. National Environmental Policy Act, PL 91-190, January, 1970.

van Breeman, N., and Wielemaker, W.G., (1974), "Buffer intensities and equilibrium pH of minerals and soils: II. Theoretical and actual pH of minerals and soils", *Soil Sci. Soc. Amer. J.*, 38:61–66.

van Olphen, H., (1977), *An Introduction to Clay Colloid Chemistry,* 2nd Ed., John Wiley & Sons, New York, 318p.

Verschueren, K., (1983), *Handbook of Environmental Data on Organic Chemicals,* 2nd Ed., van Nostrand Reinhold, New York, 1310p.

Verwey, E.J.W., and Overbeek, J.Th.G., (1948), *Theory of the Stability of Lyophobic Colloids,* Elsevier, Amsterdam, 205p.

Wagenet, R.J., (1983), "Principles of salt movement in soils", in *Chemical Mobility and Reactivity in Soil Systems*, SSSA Spec. Publ. No. 11, pp. 123–140.

Wang, B.W., (1990), "A study of the role and contribution of amorphous materials in marine soils of Eastern Canada", Ph.D. Thesis, McGill University.

Warith, M.A., (1987), "Migration of leachate through clay soil", Ph.D. Thesis, McGill University.

Warith, M.A., and Yong, R.N., (1991), "Landfill leachate attenuation by clay soil", *J. Hazardous Waste & Hazardous Materials,* 2:127–141.

Warkentin, B.P., and Schofield, R., (1962), "Swelling pressure of Na-montmorillonite in NaCl solutions", *J. Soil Sci.,* 13:98.

Warshaw, Ch., and Roy, R., (1961), "Classification and scheme for the identification of layer silicates", *Bull. Amer. Geol. Soc.,* 64:921–944.

Weber, L., (1991), "The permeability and adsorption capability of kaolinite and bentonite clays under heavy metal leaching", M. Eng. Thesis, McGill University.

West, L.J., Stewart, D.I., Duxbury, J.R., and Johnston, S.R., (1999), "Toxic metal mobility and retention at industrially contaminated sites", in *Chemical Containment of Waste in the Geosphere,* R. Metcalfe, and C.A. Rochelle (eds.), Geological Society, London, Special Publ., 157:241–264.

Wildman, W.E., Jackson, M.L., and Whittig, L.D., (1968), "Iron-rich montmorillonite formation in soils derived from serpentine", *Soil Sci. Soc. Amer. J.,* 32:787–794.

Xing, G.X., Xu, L.Y., and Hou, W.H., (1995), "Role of amorphous Fe oxides in controlling retention of heavy metal elements in soils", in Chapter 6, *Environmental Impact of Soil Component Interactions: Metals, Other Inorganics and Microbial Activities,* Vol. II, P.M. Huang, J. Berthelin, J.-M. Bollag, W.B. McGill, and A.L. Page (eds.), Lewis Publishers, Boca Raton, FL.

Yanful, E.K., Quigley R.M., and Nesbitt, H.W., (1988), "Heavy metal migration at a landfill site, Ontario, Canada — I: Thermodynamic assessment and chemical interpretations", *Appl. Geochem.,* 3:523–533.

Yanful, E.K., Quigley, R.M., and Nesbitt, H.W., (1988), "Heavy metal migration at a landfill site, Ontario, Canada — II: Metal partitioning and geotechnical implications", *Appl. Geochem.,* 3:623–629.

Yariv, S., and Cross, H., (1979), *Geochemistry of Colloid Systems,* Springer-Verlag, New York, 449p.

Yong, R.N., and Warkentin, B.P., (1959), "A physico-chemical analysis of high swelling clays subject to loading", *Proc. 1st. Pan Amer. Conf. on Soil Mech. Found. Engr.,* 2:867–888.

Yong, R.N., and Warkentin, B.P., (1975), *Soil Properties and Behaviour,* Elsevier Scientific, Amsterdam, 449p.

Yong, R.N., Sethi, A.J., Ludwig, H.P., and Jorgensen, M.A., (1979), "Interparticle action and rheology of dispersive clays", ASCE, Geotech. Eng. Div., 105:1193–1209.

Yong, R.N., (1984), "Particle interaction and stability of suspended solids", in *Sedimentation/Consolidation Models: Predictions and Validation,* R.N. Yong and F.C. Townsend (eds.), ASCE Symp. Proc. pp. 30–59.

Yong, R.N., Elmonayeri, D.A., and Chong, T.S., (1985), "The effect of leaching on the integrity of a natural clay", *J. Engr. Geology,* 21:279–299.

Yong, R.N., and Samani, H.M.V., (1987), "Modelling of contaminant transport in clays via irreversible thermodynamics", *Proc. Geotechnical Practice for Waste Disposal '87*/GT Div. ASCE, pp. 846–860.

Yong, R.N., and Samani, H.M.V., (1988), "Effect of adsorption on prediction of pollutant migration in a clay soil", *Proc. CSCE-ASCE Nat. Conf. on Env. Engr.,* pp. 510–519.

Yong, R.N., Lechowicz, Z.M., Syzmanski, A., and Wolski, W., (1988), "Consolidation of organic subsoil in terms of large strains", *Proc. ISSMFE Baltic Conf.*

Yong, R.N., and Mourato, D., (1988), "Extraction and characterization of organics from two Champlain Sea subsurface soils", *Can. Geotech. J.,* 25:599–607.

Yong, R.N., and Xu, D.M., (1988), "An identification technique for evaluation of phenomenological coefficients in unsaturated flow in soils", *Int. J. Numerical and Anal. Methods in Geomechanics,* 12:283–299.

Yong, R.N., and Hoppe, E.J., (1989), "Application of electric polarization to contaminant detection in soils", *Can. Geotech. J.,* 26:536–550.

Yong, R.N., Mohamed, A.M.O., and Cheung, S.C.H., (1990a), "Thermal behaviour of backfill material for a nuclear waste disposal vault", *Proc. Materials Res. Soc. Symp.,* 176:649–656.

Yong, R.N., and Mourato, D., (1990), "Influence of polysaccharides on kaolinite structure and properties in a kaolinite-water system", *Can. Geotech. J.,* 27:774–788.

Yong, R.N., Warkentin, B.P., Phadungchewit, Y., and Galvez, R., (1990b), "Buffer capacity and lead retention in some clay materials", *Water, Air, Soil Pollut. J.,* 53:53–67.

Yong, R.N., Mohamed, A.M.O., and El Monayeri, D.S., (1991), "Flow of surfactant fluid in nonaqueous phase liquid-saturated soils during remedial measures", ASCE GED Geotech. Engr. Congr., pp.1137–1148.

Yong, R.N., Xu, D.M. and Mohamed, A.M.O., (1991), "Analytical solution for a coupled hydraulic and subsidence model using a characteristic surface method", ASCE GED Geotech. Engr. Congr., pp.1331–1340.

Yong, R.N., (1991), "Landfilling waste management compliance with regulatory policy and guidelines", *Proc. Can. Geotech. Soc. Conf. on Env. Geotech.,* pp. 3–8.

Yong, R.N., Mohamed, A.M.O., and El-Monayeri, D.S., (1991), "A multiphase flow model for remediation of subsurface contamination by nonaqueous phase liquid", *Proc. Can. Geotech. Soc. Conf. Env. Geotech.,* pp. 293–298.

Yong, R.N., Cabral, A., and Weber, L.M., (1991a), "Evaluation of clay compatibility to heavy metals transport and containment: permeability and retention", *Proc. 1st Can. Conf. Env. Geotech.,* Montreal, pp. 314–321.

Yong, R.N., and Sheremata, T.W., (1991), "Effect of chloride ions on adsorption of cadmium from a landfill leachate", *Can. Geotech. J.,* 28:378–387.

Yong, R.N., and Rao, S.M., (1991), "Mechanistic evaluation of mitigation of petroleum hydrocarbon contamination by soil medium", *Can. Geotech. J.,* 28:84–91.

Yong, R.N., Tousignant, L., Leduc, R., Chan, E.C.S., (1991), "Disappearance of PAHs in a contaminated soil from Mascouche", in *In-Situ and On Site Bioreclamation,* R.E. Hinchee, and R.F. Olfenbuttel (eds.), pp. 377–395.

Yong, R.N., Mohamed, A.M.O., and Warkentin, B.P., (1992a), *Principles of Contaminant Transport in Soils,* Elsevier, Holland, 327p.

Yong, R.N., and Mohamed, A.M.O., (1992), "A study of particle interaction energies in wetting of unsaturated expansive clays", *Can. Geotech. J.,* 29:1060–1070.

Yong, R.N., Mohamed, A.M.O., and Wang, B.W., (1992b), "Influence of amorphous silica and iron hydroxide on interparticle action and soil surface properties", *Can. Geotech. J.,* 29:803–818.

Yong, R.N., Xu, D.M., Mohamed, A.M.O., and Cheung, S.C.H., (1992c), "An analytical technique for evaluation of coupled heat and mass flow coefficients in unsaturated soil", *Int. J. for Num. and Anal. Methods in Geomechanics,* 16:233–246.

Yong, R.N., and Galvez-Cloutier, R., (1993), "Le contrôle du pH dans les mécanismes d'accumulation lors de l'intéraction kaolinite-plomb", *Proc. Environnement et Géotechnique,* Paris, pp. 309–316.

Yong, R.N., and Prost, R., (1993), "Decontamination, treatment and rehabilitation of sites", *Proc. Environnement et Géotechnique,* Paris, pp. 1–18.

Yong, R.N., Galvez-Cloutier, R., and Phadungchewit, Y., (1993), "Selective sequential extraction analysis of heavy metal retention in soil", *Can. Geotech. J.,* 30:834–847.

Yong, R.N., and Phadungchewit, Y., (1993), "pH influence on selectivity and retention of heavy metals in some clay soils", *Can. Geotech. J.,* 30:821–833.

Yong, R.N., Tan, B.K., and Mohamed, A.M.O. (1994), "Evaluation of attenuation capability of a micaceous soil as determined from column leaching tests", ASTM STP 1142, pp. 586–606.

Yong, R.N., Mohammed, L., and Mohamed, A.M.O., (1994), "Retention and transport of refinery residual petroleum in soil", ASTM STP 1221, pp. 89–110.

Yong, R.N., (1995), "On the fate of toxic pollutants in contaminated sediments", ASTM STP 1293, pp.13–38.

Yong, R.N., (1996), "Multi-disciplinarity of Environmental Geotech.", *Proc. 2nd Int. Cong. Env. Geotech.,* Osaka, Japan, 3:1255–1273.

Yong, R.N., and Mohamed, A.M.O.,(1996), "Evaluation of coupled heat and water flow parameters in a buffer material", *J. Engr. Geol.,* 41:269–286.

Yong, R.N., MacDonald, E., and Coles, C., (1996), "Processes and partitioning of contaminants in contaminated sediments", *Proc. 2nd Int. Cong. Env. Geotech.,* Osaka, Japan, 1:429–436.

Yong, R.N., Desjardins, S., Farant, J.P., and Simon, P., (1997), "Influence of pH and exchangeable cation on oxidation of methylphenols by a montmorillonite clay", *Appl. Clay Sci.,* 12:93–110.

Yong, R.N., and MacDonald, E.M., (1998), "Influence of pH, metal concentration, and soil component removal on retention of Pb and Cu by an illitic soil", Chapter 10, in *Adsorption of Metals by Geomedia*, E.A. Jenne (ed.), Academic Press, San Diego.

Yong, R.N., Bentley, S.P., Thomas, H.R., Tan, B.K., and Yaacob, W.Z.W., (1998), "The attenuation characteristics of natural clay materials in South Wales and their potential use as engineered clay barriers", *Proc. 8th Int. Cong. IAEG,* Vancouver, pp. 2499–2504.

Yong, R.N., Tan, B.K., Bentley, S.P., Thomas, H.R., Yaacob, W.Z.W., and Hashm, A.A., (1998), "Assessment of attenuation capability of two soils via leaching column tests", *Proc. 3rd. Int. Cong. Env. Geotech.,* Lisbon, pp. 503–508.

Yong, R.N., Tan, B.K., Bentley, S.P., Thomas, H.R., and Zuhairi, W., (1999), "On the competency assessment of clay soils from South Wales as landfill liners", *Q. J. Eng. Geol.,* 32 (10), 261–270.

Yong, R.N., Bentley, S.P., Harris, C., and Yaacob, W.Z.W., (1999), "Selective sequential extraction analysis (SSE) on estuarine alluvium soils", in *Geoenvironmental Engineering — Ground Contamination: Pollutant Management and Remediation,* R.N. Yong, and H.R. Thomas (eds.), Thomas Telford , London, pp.118–126.

Yong, R.N., (1999a), "Overview of partitioning and fate of contaminants: retention, retardation and regulatory requirements", in *Chemical Containment of Waste in the Geosphere,* R. Metcalfe, and C.A. Rochelle (eds.), Geological Society, London, Spec. Publ. 157, pp. 1–20.

Yong, R.N., (1999b), "Soil suction and soil-water potentials in swelling clays in engineered clay barriers", *J. Engr. Geol.,* 84:3–14

Yong, R.N., (1999c), "Overview of modelling of clay microstructure and interactions for prediction of waste isolation barrier performance", *J. Engr. Geol.,* 84:83–92.

Zachara, J.M., (1986), "Quinoline sorption to materials: role of pH and retention of the organic cation", *Environ. Sci. Technol.,* 20:620–2627.

Zanoni, A.E., (1972), "Groundwater pollution and sanitary landfills: a critical review", *Groundwater,* 10:3–13.

Index

A

B

C

D

E

F

G

H

M

P

T

U

V